**Computational Modeling
and Simulation Examples
in Bioengineering**

Computational Modeling and Simulation Examples in Bioengineering

Edited by

Nenad D. Filipovic
Faculty of Engineering
University of Kragujevac
SERBIA

 IEEE Engineering in Medicine
and Biology Society, *Sponsor*

IEEE Press Series in Biomedical Engineering
Metin Akay, *Series Editor*

IEEE PRESS

Published by John Wiley & Sons, Inc., Hoboken, New Jersey.
Published simultaneously in Canada.

For general information on our other products and services or for technical support, please contact our Customer Care Department within the United States at (800) 762–2974, outside the United States at (317) 572-3993 or fax (317) 572-4002.

Wiley also publishes its books in a variety of electronic formats. Some content that appears in print may not be available in electronic formats. For more information about Wiley products, visit our web site at www.wiley.com.

Library of Congress Cataloging-in-Publication Data

Names: Filipovic, Nenad, D., editor.
Title: Computational modeling and simulation examples in bioengineering / edited by Nenad D. Filipovic, Faculty of Engineering, University of Kragujevac, Serbia.
Description: First edition. | Hoboken, New Jersey : Wiley, [2022] | Series: IEEE press series on biomedical engineering | Includes index.
Identifiers: LCCN 2021039449 (print) | LCCN 2021039450 (ebook) | ISBN 9781119563945 (hardback) | ISBN 9781119563921 (adobe pdf) | ISBN 9781119563914 (epub)
Subjects: LCSH: Biomedical engineering. | Computer simulation.
Classification: LCC R856 .C626 2022 (print) | LCC R856 (ebook) | DDC 610.28–dc23/eng/20211005
LC record available at https://lccn.loc.gov/2021039449
LC ebook record available at https://lccn.loc.gov/2021039450

Cover Design: Wiley
Cover Image: © archy13/Shutterstock

Set in 9.5/12.5pt STIXTwoText by Straive, Pondicherry, India

10 9 8 7 6 5 4 3 2 1

Contents

Editor Biography

 **Nenad D. Filipovic** is Rector of University of Kragujevac, Serbia, full Professor at Faculty of Engineering and Head of Center for Bioengineering at University of Kragujevac, Serbia. He was Research Associate at Harvard School of Public Health in Boston, USA. His research interests are in the area of biomedical engineering, cardiovascular disease, fluid–structure interaction, biomechanics, bioinformatics, biomedical image processing, machine learning, medical informatics, multi-scale modeling, software engineering, parallel computing, computational chemistry, and bioprocess modeling. He is author and coauthor of 11 books, over 350 publications in peer-reviewed journals, and over 10 software for modeling with finite element method and discrete methods from fluid mechanics and multiphysics. He also leads a number of national and international projects in the EU and United States in the area of bioengineering and software development.

He is Director of Center for Bioengineering at University of Kragujevac and leads joint research projects with Harvard University and University of Texas in the area of bio-nanomedicine computer simulation. He also leads a number of national and international projects in the area of bioengineering and bioinformatics. He is Editor in Chief for EAI Endorsed Transaction on Bioengineering and Bioinformatics, Managing Editor for Journal of Serbian Society for Computational Mechanics, President of Serbian Society of Mechanics and member of European Society of Biomechanics (ESB), European Society for Artificial Organs (ESAO), and IEEE member.

Author Biographies

Tijana Djukic (born on 1st April 1988) is a PhD in biomedical engineering. She finished Bachelor and Master studies at Faculty of Mechanical Engineering, University of Kragujevac, Serbia, with average mark 10. She finished PhD studies at the same faculty, with average mark 10 in 2015, for a record period of less than three years. Currently, she works as a Senior Research Associate at Institute for Information Technologies, University of Kragujevac, Serbia, and is involved in projects regarding software development of applications in bioengineering. Her main research interests are in the area of computational mechanics, solid–fluid interaction, microfluidics, multi-scale cancer modeling, parallel computing using GPU devices, and modeling using finite element and lattice Boltzmann method. She speaks English, German, and Spanish, and uses Italian and French.

Ksenija Zelic Mihajlovic is a Doctor of Dental Medicine, PhD in dental sciences, and a specialist in Periodontology and Oral Medicine with more than 12 years of clinical practice. She holds the position of a Research Associate at the School of Dentistry, University of Belgrade, Serbia, and has published 24 scientific papers in prestigious international peer-reviewed journals with more than 400 citations. Focus of her basic and clinical research for 12 years has been on nanostructure of dental tissues, nanotechnology, computational modeling and biomechanics in dental sciences and practice, forensic dentistry, periodontology, and endodontics.

Zarko Milosevic was born on 14 March 1983. He received his Master's degree from Faculty of Mechanical Engineering University of Kragujevac in 2009 and his PhD degree from the same institution in 2018. He is a Research Associate at Institute of Information Technologies in the Department of Technical Technological Sciences in the scientific field of Bioengineering and Bioinformatics. His research interests are in the area of biomedical engineering, software engineering, computer graphics, cardiovascular disease, fluid-structure interaction, biomechanics, bioinformatics, biomedical image processing, machine learning.

Dalibor D. Nikolic was born on 24 April 1985, in Smederevska Palanka, Serbia. He received his Master's degree from the Faculty of Engineering, University of Kragujevac, Serbia, in 2011, and his PhD degree from the same institution, in 2017. He is a Research Associate at Institute of Information Technologies, in Department of Technical-Technological Sciences in the scientific fields of Bioengineering and Bioinformatics. His research area covers the development and optimization of stent devices using FEA, CAD/CAE, topology, and parametric optimization methods.

Igor B. Saveljic is a Scientific Associate at the Institute for Information Technologies, University of Kragujevac and Research and Development Center for Bioengineering, Kragujevac, Serbia. He received Master's degree from Faculty of Engineering, University of Kragujevac, Serbia, in 2009, and PhD degree from the same institution, in 2016. He is author and coauthor of over 70 publications in the field of numerical modeling. His main research interests are in fluid dynamics and modeling using finite element method.

Tijana I. Šušteršič is a Teaching Assistant at the Faculty of Engineering, University of Kragujevac, Kragujevac, Serbia. She finished MSc and BSc studies in Mechanical Engineering, Department of Applied Mechanics and Automatic Control with GPA 10.00/10.00 at the same Faculty and as the best student in generation. She is also engaged with *Bioengineering* Research and Development *Center (BioIRC), Kraguje-vac. She is the* author and coauthor *of 14 papers published in peer-reviewed journals, 2 book chapters, and more than 23 papers presented at the international con-ferences.* She possesses Certificate of Proficiency in English (CPE) level C2 and Goethe Zertifikat level B2 according to the Common European Framework of Reference for Languages. Her area of expertise includes biomedical signal processing, machine learning algorithms, finite element modeling, and implementation of different algorithms on Field Programmable Gate Array (FPGA).

Arso M. Vukicevic is a Research Associate at the University of Kragujevac. His research interests are in the application of computer vision and artificial intelligence techniques for automation and digitalization in Industry and Healthcare.

Aleksandra Vulović received her Bachelor and Master's diploma from the Faculty of Engineering, University of Kragujevac. She is currently a PhD student at the same Faculty and a research assistant at BioIRC. She is working on the application of various computer methods for problems in bioengineering and bioinformatics. The focus of her PhD thesis is on the optimization of hip implant surface modification in order to promote femoral bone growth during the healing process. She is coauthor of more than 20 publications in international journals and on conferences and has participated in several national and international research projects.

Preface

The aim of this book is to provide concrete examples of applied modeling in biomedical engineering. The book is accompanied by software on the web for studying the representative bioengineering problems in more detail (www. bioengexambook.ac.rs). The primarily goal of the book is to serve as a textbook in various bioengineering university courses, as well as a support for basic and clinical research.

Different numerical examples in the area of bone, tissue, cardiovascular, augmented reality, and vertigo disease give readers an overview of typical problems that can be modeled and complete theoretical background with numerical method behind. The book can be used for lecturers of bioengineering courses at universities but also it can be very helpful for researchers, medical doctors, and clinical researchers. In comparison to the existing literature, this book will give more practical examples with supporting web platform in the area of modeling in bone, tissue, cardiovascular, cancer, lung, and vertigo diseases.

The book will be prepared mainly as a textbook for undergraduate or graduate courses in bioengineering, engineering and applied sciences in general, and medicine. There are different examples of application: bone, muscle, tissue, cardiovascular, cancer, lung, and vertigo disease. In chapters, it will be the theoretical background and basics of the computational methods used for the specific modeling. We consider that support by the software on the web should be of great help for lecturers when organizing classes. The theoretical presentations (either from the theoretical background or from bioengineering applications) can be accompanied by use of the software with menu-driven modeling and solution display. Use of the software can also aid students when studying various theoretical or bioengineering problems.

The book will be also prepared to be useful for researchers in various fields related to bioengineering and other scientific fields, including medical applications. The book will provide basic information about how a bioengineering (or medical) problem can be modeled, which computational models can be used, and what is the background of the applied computer models.

In Chapter 1, some basic theoretical and numerical examples of computational Modeling of Abdominal Aortic Aneurysms are given. Chapter 2 describes modeling the motion of rigid and deformable objects in fluid flow. Application of computational methods in dentistry with some theoretical and practical numerical examples is given in Chapter 3.

Determining Young's modulus of elasticity of cortical bone from CT scans is described in Chapter 4. Parametric modeling of blood flow and wall interaction in aortic dissection is investigated in Chapter 5. Application of AR Technology in Bioengineering is well described in Chapter 6. Augmented Reality Balance Physiotherapy is presented in Chapter 7. Modeling of the human heart – ventricular activation sequence and ECG measurement are given in Chapter 8. In Chapter 9, medical image segmentation was described using contrast stretching, edge-detection and thresholds with coupled Simulink-XSG (Xilinx System Generator) tool and FPGA (Field Programmable Gate Arrays). The book is intended for pre-graduate and postgraduate students as well as for researchers in the domains of bioengineering, biomechanics, biomedical engineering, and medicine.

Prof. Nenad D. Filipovic
Faculty of Engineering,
University of Kragujevac,
SERBIA

1

Computational Modeling of Abdominal Aortic Aneurysms

Nenad D. Filipovic[1,2]

[1]*Faculty of Engineering, University of Kragujevac, Kragujevac, Serbia*
[2]*Bioengineering Research and Development Center, BioIRC Kragujevac, Serbia*

1.1 Background

Abdominal aortic aneurysm (AAA) is a dilation of the aorta beyond 50% of the normal vessel diameter [1] that is frequently observed in the aging population [2] and it affects 6–9% of the people in the industrialized world. It is a major health problem that typically affects men after the age of 50 [3] and it is the thirteenth leading cause of death in Western societies [4]. AAAs cause about 15 000 deaths per year in the United States only [5], and 1.3% of all deaths among men aged 65–85 years in developed countries [6–8]. In the United States alone, 1.5 million have undiagnosed AAAs [3]. They can remain asymptomatic for most of their development and, if left untreated, they can enlarge and eventually rupture with catastrophic mortality rate of 80% [9] to 90% [2]. On the other hand, the mortality following elective AAA repair has significantly improved to 3–6% [10] which clearly demonstrates the need of diagnosing and monitoring AAAs on time in order to make progress in both medical and economic domain.

A myriad of different factors are established in the literature to account for AAA formation, expansion, and, eventually, rupture. Namely, a substantial amount of research on AAA expansion and rupture focuses on different biological and biomechanical factors and, lately, special attention is put to genes and chemical influences. Among biological factors and risk factors, authors usually discuss the influence of diameter, sex, blood pressure, chronic obstructive pulmonary disease, and smoking [11]. From a biomechanical point of view, major factors contributing to AAA expansion and rupture are the wall stress [12, 13] and strength [14], wall stiffness [15], vessel asymmetry [16–19], intraluminal thrombus (ILT) [20, 21], entire geometry [22, 23], etc. Namely, according to the biomechanical perspective,

Computational Modeling and Simulation Examples in Bioengineering, First Edition. Edited by Nenad D. Filipovic. © 2022 The Institute of Electrical and Electronics Engineers, Inc. Published 2022 by John Wiley & Sons, Inc.

AAA rupture is usually defined as a material failure of the degenerated AAA wall to withstand the stress exerted on it [24]. However, the relationship between rupture and biomechanics of aorta proved to be more complex [10] and the evaluation based only on one of these parameters is not sufficient.

For many years, diameter was taken to be the primary parameter associated with rupture risk estimation. Namely, the threshold of $\geq 5.5 \text{ cm}^3$ was applied as indicative for AAA repair [10]. However, continuous studies in the past showed that while 10–24% of small aneurysms (<5.5 cm) may rupture [1], aneurisms which diameter exceeds the threshold remain stable. These findings cast doubt over the suitability of surgical repair based solely on the maximum diameter criterion [16–18]. In order to refute "diameter criterion" rule, many other criteria and parameters ensued. In contrast, the survey conducted in 2006 [25] confirmed that 92% of surgeons still use maximum diameter criterion and growth rate in making decisions about the surgical intervention while 19% of them stated that they were not even aware that biomechanics may influence the rupture risk. The results of the survey suggest that cooperation of surgeons and engineers is necessary in order not just to make technical advances but to implement them in practice.

Mutual collaboration of clinicians and engineers resulted in different efforts and methodologies proposed in the last few decades, all striving to make progress in the domain of AAA expansion and rupture prediction. This paper reviews some of the most significant studies in the area of AAA modeling in the past decades which further understanding of utmost importance for making additional progress toward validation and application in the clinical setting. We firstly described computational methods applied for AAA in chronological order. Then, different experimental testing as well as our own testing is described in order to determine the mechanical properties of AAA. Mechanical–chemical including ILT modeling is separately analyzed. Finite element procedure including fluid–structure interaction (FSI) from our group is described. In the end, some of data mining (DM) approach and vision for future clinical decision support system (DSS) is given.

1.2 Clinical Trials for AAA

The known risk factors for AAA include male sex, smoking, hypertension, and a family history of AAA in a first-degree relative [26]. Many clinical studies confirmed that smoking is the most important modifiable risk factor for AAA [27–33]. Some authors discovered that the duration of smoking and daily cigarette number are also associated with a higher risk of AAA [28, 34]. One large systematic review of studies evaluating smoking and aortic aneurysm placed the relative risk of aortic aneurysm-related events in current smokers between 3 and 6 [34].

Smoking contributes significantly to the prevalence of AAA and may account for 75% of all AAA, 4 cm in diameter or larger [33].

A history of hypertension and myocardial infarction or coronary artery bypass surgery was negatively associated, whereas a body mass index $\geq$25 kg/m^2 was protective [35].

1.3 Computational Methods Applied for AAA

The study of Darling et al. [36] was one of the first studies to conclude that AAA rupture predominately occurs on posterior walls. Namely, in the autopsy study conducted almost 40 years ago on AAA patients, among whom 473 died with surgically intact arteriosclerotic AAAs, it was reported that 82% of ruptures occurred along the posterior and posteriolateral wall of AAA. Consequently, it was important to examine the wall stress acting on these regions in AAA, so that many studies were influenced to direct their research toward examining posterior wall stress for clinical relevance. One of the first studies to apply the finite element analysis (FEA) to determine the AAA wall stress was the study of [37] who concluded that FEA has a potential of becoming a crucial tool in the study of vascular mechanics. Their study was followed by extensive research in the area of numerically predicted AAA wall stress that continues even at present. However, limitations of this study include the use of idealized models with regular structures and evenly distributed wall stress.

In 1998, Vorp et al. [38] published a study on the influence of maximum diameter and aortic aneurysm asymmetry on mechanical wall stress. They investigated the effects of asymmetry on 3D stress distribution in the wall of AAA and refuted the critical diameter criterion suggesting that all AAAs with the same diameter have the same risk of rupture. They generated 10 virtual computer models with commercial software (Pro-Engineer v. 16.0; Parametric Technology Waltham, Mass) according to two protocols. In the first protocol, five models were generated with constant maximum diameter parameter (6 cm) while asymmetry β varied from 0.3 to 1.0. In the second protocol, asymmetry was kept constant $\beta = 0.4$ while maximum diameter parameter varied from 4 to 8 cm. The results confirmed that both parameters had the influence on the increase and decrease in the wall stress in the different sections of the aneurisms and that aneurysm rupture was caused by a gross mechanical failure of the aortic wall which occurs when wall stress exceeds the strength of the tissue. It was also concluded that maximum stress occurs on the posterior wall for small AAAs ($\leq$5), while for larger AAAs peak stress is on the anterior surface. Although it pioneered in proving the effects of asymmetry, the study was performed on virtual models, so potential limitations assuming that

AAA wall is homogenous, isotropic, and linearly elastic with small strains and uniform thickness have to be taken into account when analyzing real AAA models.

Venkatasubramaniam et al. [10] refuted traditional views that relate aneurysm size to the risk of rupture. They conducted a comparative study of aortic wall stress in ruptured and non-ruptured aneurysms with an aim to prove the importance of wall stress when predicting the risk of rupture in individual patients. Namely, the study included computed tomography (CT) scans of 27 patients (12 ruptured and 15 non-ruptured AAA), predominantly males. Using the finite element method, they calculated wall stress using the geometry of AAA, the material properties of the aortic wall, and the forces and constraints acting on the wall. The material properties were used from a previously validated mathematical model by [39–41]. ANSYS 6.1 program (ASN Systems Ltd, Cannonsburg, USA) was utilized for the analysis and post-processing while the von Mises stress was used to evaluate the state of the aneurysms. There were no important differences in the mean diameter between two groups (6.8 cm for non-ruptured and 7.6 cm for ruptured, $P > 0.1$) and there were two aneurysms that ruptured at small diameters of 5.0 and 5.7 cm. The authors concluded that AAA that ruptured or went on to rupture had significantly higher peak stress (mean 1.02 MPa) compared with non-ruptured (mean 0.62 MPa). Moreover, systolic blood pressure was also significantly higher in ruptured AAA. Noting that 45 and 65 mm diameter AAAs can have the same stress, they emphasized the role of the shape and asymmetry of the aneurism including the anterior and superior limits. They also demonstrated that wall stress can be calculated from a routine CT scan and that it may be a better predicator of AAA risk of rupture than diameter alone on the individual basis. On the other hand, the study assumed a uniform AAA wall thickness of 2 mm and did not take into account the effect of thrombus on wall stress.

In 2006, Vande Geest et al. [42] developed a biomechanics-based rupture potential index (RPI) that became a useful rupture prediction tool. Namely, the RPI predetermined the wall strength on a patient-specific basis by utilizing experimental tensile testing and statistical modeling. The tissue strength was calculated by taking parameters such as age, sex, smoking status, family history of AAA, normalized diameter, and the maximum thickness of the ILT into account. Then, the wall stress was predicted with FEA. Although the authors reported that the RPI has a potential to identify high rupture risk of AAA better than diameter or peak wall stress (PWS) alone, their approach still requires validation before it can be introduced into clinical setting.

In the study conducted by Scotti et al. [43], 10 idealized models of AAA were used, generated with the CAD software ProEngineer Wildfire (Parametric Technology Corporation, Needham, MA) [44] together with an additional non-aneurismal model as control in order to assess the significance of an arbitrary estimated peak fluid pressure (117 mmHg) compared with nonuniform pressure resulting from a coupled FSI. The models differed in the degree of asymmetry and wall heterogeneity and the FEA was used to estimate the effects of asymmetry and wall thickness on the wall stress and fluid dynamics. Each model contained fluid and solid domain and was analyzed with static pressure-deformation analysis together with FSI. The Navier–Stokes equations were used as governing equations for homogeneous blood flow, while for computational solid stress (CSS) analysis, only the solid domain was considered. The results exhibited that a realistic fluid pressure distribution to the inner AAA wall (instead of an arbitrary peak systolic pressure) resulted in at least 20% higher wall stress, despite wall thickness heterogeneity. Additionally, maximum AAA wall stress increased with asymmetry although the computational model included blood flow.

Finol and Ender [45, 46] used a Spectral Element Method with three-step time splitting scheme for the semidiscrete formulation of the time-dependent terms in the momentum equations for axisymmetric two-aneurysm abdominal model. This methodology has been widely used for the Direct Numerical Simulation of transitional flows with fast-evolving temporal phenomenon and complex geometries.

Multi-scale models for AAA are considered constitutive models for vascular tissue, where collagen fibers are assembled by proteoglycan cross-linked collagen fibrils (CFPG-complex) and reinforce an otherwise isotropic matrix (elastin). There is multiplicative kinematics for the straightening and stretching of collagen fibrils. Mechanical and structural assumptions at the collagen fibril level define a piece-wise analytical stress–stretch response of collagen fibers.

The concept of multi-scale constitutive model performs integration at the material point for macroscopic stress which takes into account micro plane concept incorporated in the finite element modeling (FEM) [47, 48].

Zhang et al. [49] used multi-scale and multi-physical models for understanding disease development and progression, and for designing clinical interventions. They investigated multi-scale models of cardiac electrophysiology and mechanics for diagnosis, clinical decision support, and personalized and precision medicine in cardiology with examples in arrhythmia and heart failure.

FSI describes the wave propagation in arteries driven by the pulsatile blood flow. These problems are complex and challenging due to the high nonlinearity of the problem. The nonlinearity exists in the fluid equation but also in the structure displacement which modifies the fluid domain and generates geometrical nonlinearities as well [50].

Some authors used the generalized string model as the structure of blood flow in compliant vessels and arteries [50–56]. Causin et al. [57] described this string model as a structural model derived from the theory of linear elasticity for a cylindrical tube with small thickness. Nobile and Vergara [56] emphasized that the generalized string model neglects bending as well. Čanić et al. [58–61] claimed that there are no analytical results which are able to prove the well posedness of FSI problems without assuming the structure model that includes the higher-order derivative terms, capturing the viscoelastic behavior. For blood flow, there is a strong added mass effect issue in which the fluid and structure have comparable densities.

All the above studies demonstrate the importance of biomechanical modeling of AAA by using FEM approach with and without FSI and nonlinear wall deformation. This mechanical approach provides an additional understanding of potential indicators of rupture risk.

1.4 Experimental Testing to Determine Material Properties

Experimental tests are used to determine the mechanical properties of AAA. In-vivo measurements are based on the imaging modality. For in-vivo measurements, the main difficulty is to accurately determine the true force and the displacement distribution for the aorta wall. For isolating samples often unknown changes of their behavior affecting the results of such tests. MacSweeney et al. [62] found that elastic modulus was higher in aneurysmal abdominal aorta compared with controls. More recently, van't Veer et al. [63] estimated the compliance and distensibility of AAA by means of simultaneous instantaneous pressure and volume measurements obtained with the magnetic resonance imaging (MRI). Ganten et al. [64] by using time resolved electrocardiography (ECG)-gated CT imaging data from 67 patients, found that the compliance of AAA did not differ between small and large lesions. Molacek et al. [65] did not find any correlation between aneurysm diameter and distensibility of AAA wall and of normal aorta. Uniaxial extension testing is the simplest and most common of *ex-vivo* testing methods. The recorded force-extension data are converted to stress/strain. Di Puccio et al. [66] provided a recent review of the incompressibility assumption on soft biological tissue. Biaxial test was used as an initial square thin sheet of material which is stress normally to both edges. Even this test is not sufficient to fully characterize anisotropic materials [67, 68], although it can capture additional information regarding the mechanical behavior of the specimens with respect to uniaxial one.

One of the most complete data for biaxial mechanical behavior of aorta and AAAs is described in Vande Geest et al. [69, 70]. They reported on biaxial mechanical data for AAA (26 samples) and normal human AA as a function of age: less than 30, between 30 and 60, and over 60 years of age. They found that the aortic tissue becomes less compliant with age and that AAA tissue is significantly stiffer than normal abdominal aortic tissue.

At Clinical Center in Belgrade, Serbia, we developed an experimental procedure for bubble inflation test. Intraoperatively, specimens of anterior wall of AAA are harvested, stored in saline at 4° C immediately, and placed in the laboratory setup that simulates natural forces by exposing aortic tissue specimen to inflation with pressurized solution [71] (Figure 1.1). The model consists of the mechanical pump with Crebs–Ringer solution, heater and heat exchanger, tissue container, pressure sensor, and transducer with camera. Camera and pressure transducer system were connected by USB connection with the laptop serving as a control unit and data collector. After heating the Crebs–Ringer solution in the system to 37 °C, the pump was turned on and the pressure was gradually increased exposing aortic tissue to the maximum pressure until the moment of tissue rupture that was recorded by a webcam placed above the tissue. Pressure value in the moment of rupture was known due to the dedicated software defining tissue (failure) strength.

Ex-vivo testing is mostly based on uniaxial or biaxial stretching of the intraoperatively harvested tissue specimens. These tests have the ability to characterize intrinsic properties of the tissue itself with independent physical meanings, stiffness, failure stress, and strain. Tissue sample is exposed to extension along its length at a constant displacement rate, while the force is recorded during extension and until failure of the tissue. Uniaxial extension testing is the simplest and most common. Van de Geest et al. [72] reported uniaxial extension testing of 69 AAA specimens, from 21 patients. A novel mathematical model to estimate physically meaningful measures such as stiffness varied from 21.2 to 19.3 N/cm^2, mean 80.5 N/cm^2.

Improvement of uniaxial testing was achieved by performing biaxial testing, when specimen is exposed to determined forces between the two orthogonal directions. Using biaxial testing, the same authors compared the tissue of AAA and normal aorta and compared their behavior in longitudinal and circumferential direction. They found that aneurysmal degeneration of the abdominal aorta is associated with an increase in mechanical anisotropy, with preferential stiffening in the circumferential direction. Information related to stiffness and strain of aortic tissue was gained. Thubrikar et al. [73] were testing different regions of aneurysm in terms of yield stress, yield strains, and other mechanical properties, and found that the anterior wall of AAA is the weakest. In the circumferential direction, the yield stress of the lateral region was greater than that of the anterior or posterior region ($73 + 22 \, N/cm^2$ versus $52 + 20 \, N/cm^2$ or $45 + 14$ N/cm^2, respectively).

Our results showed rupture of the tissue at the inflation pressure of 9.8 N/cm^2. The analytical calculated wall strength for wall thickness 0.2 cm and aorta radius 1 cm was 24.5 N/cm^2. FEA showed PWS of 57 N/cm^2 and wall stress of 21.2 N/cm^2 at the anterolateral wall of AAA in the area of harvested tissue [71].

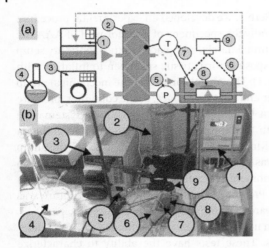

Figure 1.1 (a) Cartoon schema. (b) Real laboratory model. Laboratory model consists of the mechanical pump (3) with Crebs–Ringer solution (4), heater (1) and heat exchanger (2), tissue container (6), pressure sensor (5), and transducer with camera (9). Camera and pressure transducer system were connected by USB connection with the laptop serving as a control unit and data collector. After heating the Crebs–Ringer solution in the system to 37 °C (7 – temperature sensor), the pump was turned on and the pressure was gradually increased exposing aortic tissue (6) to the maximum pressure until the moment of tissue rupture that was recorded by a webcam placed above the tissue. Pressure value in the moment of rupture was known due to the dedicated software defining tissue (failure) strength.

1.5 Material Properties of the Aorta Wall

It is very important to use arterial wall with proper material model because the wall stiffness increases when lumen diameter increases and calcification and medial sclerosis occur with aging and disease. The aneurysmal wall has been found to be mechanically anisotropic [74]. Isotropic properties assumption is a reasonable assumption in most cases. A more accurate constitutive model is needed to describe the properties of the wall.

The flat arterial tissue layer is assumed to be a fiber-reinforced material with relatively stiff collagenous fibers embedded in a homogeneous isotropic (soft) ground matrix [75]. The assumption of strain energy functions holds good only for single continuous medium tissues which is not the case with arterial tissue [76]. The mechanical properties of soft biological tissues depend greatly on their microstructure integration attained in the constitutive model [49]. Taghizadeh et al. [77] proposed a new biaxial constitutive model based on microstructural properties as opposed to the simple uniaxial tests carried out by Sokolis et al. [78] and Karimi et al. [79]. Wall thickness is also very important. Since aneurysmal rupture occurs at a specific site of the aortic lumen, the properties of the wall affect the computed solution. The aortic wall thickness in computational methods is assumed to be in the range of 1.5–2 mm. While this may be largely accurate for most simulations, it has been acknowledged as a major limitation in the completeness of the prediction solution [80]. Raut et al. [81] suggested that the wall thickness as a constant value is not accurate and described a novel method that incorporated the regionally varying wall thickness, especially in the area of rupture.

1.6 ILT Modeling

Most aneurysms have ILT within their lumen. Stenbaek et al. [82] described that the development of ILT may be a better predictor than the maximum diameter of the AAA as a rupture risk parameter. Li et al. [83] calculated that the non-ILT models had higher stress development than the ILT models. Di Martino and Vorp [84] found that ILT might protect the AAA wall from the pressure applied by blood flow. O'Leary et al. [85] performed mechanical tests on 356 samples and classified them into 3 morphologies, type 1 which was a multilayered ILT which strength and stiffness decreased gradually, type 2 which strength decreased abruptly, and a single-layered ILT with lower strength and stiffness compared with the other two types. Tong et al. [86] carried out biomechanical behavioral studies on 90 AAA samples (78 men and 12 women). They found that the female ILT luminal layer showed a lower stiffness in the longitudinal direction than male and, consequently, the thrombi may have different wall weakening effects in males and females. Speelman et al. [87–89] also showed that AAA wall stress is closely related to the AAA diameter. Namely, they investigated whether wall stress can be used to predict stable and progressive AAAs as well as the effect of ILT on wall stress. The study was conducted on the finite element models of 30 patients with wall stresses computed with and without ILT for stable and progressive AAAs. The results showed that ILT reduced AAA wall stress, progressive AAA growth was not related to the diameter but AAA volume and relative ILT volume, and that higher wall stress was related to AAA growth only when ILT was not included in the simulations.

Kontopodis et al. [90] analyzed a single uncommon case of a 75-year-old Caucasian male diagnosed with fast-growing AAA without family history but with other medical conditions such as smoking, hypertension, diabetes, and COPD. The aneurism which was initially small (45 mm in diameter) presented a growth of 1 cm in only six months. The focus of research was on thrombus and lumen volumes, thrombus maximum thickness, maximum centerline curvature together with biomechanical aspects such as PWS and areas with high stress and stress redistribution. Two 3D AAA models (initial and follow up) were reconstructed from 2D CT angiography images with manual and automatic segmentation process performed with open source software ITK-SNAP [91]. The results indicated that the total volume increased from 85 to 120 ml for six months while lumen volume remained constant (72 ml for initial and 71 for follow up examination). Moreover, ILT increased from 14 to 50 ml and maximum thickness increased from 0.3 to 1.6 cm in the anterior part with maximum curvature of the centerline increased from 0.4 to 0.5 cm^{-1} suggesting an anterior bulging. Additionally, area under high wall stress remained almost constant while there was a marked redistribution of wall stress from anterior site and AAA neck to posterior wall. Total posterior wall area exposed to stress was 0 cm^2 at the initial stage and 9.7 cm^2 at the follow up. It was interesting to note that PWS did not increase as was the case with previous studies (e.g. [89]). Authors suggested that although large-scale studies are needed, information gathered by the analysis could be of use in determining AAA natural history and patient-specific rupture risk estimation. Anton et al. [92] provided an insight into a comparative study, numerical and experimental, to analyze the pressure field in AAA models and validate numerical model's ability to predict experimental pressure field. They used patient-specific AAA geometry with and without ILT with iliac arteries included. Iliac arteries were taken into account because numerical observations suggest that their inclusion is critical for accurate prediction of pressure, flow pattern, and wall stress [93]. Four CFD mathematical models were employed for ILT models and three CFD models were used for model without ILT because Reynolds Stress Model did not converge satisfactorily. CFD models predicted pressure fields substantially well with average difference of 1.1% for the model without ILT and 15.4% for ILT model. Moreover, they aimed to compare the spatial pressure drop before and after the formation of ILT although only few studies before performed spatial pressure drop measurements in AAA (e.g. [94, 95]). The special pressure drop in the ILT model was 5000 Pa while in the model without ILT, it was around 1500 Pa which is explained by the fact that with smaller lumen, flow velocities are higher. The study of Polzer et al. [96] broadened finite element single-phase AAA models (e.g. [69, 97–100]) with poroelastic description of ILT. The model was loaded by a pressure step and a cyclic pressure wave. The numerical results of the studied idealized axisymmetric two-phase AAA models proved that the entire blood pressure was transmitted to the AAA wall in spite of the existence of the ILT, which was also previously confirmed by [101]. The study found that the stress in the AAA wall, at steady state, did not depend on the permeability of ILT and did not differ from single-phase conventional models. Consequently, the results of the study suggested that single-phase description could be used reliably to predict the stress in AAA wall instead of biphasic one which is less computationally efficient. The study also showed that ILT reduced wall tensile stress by a value between 46 and 62%. Additionally, the analysis helped the explanation of differences between in-vivo and in-vitro measurements by demonstrating that pressure under ILT depends on the local geometry of the AAA. On the other hand, poroelastic description cannot be replaced when investigating transport phenomena through AAA tissue. Baek et al. [102] used a set of 39 patients' CT images data from 9 different patients. The images were divided into two groups – low and high ILT group, so that the effect of ILT on the AAA expansion could be established. The results indicate that the relationship between AAA expansion rate and maximum diameter can be changed by inhomogeneous distribution of ILT thickness. Although generally ILT presence is associated with aneurysm expansion rate, a slowdown of expansion was established in areas of thick ILT.

By using an inverse optimization method, Zeinali and Baek [103] created a computational framework toward patient-specific AAA modeling. Namely, using a 3D geometry from medical images, they identified initial material parameters for healthy aorta to satisfy homeostatic condition and then created different computational shapes and considered multiple spatiotemporal forms of elastin degradation and stress-mediated collagen turnover. The results exhibited the importance of the role of elastin damage extent, geometric complexity of an enlarged AAA, and sensitivity of stress-mediated collagen turnover on the wall stress distribution and the rate of expansion. Also, the study showed that the distributions of stress and local expansion initially correspond to the extent of elastin damage, but change because of stress-mediated tissue growth and remodeling dependent on the aneurysm shape. The specificity of their study lies in the fact that the authors did not use AAA patient-specific model, but medical images of a healthy subject. On the other hand, they suggest that in spite of the model used for the present study, their computational framework could be used in a patient-specific modeling to predict AAA shape and mechanical properties if improved in the domain of boundary conditions, description of aortic tissue, growth and remodeling, and the development of inverse scheme using AAA patients' longitudinal images.

The studies continuously prove that ILT has the potential to influence AAA both biochemically and biomechanically. Namely, ILT induces localized hypoxia, possibly leading to increased neovascularization, inflammation, and local wall weakening [104]. On the other hand, ILT development is strongly influenced by agonists and antagonists of platelets activation, aggregation, adhesion, and the proteins involved in the coagulation cascade which are not thoroughly discussed in the present literature. The study of Biasetti et al. [105] analyzed the evolution of chemical species involved in the coagulation cascade, their relation to coherent vertical structures, and the possible effect on ILT development. The authors developed a fluid–chemical model that simulates the coagulation cascade through a series of convection–diffusion–reaction equations. They followed the coagulation cascade model ([106]) which consisted of 18 species and involved plasma-phase and surface-bound enzymes and zymogens, with both plasma-phase and membrane-phase reactions. Blood was modeled as a non-Newtonian incompressible fluid. The coupling was achieved by a set of convection–diffusion–reaction equations that were added to the computed blood flow field in order to predict the distribution of chemicals. Since the model was able to couple the fluid and chemical domains, it represents an integrated mechanochemical potential for further understanding of ILT formation and development mechanisms.

Meaningful limitation of the study is mirrored in the fact that authors assumed a rigid wall which may alter the results since ILT tissue can deform during a cardiac cycle and consequently influence the fluid dynamics and the distribution of chemicals.

1.7 Finite Element Procedure and Fluid–Structure Interaction

Blood flow in aorta has a time-dependent 3D flow, so the time-dependent and full three-dimensional Navier–Stokes equations were solved. The laminar flow condition appropriate for this type of analysis [107] was used. The finite element code was validated using the analytical solution for shear stress and velocities through the curved tube [108]. A penalty formulation will be used [109]. The incremental–iterative form of the equations for time step and equilibrium iteration "i" is:

$$
\left[\begin{array}{cc} \frac{1}{\Delta t}\mathbf{M_v} + {}^{t+\Delta t}\mathbf{K}_{vv}^{(i-1)} + {}^{t+\Delta t}\mathbf{K}_{\mu v}^{(i-1)} + {}^{t+\Delta t}\mathbf{J}_{vv}^{(i-1)} & \mathbf{K_{vp}} \\ \mathbf{K_{vp}^T} & 0 \end{array} \right] \left\{ \begin{array}{c} \Delta \mathbf{v}^{(i)} \\ \Delta \mathbf{p}^{(i)} \end{array} \right\}
$$

$$
= \left\{ \begin{array}{c} {}^{t+\Delta t}\mathbf{F}_v^{(i-1)} \\ {}^{t+\Delta t}\mathbf{F}_p^{(i-1)} \end{array} \right\}
$$

$$(1.1)$$

The left upper index "$t + \Delta t$" denotes that the quantities are evaluated at the end of time step. The matrix $\mathbf{M_v}$ is mass matrix, $\mathbf{K_{vv}}$ and $\mathbf{J_{vv}}$ are convective matrices, $\mathbf{K_{\mu v}}$ is the viscous matrix, $\mathbf{K_{vp}}$ is the pressure matrix, and $\mathbf{F_v}$ and $\mathbf{F_p}$ are forcing vectors. The pressure is eliminated at the element level through the static condensation. For the penalty formulation, the incompressibility constraint is defined in the following manner:

$$
\mathrm{div}\mathbf{v} + \frac{p}{\lambda} = 0 \tag{1.2}
$$

where λ is a relatively large positive scalar so that p/λ is a small number (practically zero).

1.7.1 Displacement Force Calculations

Displacement force was calculated from direct integration of the pressure and wall shear stress distributions on the surfaces of the aortic wall:

$$\sum F = \int p \mathrm{d}A + \int \tau \mathrm{d}A + F_\mathrm{g} \tag{1.3}$$

where the integrals are surface finite elements along the surface of the aortic wall. Weight of blood in standing or supine position was also considered. It was assumed that all of the weight of the blood is supported by the endograft.

1.7.2 Shear Stress Calculation

At the regions of stasis and flow reversals, there is an important role of motion and deformation in the shear stress calculation [110, 111], concept of the rigid wall condition was assumed due to the negligible effect of shear stress on total DF [112]. The distribution of stresses within the blood was estimated. The stresses ${}^t\sigma_{ij}$ at time "t" is equal

$$ {}^t\sigma_{ij} = - {}^tp\delta_{ij} + {}^t\sigma_{ij}^\mu \tag{1.4}$$

where

$$ {}^t\sigma_{ij}^\mu = {}^t\mu^t \left(v_{i,j} + v_{j,i} \right) \tag{1.5}$$

is the viscous stress, ${}^t\mu$ is viscosity corresponding to the velocity vector ${}^t\mathbf{v}$ at a spatial point within the blood domain. The viscous stresses are represented by (1.5). The wall shear stress is calculated as:

$$ {}^t\tau = {}^t\mu \frac{\partial {}^tv_t}{\partial n} \tag{1.6}$$

where ${}^tv_t$ denotes the tangential velocity, and n is the normal direction at the vessel wall. At the integration points near the wall surface, the tangential velocity was estimated first, and then the velocity gradient $\partial {}^tv_t/\partial n$ was calculated. Finally, the viscosity coefficient ${}^t\mu$ using the average velocity was evaluated. Blood was taken as an incompressible Newtonian fluid, appropriate for larger arteries [110]. The kinematic viscosity was $\nu = 3.5\mathrm{e}{-6}\,\mathrm{m}^2/\mathrm{s}$ and the blood density was $\rho = 1050\,\mathrm{kg/m}^3$.

1.7.3 Modeling the Deformation of Blood Vessels

In modeling blood flow in large blood vessels, we have recognized two distinct cases: (i) rigid walls and (ii) deformable walls. The assumption (i) is mostly adopted in practical applications. It is very important to determine the stress–strain state in tissue, when blood and blood vessel system is analyzed, as well as the effects of the wall deformation on the blood flow characteristics.

There are complex mechanical characteristics of blood vessel tissue. The tissue can be modeled from linear elastic to nonlinear viscoelastic model. The governing finite element equations used in modeling wall tissue deformation with emphasis on implementation of nonlinear constitutive models are summarized.

If the principle of virtual work is applied, the differential equations of motion of a finite element are

$$\mathbf{M\ddot{U}} + \mathbf{B}^w\mathbf{\dot{U}} + \mathbf{KU} = \mathbf{F}^{\text{ext}} \tag{1.7}$$

where the element matrices are: $\mathbf{M}$ is mass matrix; $\mathbf{B}^w$ is the damping matrix, in case when the material has a viscous resistance; $\mathbf{K}$ is the stiffness matrix; and $\mathbf{F}^{\text{ext}}$ is the external nodal force vector which includes body and surface forces acting on the element. The dynamic differential equations of motion by the standard assembling procedure are obtained. These differential equations are integrated with a selected time step size Δt. The displacements $^{n+1}\mathbf{U}$ at end of time step are finally obtained according to equation:

$$\mathbf{\hat{K}}_{\text{tissue}}{}^{n+1}\mathbf{U} = {}^{n+1}\mathbf{\hat{F}} \tag{1.8}$$

where the tissue stiffness matrix $\mathbf{\hat{K}}_{\text{tissue}}$ and vector $^{n+1}\mathbf{\hat{F}}$ are expressed in terms of the matrices and vector in (1.7). Above equation is obtained under the assumption that the problem is linear: the viscous resistance is constant, displacements are small, and the material is linear elastic.

In many real examples as in case of aneurism or heart ventricle motion, the wall displacements can be large, hence the problem becomes geometrically nonlinear. Also, the tissue of blood vessels has nonlinear constitutive law which has to be expressed with materially nonlinear finite element formulation. Therefore, the linear formulation of the equation may not be appropriate. For a nonlinear problem, there is incremental–iterative equation

$$^{n+1}\mathbf{\hat{K}}_{\text{tissue}}^{(i-1)}\Delta\mathbf{U}^{(i)} = {}^{n+1}\mathbf{\hat{F}}^{(i-1)} - {}^{n+1}\mathbf{F}^{\text{int}(i-1)} \tag{1.9}$$

Here, $\Delta\mathbf{U}^{(i)}$ are the nodal displacement increments for the iteration "i," and the system matrix $^{n+1}\mathbf{\hat{K}}_{\text{tissue}}^{(i-1)}$, the force vector $^{n+1}\mathbf{\hat{F}}^{(i-1)}$, and the vector of internal forces $^{n+1}\mathbf{F}^{\text{int}(i-1)}$ correspond to the previous iteration.

We described the material nonlinearity of blood vessels which is used in further applications. The geometrically linear part of the stiffness matrix, $\left(^{n+1}\mathbf{K}_{\text{L}}\right)_{\text{tissue}}^{(i-1)}$, and nodal force vector, $^{n+1}\mathbf{F}^{\text{int}(i-1)}$, are defined:

$$\left(^{n+1}\mathbf{K}_{\text{L}}\right)_{\text{tissue}}^{(i-1)} = \int_V \mathbf{B}_{\text{L}}^{T}{}^{n+1}\mathbf{C}_{\text{tissue}}^{(i-1)}\mathbf{B}_{\text{L}}\mathrm{d}V, \quad \left(^{n+1}\mathbf{F}^{\text{int}}\right)^{(i-1)} = \int_V \mathbf{B}_{\text{L}}^{T}{}^{n+1}\boldsymbol{\sigma}^{(i-1)}\mathrm{d}V$$

$$\tag{1.10}$$

Here, the consistent tangent constitutive matrix $^{n+1}\mathbf{C}_{\text{tissue}}^{(i-1)}$ of tissue and the stresses at the end of time step $^{n+1}\boldsymbol{\sigma}^{(i-1)}$ depend on the material model used.

1.7.4 FSI Interaction

In many models of cardiovascular examples where deformation of blood vessel walls was taken into account, we can implement the loose coupling approach for the FSI [113–116]. The overall algorithm consists of the following steps:

1) For the current geometry of the blood vessel, determine blood flow (with Arbitrary Lagrangian–Eulerian (ALE) formulation). The boundary conditions for the fluid are wall velocities at the common blood–blood vessel surface.
2) Calculate the loads, which act on the walls from fluid domain (blood).
3) Determine deformation of the walls taking the current loads from the fluid domain (blood).
4) Check for the overall convergence which includes fluid and solid domain. If convergence is reached, go to the next time step. Otherwise go to step (1).
5) Update blood domain geometry and velocities at the common solid–fluid boundary for the new calculation of the fluid domain. In case of large wall displacements, update the finite element mesh for the fluid domain. Go to step (1).

The shear stress and drag force distribution have been presented in Figures 1.2 and 1.3 for two different patients for proximal and distal AAA.

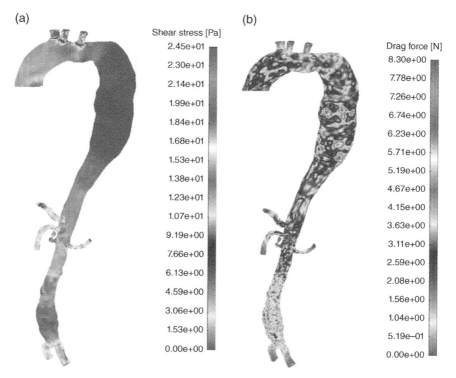

Figure 1.2 (a) Shear stress distribution. (b) Drag force distribution.

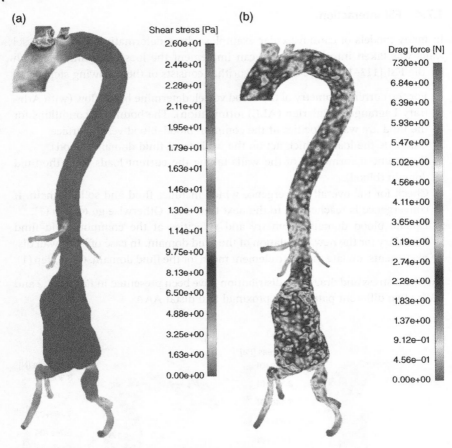

Figure 1.3 (a) Shear stress distribution. (b) Drag force distribution.

1.8 Data Mining and Future Clinical Decision Support System

Together with CFD simulation, there are numerous statistics-based machine learning methods that can be used to give more accurate and faster conclusions for clinicians [117].

Kolachalama et al. [118] proposed a DM technique that accounted for the geometric variability in patients for predicting cardiovascular flows. A Bayesian network-based algorithm was used to understand the influence of key parameters through a sensitivity analysis. Martufi et al. [119] investigated a geometrical characterization of the wall thickness distribution in AAA. They were able to train a model to differentiate the wall thicknesses in ruptured and un-ruptured AAA. Shum et al. [120] developed a model from 66 ruptured data sets and 10 non-ruptured data sets and their geometric indices and wall thickness variations. The results of this study showed that, in addition to maximum diameter, sac length, sac height, volume, surface area, bulge height, and ILT volume were all highly correlated with rupture status. In this study, the overall classification accuracy was 86.6%. It used a decision tree algorithm that is one of the possible machine learning methods that can be used for large data sets requiring a decision output.

Filipovic et al. [121] combined DM techniques and CFD for the estimation of the wall shear stresses in AAA under prescribed geometry changes. They performed large-scale CFD runs for creating machine learning data on the Grid infrastructure and their results showed that DM models provide good prediction of the shear stress at the AAA in comparison with full CFD model results on real patient data.

The abovementioned studies have been limited by the use of geometric parameters and, in particular, the maximum diameter of lumen alone as factors contributing to the rupture of an AAA. But other parameters such as patient history and comorbidities and presence of stents or other geometric parameters such as the aneurysm neck angles, tortuosity, and genetics factors [122–126] should be included. Despite the fact that state-of-the-art FEM approaches represent powerful tool for estimation of AAA, their application in clinical practice remains limited due to several reasons. Firstly, every patient has specific and complex anatomy and it is not possible to create a general or parametric human model. Moreover, accuracy of simulations depends on the considered level of details, meaning that increasing required computation time and power will be necessary for obtaining precise results. As a consequence, performing patient-specific simulation may take a few hours (if patient scans are available in the first place). This makes current FEM approaches inadequate for urgent situations such as alerting patient in case of AAA rupture.

In order to avoid described limitations, the patient-specific DSS could be proposed. The main idea is to perform patient-specific forward simulations in advance. AAA of different types (sizes and positions) will be simulated and calculated stress analysis will be used for training of intelligent model. But, in order to perform forward FEM simulations patient geometry is required. For this reason, during the registration into our system, in local workstation (user hospital), patient or his/her medical institution will be asked to provide us patients' medical scans (CT or MRI for example), if there are any. However, it is assumed that some patients will not have medical scans.

For future clinical DSS, we suggest the following scenario: new user has already performed medical scanning (CT or MRI) and these data are uploaded to the central server (or they are available over institution existing connection). In that case, we will generate 3D patient-specific FEM model by using image-based modeling software developed on the server. It is important to mention that, for the project purposes, accurate reconstructions will be necessary only for torso and aorta. We will reconstruct only aorta (by using statistical shape models) and torso interface (by using level set and marching cubes algorithms) because they are needed for setting FEM boundary conditions and because, on the other hand, accurate reconstruction of all organs represents time-consuming and computationally expensive problem with current medical imaging methods. In this way, highly accurate geometry, materials, and boundary conditions could be obtained for every patient (Figure 1.4).

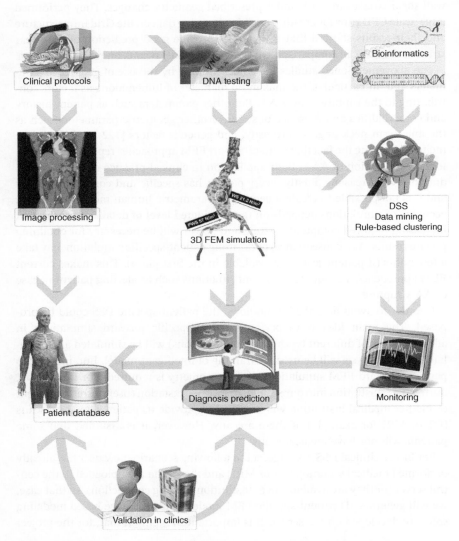

Clinical protocols

DNA testing

Bioinformatics

Image processing

3D FEM simulation

DSS
Data mining
Rule-based clustering

Patient database

Diagnosis prediction

Monitoring

Validation in clinics

Figure 1.4 Description of clinical decision support system for AAA disease.

1.9 Conclusions

In the first part of this paper, literature survey on AAA biomechanics is reported including several aspects from experimental test, constitutive model, and ILT. Further, we presented our FE models, aimed at simulating and enhancing the computational study of the aneurismatic pathology. The combination of fluid and nonlinear structure modeling can give better understanding of the flow, pressure distribution, wall shear stress quantification, and effect of material properties and geometrical parameters. Computational methods have made patient-specific analyses possible, a feature essential for understanding the progression of AAA in a particular patient. Finally, future clinical DSS is suggested by using DM approach. The main aim is to run predictive FSI model in order to estimate the risk of rupture and to use patient-specific wall properties with calcium, tissue disease, and thrombus to overcome multiple level of uncertainties.

During the last two decades, significant efforts have been made in order to define a computational model which includes biomechanical and biological approach, but still a lot of clinical studies are necessary in order to make these computational studies real in everyday clinical practice.

Exercise 1.1   Modeling of Blood Flow Within the AAA

The shape of AAA is very important. The severity of AAA is commonly estimated in clinical practice by considering the AAA maximal diameter. However, from the mechanical point of view, the hemodynamic effects and the mechanical stresses within the AAA tissue certainly are important in the process of the AAA rupture. Bulge diameter alone may not be a sufficient criterion for determination of rupture risk; therefore, an insight into the hemodynamic effects and the stress–strain quantification and distribution within the vessel wall are of great significance even in medical practice.

Generation of the Finite Element Model

A simplified geometry of an aneurism is shown in Figure 1.5. With the on-web software, the 3D finite element model for the blood flow domain can be parametrically generated. A transition smoothness between the surfaces is achieved by using Bezier's curves. Also, the results can be displayed with a user-friendly menu in a way suitable for an insight into medical aspects of the blood flow conditions. A detailed description of the software use is given on the web (within Tutorial of each example), while the description of geometric parameters is given in the caption of Figure 1.5.

(Continued)

Exercise 1.1   (Continued)

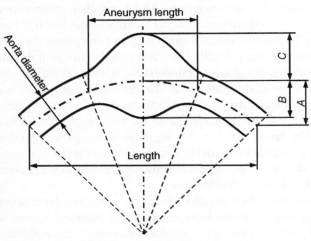

Figure 1.5 Geometrical parameters of AAA: "Length" is the parameter which defines the total horizontal projection of the generated aneurysm model; "*A*" is the height of the arc of central line; "Aorta diameter" is the abdominal aorta diameter; "*B*" is the radius from the central line to the inner wall of the aneurysm; "*C*" is the radius from the central line to the outer wall of the aneurysm; "Aneurysm length" is an average length of the AAA.

Boundary Conditions

At the inflow aorta cross-section, a fully developed parabolic flow is assumed, determined by a selected volume flux. The normal stress and tangential stress are set to be equal to zero (stress-free condition) or they are prescribed at the outlet cross-section.

It is assumed that the entering flow is pulsatile, with a typical waveform shown in Figure 1.6 [127]. As described in the software menu, the waveform can be changed.

Results

Results for two examples of the symmetric AAA are given here: (i) case with rigid walls and (ii) AAA with deformable walls. Results not shown here and solutions for other model parameters can be obtained using Software on the web.

(Continued)

Exercise 1.1   (Continued)

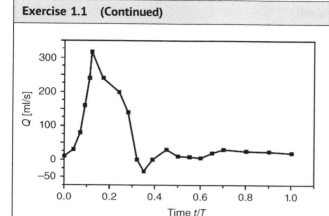

Figure 1.6   A typical in-flow waveform at the aorta entry. Q is the volumetric in-flux and t/T is the relative time with respect to the cycle period T.

Modeling of AAA Assuming Rigid Walls

We analyze an aneurism at the straight aorta domain, where aorta proximal and distal to the AAA bulge is idealized as straight rigid tube and branching arteries are excluded. The model has ratio $D/d = 2.75$ and geometry generated according to Figure 1.5 (D and d are diameters of the bulge and aorta, respectively). The data are: blood density is $\rho = 1.05$ g/cm^3; kinematic viscosity (Newtonian fluid) $\nu = 0.035$ cm^2/s, $d = 12.7$ mm. The inflow velocity is defined by the flux function given in Figure 1.6. The FE mesh consisted of approximately 8000 3D 8-node brick elements.

The results for the velocity and pressure at peak systole $t/T = 0.16$ are shown in Figure 1.7. The velocity disturbance in the region of the aneurism is notable. Also, the region of maximum pressure is located inside AAA.

Modeling AAA with Deformable Walls

Here, an aneurysm of the straight aorta with deformable walls is modeled according to the FSI algorithm. Blood flow is calculated using 2112 eight-node 3D elements, and 264 four-node shell elements used to model the aorta wall, with the wall thickness $\delta = 0.2$ cm. The material constants for blood as in the previous example, while data for the vessel wall are: Young's modulus $E = 2.7$ MPa, Poisson's ratio $\nu = 0.45$, wall thickness $\delta = 0.2$ cm, and tissue density $\rho = 1.1$ g/cm^3. Boundary conditions for the model are prescribed velocity profile (see Figure 1.8a) and output pressure profile as given in Figure 1.8b.

(Continued)

Exercise 1.1 (Continued)

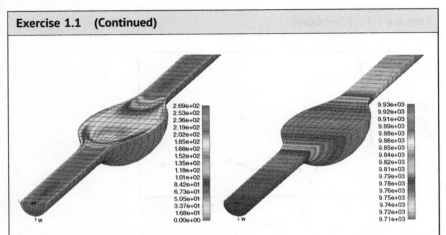

Figure 1.7 Velocity field (left panel) and pressure distribution (right panel) for peak systole $t/T = 0.16$ of AAA for the model with $D/d = 2/75$, $d = 12.7$ mm.

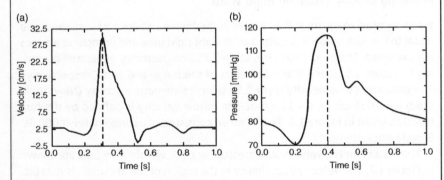

Figure 1.8 Input velocity and output pressure profiles for the AAA on a straight vessel. Inlet peak systolic flow is at $t = 0.305$ s and outlet peak systolic pressure is at $t = 0.4$ s. (a) Velocity waveform; (b) pressure waveform. *Source*: Modified from Scotti et al. [44].

The results for velocity magnitude distribution at $t = 0.305$ s are shown in Figure 1.9a. The von Mises wall stress distributions at $t = 0.4$ s is given in Figure 1.9b. It can be seen that the velocities are low in the domain of the aneurism, while the larger values of the wall stress are at the proximal and distal aneurism zones.

(Continued)

Exercise 1.1   (Continued)

(a)   (b)

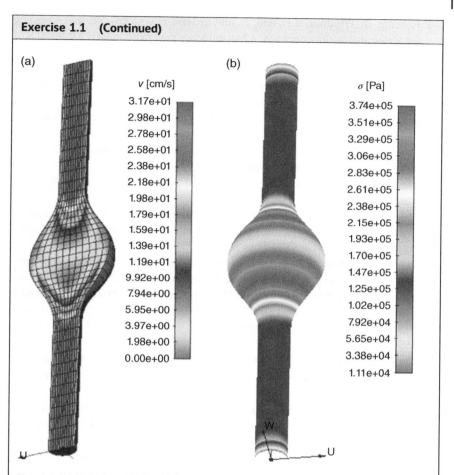

Figure 1.9 Velocity magnitude field and von Mises wall stress distribution for symmetric AAA on the straight vessel. (a) Velocity field distribution for peak at $t = 0.305$ s; (b) von Mises wall stress distributions for blood pressure peak at $t = 0.4$ s.

References

1 McGloughlin, T.M. and Doyle, B.J. (2010). New approaches to abdominal aortic aneurysm rupture risk assessment: engineering insights with clinical gain. *Arterioscler. Thromb. Vasc. Biol.* **30**: 1687–1694. https://doi.org/10.1161/ATVBAHA.110.204529.

2 Fleming, C., Whitlock, E.P., Beil, T., and Lederle, F.A. (2005). Review: screening for abdominal aortic aneurysm: a best-evidence systematic review for the U.S. Preventive Services Task Force. *Ann. Intern. Med.* **142**: 203–211.

3 O'Gara, T.P. (2003). Aortic aneurysm. *Circulation* **107**: e43–e45.

4 Minino, A.M., Heron, M.P., Murphy, S.L., and Kochanek, K.D. (2007). Deaths: final data for 2004. *Nat. Vital. Stat. Rep.* **55**: 1–119.

5 Sakalihasan, N., Limet, R., and Defawe, O.D. (2005). Abdominal aortic aneurysm. *Lancet* **365** (9470): 1577–1589.

6 Thompson, M.M. (2003). Controlling the expansion of abdominal aortic aneurysms. *Br. J. Surg.* **90**: 897–898.

7 Thompson, M.M. (2003). Infrarenal abdominal aortic aneurysms. *Curr. Treat Opt. Cardiovasc. Med.* **5** (2): 137–146.

8 Thompson, R.W. and Baxter, B.T. (1999). MMP inhibition in abdominal aortic aneurysms. Rationale for a prospective randomized clinical trial. *Ann. N. Y. Acad. Sci.* **878**: 159–178.

9 Bown, M.J., Sutton, A.J., Bell, P.R., and Sayers, R.D. (2002). A meta-analysis of 50 years of ruptured abdominal aortic aneurysm repair. *Br. J. Surg.* **89**: 714–730.

10 Venkatasubramaniam, A.K., Fagan, M.J., Mehta, T. et al. (2004). A Comparative Study of Aortic Wall Stress Using Finite Element Analysis for Ruptured and Non-ruptured Abdominal Aortic Aneurysms. *Eur. J. Vasc. Endovasc. Surg.* **28** (2): 168–176.

11 Choke, E., Cockerill, G., Wilson, W.R.W. et al. (2005). A review of biological factors implicated in abdominal aortic aneurysm rupture. *Eur. J. Vasc. Endovasc. Surg.* **30**: 227–244.

12 Fillinger, M.F., Marra, S.P., Raghavan, M.L., and Kennedy, F.E. (2003). Prediction of rupture risk in abdominal aortic aneurysm during observation: wall stress versus diameter. *J. Vasc. Surg.* **37**: 724–732.

13 Fillinger, M.F., Raghavan, M.L., Marra, S.P. et al. (2002). In vivo analysis of mechanical wall stress and abdominal aortic aneurysm rupture risk. *J. Vasc. Surg.* **36**: 589–597.

14 Upchurch, G.R. Jr. and Schaub, T.A. (2006). Abdominal aortic aneurysm. *Am. Fam. Phys.* **73**: 1198–1204.

15 Sonesson, B., Sandgren, T., and Lanne, T. (1999). Abdominal aortic aneurysm wall mechanics and their relation to risk of rupture. *Eur. J. Vasc. Endovasc. Surg.* **18**: 487–493.

16 Doyle, B.J., Callanan, A., Walsh, M.T. et al. (2009). A finite element analysis rupture index (FEARI) as an additional tool for abdominal aortic aneurysm rupture prediction. *Vasc. Dis. Prev.* **6**: 114–121.

17 Doyle, B.J., Cloonan, A.J., Walsh, M.T. et al. (2010). Identification of rupture locations in patient-specific abdominal aortic aneurysms using experimental and computational techniques. *J. Biomech.* **43**: 1408–1416.

18 Doyle, B.J., Corbett, T.J., Callanan, A. et al. (2009). An experimental and numerical comparison of the rupture locations of an abdominal aortic aneurysm. *J. Endovasc. Ther.* **16**: 322–335.

19 Doyle, B.J., Grace, P.A., Kavanagh, E.G. et al. (2009). Improved assessment and treatment of abdominal aortic aneurysms: the use of 3D reconstructions as a surgical guidance tool in endovascular repair. *Ir. J. Med. Sci.* **178**: 321–328.

20 Biasetti, J. and Gasser, T.C. (2012). A fluido-chemical model to predict the growth of intra-luminal thrombus in abdominal aortic aneurysms, ECCOMAS 2012. *6th European Congress on Computational Methods in Applied Sciences and Engineering*, Vienna, Austria (10–14 September 2012).

21 Biasetti, J., Gasser, T.C., Auer, M. et al. (2010). Hemodynamics of the normal aorta compared to fusiform and saccular abdominal aortic aneurysms with emphasis on a potential thrombus formation mechanism. *Ann. Biomed. Eng.* **38** (2): 380–390.

22 Georgakarakos, E., Ioannou, C., Kamarianakis, Y. et al. (2010). The role of geometric parameters in the prediction of abdominal aortic aneurysm wall stress. *Eur. J. Vasc. Endovasc. Surg.* **39**: 42–48.

23 Georgakarakos, E., Ioannou, C.V., Papaharilaou, Y. et al. (2013). Peak wall stress does not necessarily predict the location of rupture in abdominal aortic aneurysms. *Eur. J. Vasc. Endovasc. Surg.* **39** (3): 302–304.

24 Vorp, D.A. (2007). Biomechanics of abdominal aortic aneurysm. *J. Biomech.* **40**: 1887–1902.

25 VASCOPS Vascular Diagnostics (2007). On-line survey: clinical assessment of AAA rupture risk: are biomechanical predictors needed? http://www.vascops.com (accessed 12 September 2021).

26 Alcorn, H.G., Wolfson, S.K. Jr., Sutton-Tyrrell, K. et al. (1996). Risk factors for abdominal aortic aneurysms in older adults enrolled in The Cardiovascular Health Study. *Arterioscler. Thromb. Vasc. Biol.* **16**: 963–970.

27 Derubertis, B.G., Trocciola, S.M., Ryer, E.J. et al. (2007). Abdominal aortic aneurysm in women: prevalence, risk factors, and implications for screening. *J. Vasc. Surg.* **46**: 630–635.

28 Forsdahl, S.H., Singh, K., Solberg, S. et al. (2009). Risk factors for abdominal aortic aneurysms: a 7-year prospective study: the Tromso Study, 1994–2001. *Circulation* **119**: 2202–2208.

29 Svensjo, S., Bjorck, M., and Wanhainen, A. (2013). Current prevalence of abdominal aortic aneurysm in 70-year-old women. *Br. J. Surg.* **100**: 367–372.

30 Lederle, F.A., Johnson, G.R., Wilson, S.E. et al. (2001). Abdominal aortic aneurysm in women. *J. Vasc. Surg.* **34**: 122–126.

31 Singh, K., Bonaa, K.H., Jacobsen, B.K. et al. (2001). Prevalence of and risk factors for abdominal aortic aneurysms in a population-based study: the Tromso Study. *Am. J. Epidemiol.* **154**: 236–244.

32 Kent, K.C., Zwolak, R.M., Egorova, N.N. et al. (2010). Analysis of risk factors for abdominal aortic aneurysm in a cohort of more than 3 million individuals. *J. Vasc. Surg.* **52**: 539–548.

33 Lederle, F.A., Johnson, G.R., Wilson, S.E. et al. (2000). The aneurysm detection and management study screening program: validation cohort and final results. Aneurysm Detection and Management Veterans Affairs Cooperative Study Investigators. *Arch. Intern. Med.* **160**: 1425–1430.

34 Lederle, F.A., Nelson, D.B., and Joseph, A.M. (2003). Smokers' relative risk for aortic aneurysm compared with other smoking-related diseases: a systematic review. *J. Vasc. Surg.* **38**: 329–334.

35 Jahangir, E., Lipworth, L., Edwards, T.L. et al. (2015). Smoking, sex, risk factors and abdominal aortic aneurysms: a prospective study of 18 782 persons aged above 65 years in the Southern Community Cohort Study. *J. Epidemiol. Community Health* **69** (5): 481–488.

36 Darling, R.C., Messina, C.R., Brewster, D.C., and Ottinger, L.W. (1977). Autopsy study of unoperated abdominal aortic aneurysms. The case for early resection. *Circulation* **56** (Suppl. 3): 161–164.

37 Stringfellow, M.M., Lawrence, P.F., and Stringfellow, R.G. (1987). The influence of aorta-aneurysm geometry upon stress in the aneurysm wall. *J. Surg. Res.* **42**: 425–433.

38 Vorp, D.A., Raghavan, M.L., and Webster, M. (1998). Mechanical wall stress in abdominal aortic aneurysm: inuence of diameter and asymmetry. *J. Vasc. Surg.* **27**: 632–639.

39 Raghavan, M.L. and Vorp, D.A. (2000). Toward a biomechanical tool to evaluate rupture potential of abdominal aortic aneurysm: identification of a finite strain constitutive model and evaluation of its applicability. *J. Biomech.* **33** (4): 475–482.

40 Raghavan, M.L., Kratzberg, J., Castro de Tolosa, E.M. et al. (2006). Regional distribution of wall thickness and failure properties of human abdominal aortic aneurysm. *J. Biomech.* **39**: 3010–3016.

41 Raghavan, M.L., Vorp, D.A., Federle, M.P. et al. (2000). Wall stress distribution on three-dimensionally reconstructed models of human abdominal aortic aneurysm. *J. Vasc. Surg.* **31**: 760–769.

42 Vande Geest, J.P., Sacks, M.S., and Vorp, D.A. (2006). A planar biaxial constitutive relation for the luminal layer of intra-luminal thrombus in abdominal aortic aneurysms. *J. Biomech.* **39**: 2347–2354.

43 Scotti, C.M., Jimenez, J., Muluk, S.C., and Finol, E.A. (2008). Wall stress and flow dynamics in abdominal aortic aneurysms: finite element analysis vs. fluid–structure interaction. *Comput. Methods Biomech. Biomed. Eng.* **11** (3): 301–322.

44 Scotti, C.M., Shkolnik, A.D., Muluk, S., and Finol, E.A. (2005). Fluid–structure interaction in abdominal aortic aneurysms: effects of asymmetry and wall thickness. *Biomed. Eng. Online* **4** (4): 64.

45 Finol, E.A. and Amon, C.H. (2001). Blood flow in abdominal aortic aneurysms: pulsatile flow hemodynamics. *J. Biomech. Eng.* **123**: 474–484.

46 Finol, E.A. and Amon, C.H. (2002). Flow-induced wall shear stress in abdominal aortic aneurysms: part I – steady flow hemodynamics. *Comput. Methods Biomech. Biomed. Eng.* **5** (4): 309–318.

47 Federico, S. and Gasser, T.C. (2010). Nonlinear elasticity of biological tissues with statistical fiber orientation. *J. R. Soc. Interf.* **7** (47): 955–966.

48 Hardin, R.H. and Sloane, N.J.A. (1996). McLaren's improved snub cube and other new spherical designs in three dimentions. *Discret. Comput. Geom.* **15**: 429–441.

49 Zhang, Y., Barocas, V.H., Berceli, S.A. et al. (2016). Multi-scale modeling of the cardiovascular system: disease development, progression, and clinical intervention. *Ann. Biomed. Eng.* **44** (9): 2642–2660.

50 Guidoboni, G., Glowinski, R., Cavallini, N. et al. (2009). A kinematically coupled time-splitting scheme for fluid–structure interaction in blood flow. *Appl. Math. Lett.* **22** (5): 684–688.

51 Guidoboni, G., Glowinski, R., Cavallini, N., and Čanić, S. (2009). Stable loosely-coupled-type algorithm for fluid–structure interaction in blood flow. *J. Comput. Phys.* **228** (18): 6916–6937.

52 Bukač, M., Čanić, S., Glowinski, R. et al. (2013). Fluid–structure interaction in blood flow capturing non-zero longitudinal structure displacement. *J. Comput. Phys.* **235**: 515–541.

53 Quarteroni, A. and Formaggia, L. (2004). Mathematical modelling and numerical simulation of the cardiovascular system in modelling of living systems. In: *12 of Handbook of Numerical Analysis*, 3–127. Amsterdam: North-Holland.

54 Formaggia, L., Gerbeau, J.F., Nobile, F., and Quarteroni, A. (2001). On the coupling of 3D and 1D Navier–Stokes equations for flow problems in compliant vessels. *Comput. Methods Appl. Mech. Eng.* **191** (6–7): 561–582.

55 Formaggia, L., Lamponi, D., and Quarteroni, A. One-dimensional models for blood flow in arteries. *J. Eng. Math.* **47** (3–4): 251–276.

56 Nobile, F. and Vergara, C. (2008). An effective fluid–structure interaction formulation for vascular dynamics by generalized Robin conditions. *SIAM J. Sci. Comput.* **30** (2): 731–763.

57 Causin, P., Gerbeau, J.F., and Nobile, F. (2005). Added-mass effect in the design of partitioned algorithms for fluid–structure problems. *Comput. Methods Appl. Mech. Eng.* **194** (42–44): 4506–4527.

58 Čanić, S., Mikelić, A., and Tambača, J. (2005). A two-dimensional effective model describing fluid–structure interaction in blood flow: analysis, simulation and experimental validation. *Comptes Rendus.* **333** (12): 867–883.

59 Čanić, S., Tambača, J., Guidoboni, G. et al. (2006). Modeling viscoelastic behavior of arterial walls and their interaction with pulsatile blood flow. *SIAM J. Appl. Math.* **67** (1): 164–193.

60 Čanić, S., Hartley, C.J., Rosenstrauch, D. et al. (2006). Blood flow in compliant arteries: an effective viscoelastic reduced model, numerics, and experimental validation. *Ann. Biomed. Eng.* **34** (4): 575–575.

61 Heil, A., Hazel, L., and Boyle, J. (2008). Solvers for large-displacement fluid–structure interaction problems: segregated versus monolithic approaches. *Comput. Mech.* **43** (1): 91–101.

62 MacSweeney, S.T.R., Powell, J.T., and Greenhalgh, R.M. (1994). Pathogenesis of abdominal aortic aneurysm. *Br. J. Surg.* **81**: 935–941.

63 van't Veer, M., Buth, J., Merkx, M. et al. (2008). Biomechanical properties of abdominal aortic aneurysms assessed by simultaneously measured pressure and volume changes in humans. *J. Vasc. Surg.* **48** (6): 1401–1407.

64 Ganten, M.K., Krautter, U., von Tengg-Kobligk, H. et al. (2008). Quantification of aortic distensibility in abdominal aortic aneurysm using ecg-gated multi-detector computed tomography. *Vasc. Intervent.* **18** (5): 966–973.

65 Molacek, J., Baxa, J., Houdek, K. et al. (2011). Assessment of abdominal aortic aneurysm wall distensibility with electrocardiography-gated computed tomography. *Ann. Vasc. Surg.* **25** (8): 1036–1042.

66 Di Puccio, F., Celi, S., and Forte, P. (2012). Review of experimental investigations on compressibility of arteries and the introduction of a new apparatus. *Exp. Mech.* **52** (7): 1–8. https://doi.org/10.1007/s11340-012-9614-4.

67 Humphrey, J.D. and Yin, F.C. (1987). A new constitutive formulation for characterizing the mechanical behavior of soft tissues. *Biophys. J.* **52** (4): 563–570.

68 Ogden, R.W. (2009). Anisotropy and nonlinear elasticity in arterial wall mechanics. In: *Biomechanical Modelling at the Molecular, Cellular and Tissue Levels. CISM Courses and Lectures*, vol. **508** (eds. G.A. Holzapfel, R.W. Ogden, F. Pfeiffer, et al.), 179–258. Vienna: Springer.

69 Vande Geest, J.P., Sacks, M.S., and Vorp, D.A. (2004). Age dependency of the biaxial biomechanical behavior of human abdominal aorta. *J. Biomech. Eng.* **12**: 815–822.

70 Vande Geest, J.P., Sacks, M.S., and Vorp, D.A. (2006). The effects of aneurysm on the biaxial mechanical behavior of human abdominal aorta. *J. Biomech.* **39**: 1324–1334.

71 Koncar, I., Nikolic, D., Pantovic, S. et al. (2013). Modeling of abdominal aortic aneurysm rupture by using bubble inflation test. *Bioinform. Bioeng. (BIBE)* https://doi.org/10.1109/BIBE.2013.6701612.

72 Vande Geest, J.P., Di Martino, E.S., Bohra, A. et al. (2006). A biomechanics-based rupture potential index for abdominal aortic aneurysm risk assessment. *Ann. N. Y. Acad. Sci.* **1085**: 11–21.

73 Thubrikar, M.J., Labrosse, M., Robicsek, F. et al. (2001). Mechanical properties of abdominal aortic aneurysm wall. *J. Med. Eng. Techn.* **25** (4): 133–142.

74 Stamatopoulos, C., Mathioulakis, D.S., Papaharilaou, Y., and Katsamouris, A. (2011). Experimental unsteady flow study in a patientspecific abdominal aortic aneurysm model. *Exp. Fluids* **50** (6): 1695–1709.

75 Holzapfel, G.A. (2006). Determination of material models for arterial walls from uniaxial extension tests and histological structure. *J. Theor. Biol.* **238** (2): 290–302.

76 Simsek, F.G. and Kwon, Y.W. (2015). Investigation of material modeling in fluid–structure interaction analysis of an idealized three layered abdominal aorta: aneurysm initiation and fully developed aneurysms. *J. Biol. Phys.* **41** (2): 173–201.

77 Taghizadeh, H., Tafazzoli-Shadpour, M., Shadmehr, M., and Fatouraee, N. (2015). Evaluation of biaxial mechanical properties of aortic media based on the lamellar microstructure. *Materials* **8** (1): 302–316.

78 Sokolis, D.P., Kefaloyannis, E.M., Kouloukoussa, M. et al. (2006). A structural basis for the aortic stress–strain relation in uniaxial tension. *J. Biomech.* **39** (9): 1651–1662.

79 Karimi, A., Navidbakhsh, M., Shojaei, A., and Faghihi, S. (2013). Measurement of the uniaxial mechanical properties of healthy and atherosclerotic human coronary arteries. *Mater. Sci. Eng. C* **33** (5): 2550–2554.

80 Taylor, C.A. and Humphrey, J.D. (2009). Open problems in computational vascular biomechanics: hemodynamics and arterial wall mechanics. *Comput. Methods Appl. Mech. Eng.* **198** (45–46): 3514–3523.

81 Raut, S.S., Chandra, S., Shum, J., and Finol, E.A. (2013). The role of geometric and biomechanical factors in abdominal aortic aneurysm rupture risk assessment. *Ann. Biomed. Eng.* **41** (7): 1459–1477.

82 Stenbaek, J., Kalin, B., and Swedenborg, J. (2000). Growth of thrombus may be a better predictor of rupture than diameter in patients with abdominal aortic aneurysms. *Eur. J. Vasc. Endovasc. Surg.* **20** (5): 466–469.

83 Li, Z.-Y., U-King-Im, J., Tang, T.Y. et al. (2008). Impact of calcification and intraluminal thrombus on the computed wall stresses of abdominal aortic aneurysm. *J. Vasc. Surg.* **47** (5): 928–936.

84 Di Martino, E.S. and Vorp, D.A. (2003). Effect of variation in intraluminal thrombus constitutive properties on abdominal aortic aneurysm wall stress. *Ann. Biomed. Eng.* **31** (7): 804–809.

85 O'Leary, S.A., Kavanagh, E.G., Grace, P.A. et al. (2014). The biaxial mechanical behaviour of abdominal aortic aneurysm intraluminal thrombus: classification of morphology and the determination of layer and region specific properties. *J. Biomech.* **47** (6): 1430–1437.

86 Tong, J., Schriefl, A.J., Cohnert, T., and Holzapfel, G.A. (2013). Gender differences in biomechanical properties, thrombus age, mass fraction and clinical factors of abdominal aortic aneurysms. *Eur. J. Vasc. Endovasc. Surg.* **45** (4): 364–372.

87 Speelman, L., Bosboom, E.M.H., Schurink, G.W.H. et al. (2008). Patient-specific AAA wall stress analysis: 99-percentile versus peak stress. *Eur. J. Vasc. Endovasc. Surg.* **36**: 668–676.

88 Speelman, L., Bosboom, E.M.H., Schurink, G.W.H., Jacobs, M.J.H.M., and van de Vosse, F.N. (2008). *AAA Growth Predicted with Wall Stress. Poster Session Presented at Conference.* Mate Poster Award 2008: 13th Annual Poster Contest.

89 Speelman, L., Hellenthal, F.A., Pulinx, B. et al. (2010). The influence of wall stress on AAA growth and biomarkers. *Eur. J. Vasc. Surg.* **39**: 410–416.

90 Kontopodis, N., Metaxa, E., Papaharilaou, Y. et al. (2013). Changes in geometric configuration and biomechanical parameters of a rapidly growing abdominal aortic aneurysm may provide insight in aneurysms natural history and rupture risk. *Theor. Biol. Med. Model.* **10**: 67.

91 Yushkevich, P.A., Piven, J., Hazlett, H.C. et al. (2006). User-guided 3D active contour segmentation of anatomical structures: significantly improved efficiency and reliability. *NeuroImage* **31**: 1116–1128.

92 Anton, R., Chen, C.Y., Hung, M.Y. et al. (2015). Experimental and computational investigation of the patient-specific abdominal aortic aneurysm pressure field. *Comput. Methods Biomech. Biomed. Eng.* **18** (9): 981–992. https://doi.org/10.1080/10255842.2013.865024.

93 Frauenfelder, T., Lotfey, M., Boehm, T., and Wildermuth, S. (2006). Computational fluid dynamics: hemodynamic changes in abdominal aortic aneurysm after stent-graft implantation. *Cardiovasc. Intervent. Radiol.* **29**: 613–623.

94 Peattie, R.A., Riehle, T.J., and Bluth, E.I. (2004). Pulsatile flow in fusiform models of abdominal aortic aneurysms: flow fields, velocity patterns and flow-induced wall stresses. *J. Biomech. Eng.* **126**: 438–446.

95 Dorfmann, A., Wilson, C., Edgar, E.S., and Peattie, R.A. (2010). Evaluating patient-specific abdominal aortic aneurysm wall stress based on flow-induced loading. *Biomech. Model. Mechanobiol.* **9**: 127–139.

96 Polzer, S., Gasser, T.C., Markert, B. et al. (2012). Impact of poroelasticity of intraluminal thrombus on wall stress of abdominal aortic aneurysms. *Biomed. Eng. Online* **11**: 62.

97 Gasser, T.C., Gorgulu, G., Folkesson, M., and Swedenborg, J. (2008). Failure properties of intraluminal thrombus in abdominal aortic aneurysm under static and pulsating mechanical loads. *J. Vasc. Surg.* **48**: 179–188.

98 Gasser, T.C., Auer, M., Labruto, F. et al. (2010). Biomechanical rupture risk assessment of abdominal aortic aneurysms: model complexity versus predictability of finite element simulations. *Eur. J. Vasc. Endovasc. Surg.* **40**: 176–185.

99 Wang, D.H.J., Makaroun, M.S., Webster, M.W., and Vorp, D.A. (2002). Effect of intraluminal thrombus on wall stress in patient specific models of abdominal aortic aneurysm. *J. Vasc. Surg.* **I36**: 598–604.

100 Wang, D.H., Makaroun, M.S., Webster, M.W., and Vorp, D.A. (2001). Mechanical properties and microstructure of intraluminal thrombus from abdominal aortic aneurysm. *J. Biomech. Eng.* **123**: 536–539.

101 Ayyalasomayajula, A., Vande Geest, J.P., and Simon, B.R. (2010). Porohyperelastic finite element modeling of abdominal aortic aneurysms. *J. Biomech. Eng.* **132**: 104502.

102 Baek, S., Zambrano, B.A., Choi, J., and Lim, C.-Y. (2014). Growth prediction of abdominal aortic aneurysms and its association of intraluminal thrombus. *11th World Congress on Computational Mechanics (WCCM XI); 5th European Conference on Computational Mechanics (ECCM V); 6th European Conference on Computational Fluid Dynamics (ECFD VI)*, Barcelona, Spain (20–25 July 2014).

103 Zeinali-Davarani, S. and Baek, S. (2012). Medical image-based simulation of abdominal aortic aneurysm growth. *Mech. Res. Commun.* **42**: 107–117.

104 Vorp, D.A., Lee, P.C., Wang, D.H. et al. (2001). Association of intraluminal thrombus in abdominal aortic aneurysm with local hypoxia and wall weakening. *J. Vasc. Surg.* **34**: 291–299.

105 Biasetti, J. (2013). Physics of blood flow in arteries and its relation to intra-luminal thrombus and atherosclerosis. Doctoral dissertation no. 84. KTH School of Engineering Sciences, Department of Solid Mechanics – vascuMECH, KTH Royal Institute of Technology, SE-100 44 Stockholm, Sweden.

106 Jones, K.C. and Mann, K.G. (1994). A model for the tissue factor pathway to thrombin.II. A mathematical simulation. *J. Biol. Chem.* **269**: 23367–23373.

107 Filipovic, N., Milasinovic, D., Zdravkovic, N. et al. (2011). Impact of aortic repair based on flow field computer simulation within the thoracic aorta. *Comput. Methods Prog. Biomed.* **101** (3): 243–252.

108 Filipovic, N., Mijailovic, S., Tsuda, A., and Kojic, M. (2006). An implicit algorithm within the arbitrary Lagrangian–Eulerian formulation for solving incompressible fluid flow with large boundary motions. *Comp. Meth. Appl. Mech. Engrg.* **195**: 6347–6361.

109 Filipovic, N., Kojic, M., Ivanovic, M. et al. (2006). *MedCFD, Specialized CFD Software for Simulation of Blood Flow Through Arteries*. Serbia: University of Kragujevac.

110 Perktold, K. and Rappitsch, G. (1995). Computer simulation of local blood flow and vessel mechanics in a compliant carotid artery bifurcation model. *J. Biomech.* **28**: 845–856.

111 Figueroa, C.A., Taylor, C.A., Yeh, V. et al. (2009). Effect of curvature on displacement forces acting on aortic endografts: a 3-dimensional computational analysis. *J. Endovasc. Ther.* **16**: 284–294.

112 Figueroa, C.A., Taylor, C.A., Chiou, A.J. et al. (2009). Magnitude and direction of pulsatile displacement forces acting on thoracic aortic endografts. *J. Endovasc. Ther.* **16**: 350–358.

113 Filipovic, N., Rosic, M., Tanaskovic, I. et al. (2012). ARTreat project: Three-dimensional numerical simulation of plaque formation and development in the arteries. *IEEE Trans. Inf. Technol. Biomed.* **16** (2): 272–278.

114 Filipovic, N. and Schima, H. (2011). Numerical simulation of the flow field within the aortic arch during cardiac assist. *Artif. Organs* **35** (4): 73–83.

115 Veljkovic, D., Filipovic, N., and Kojic, M. (2012). The effect of asymmetry and axial prestraining on the amplitude of mechanical stresses in abdominal aortic aneurysm. *J. Mech. Med. Biol.* **12** (5): 1250089.

116 Krsmanovic, D., Koncar, I., Petrovic, D. et al. (2012). Computer modelling of maximal displacement forces in endoluminal thoracic aortic stent graft. *Comput. Methods Biomech. Biomed. Eng.* **17** (9): 1012–1020.

117 Hastie, T., Tibshirani, R., and Friedman, J. (2008). *The Elements of Statistical Learning: Data Mining, Inference, and Prediction*, Springer Series in Statistics, 2e. New York, NY, USA: Springer.

118 Kolachalama, V.B., Bressloff, N.W., and Nair, P.B. (2007). Mining data from hemodynamic simulations via Bayesian emulation. *Biomed. Eng. Online* **6**: 47.

119 Martufi, G., DiMartino, E.S., Amon, C.H. et al. (2009). Three-dimensional geometrical characterization of abdominal aortic aneurysms: image-based wall thickness distribution. *J. Biomech. Eng.* **131** (6): 061015.

120 Shum, J., Martufi, G., di Martino, E. et al. (2011). Quantitative assessment of abdominal aortic aneurysm geometry. *Ann. Biomed. Eng.* **39** (1): 277–286.

121 Filipovic, N., Ivanovic, M., Krstajic, D., and Kojic, M. (2011). Hemodynamic flow modeling through an abdominal aorta aneurysm using data mining tools. *IEEE Trans. Inf. Technol. Biomed.* **15** (2): 189–194.

122 Pannu, H., Fadulu, V.T., Chang, J. et al. (2005). Mutations in transforming growth factor-beta receptor type II cause familial thoracic aortic aneurysms and dissections. *Circulation* **112**: 513–520.

123 Zhu, L., Vranckx, R., Van Kien, P.K. et al. (2006). Mutations in myosin heavy chain 11 cause a syndrome associating thoracic aortic aneurysm/aortic dissection and patent ductus arteriosus. *Nat. Genet.* **38**: 343–349.

124 Renard, M., Callewaert, B., Baetens, M. et al. (2013). Novel MYH11 and ACTA2 mutations reveal a role for enhanced TGFß signaling in FTAAD. *Int. J. Cardiol.* **165** (2): 314–321.

125 Guo, D.C., Pannu, H., Tran-Fadulu, V. et al. (2007). Mutations in smooth muscle alpha-actin (ACTA2) lead to thoracic aortic aneurysms and dissections. *Nat. Genet.* **39**: 1488–1493.

126 van de Laar, I.M.B.H., Oldenburg, R.A., Pals, G. et al. (2011). Mutations in SMAD3 cause a syndrome form of aortic aneurysms and dissections with early-onset osteoarthritis. *Nat. Genet.* **43**: 121–126.

127 Ku, D.N. (1997). Blood flow in arteries. *Annu. Rev. Fluid Mech.* **29**: 399–434.

2

Modeling the Motion of Rigid and Deformable Objects in Fluid Flow

Tijana Djukic and Nenad D. Filipovic

Bioengineering Research and Development Center, BIOIRC, Kragujevac, Serbia

2.1 Introduction

Fluid flow can be simulated using a wide range of methods. There are standard methods used in computational fluid dynamics (CFD), such as finite element (FE), finite difference, finite volume method, where the Navier–Stokes equations are numerically solved in order to obtain the velocity and pressure field over the entire domain. Besides these standard methods that are based on equations derived for continuum domain, there are also discrete methods, such as lattice Boltzmann (LB) method, dissipative particle dynamics (DPD), and smoothed-particle hydrodynamics (SPH). In this Chapter, the LB method was used for the simulation of fluid flow. This method observes the fluid as a set of fictitious particles and by studying the dynamics of motion of these particles, the fluid flow is modeled on the macroscopic level. The greatest advantages of the LB method are the simplicity of implementation and the possibility of parallelization. The LB method has been successfully applied in simulating blood flow, including simulation of cancer invasion and tumor cell migration [1], motion of leukocytes under shear-thinning blood flow [2], analysis of leukocytes–erythrocytes interaction [3–5] as well as in many other areas in biomedicine.

Accurate simulation of fluid flow, with solid objects (whether rigid or deformable) moving through the fluid domain, represents nowadays one of the greatest challenges in CFD. There are several approaches in modeling the motion of solid bodies through a fluid domain, including the usage of the appropriate coordinate transformation [6], the usage of a moving FE mesh [7], or the application of so-called arbitrary Lagrangian Eulerian (ALE) formulation [8]. Each of these approaches has its advantages and disadvantages. There is another approach in

Computational Modeling and Simulation Examples in Bioengineering, First Edition. Edited by Nenad D. Filipovic. © 2022 The Institute of Electrical and Electronics Engineers, Inc. Published 2022 by John Wiley & Sons, Inc.

modeling solid–fluid interaction that has been proposed in literature and that is applied in this chapter – the immersed boundary method (IBM).

Combination of the LB method and IBM approach has proved to be an efficient way to simulate solid–fluid interaction. IBM was first developed by Peskin [9] and this method treats the immersed solid object as a part of the fluid that is separated by a deformable membrane. The fluid domain is represented with a fixed mesh and a set of Lagrange points are used to model the boundary of the solid domain. Many papers can be found in literature that use the IBM approach to model the solid–fluid interaction and to simulate motion of both in case of rigid bodies [10–12] and in case of deformable bodies immersed in fluid [13, 14].

Numerical simulations that predict the motion and behavior of rigid and deformable objects in fluid flow can be very useful for the understanding of numerous phenomena in biomedicine, including the delivery of intravenous drugs, stem cells and genetic material to target organs, the circulation of blood constituents through blood vessels, most importantly through stenotic arteries, the circulation of cancer cells through the bloodstream, etc.

Modern drug delivery systems that are used in cardiovascular and oncological treatments consist of nanoparticles that can be loaded with contrast agents or drug molecules. These particles are small and can be transported through the blood circulation to specific organs and tissues, where they should release their load, most commonly drug molecules. In this type of drug delivery systems, it is very important to control the size, shape, and density of these particles [15]. Numerical modeling of motion of these particles provides a useful tool to analyze the influence of various sizes, shapes, and other parameters during the design of nanoparticles.

Human blood is a complex fluid that consists of cells (formed cellular elements) and blood plasma. There are several types of blood cells, and those are erythrocytes, leukocytes, and platelets. The erythrocytes (red blood cells - RBCs) constitute around 40–45% of overall blood volume and they have a great impact on the characteristics of blood flow at the microscale level. Their role in human organism is to take the oxygen during the passage of blood through the lungs, to transport the oxygen to all other parts of the body, and to enable the exchange of oxygen and carbon dioxide in the surrounding tissue through the capillary blood vessels. Due to the important role of these cells in human organism, the analysis of dynamics of motion of RBCs separately is one of the most important problems in physiology and biomechanics. Also, RBCs are highly deformable cells and hence they represent a very demanding challenge when it comes to the simulation of deformable bodies in fluid flow.

In the field of numerical modeling of behavior of RBCs in fluid domain, several approaches are presented in literature. Pozrikidis [16] applied the boundary element method (BEM) to model the motion of RBCs. Ramanujan and Pozrikidis [17] extended this approach and considered various material models to define the relationship between force and deformation of particle. To model the influence of solid on fluid and vice versa, Eggleton and Popel [18] used the IBM. Sui et al. [19] improved this approach by adding the refinement of the mesh, in order to be able to model the motion more precisely with higher mesh density near the particle. Discrete particle methods have also been applied to model this type of phenomena. Boryczko et al. [20] modeled the solid using classic continuum mechanics, while the fluid was modeled as a cluster of particles.

In this chapter, three numerical methods are used (the first one to model the fluid flow, the second one to model the behavior of particles, and the third one to model the interaction between these two entities) and these methods are linked together into a single numerical model. Numerical simulations using the presented numerical model have been successfully applied in many areas of biomedicine. The simulations involving rigid particles are applied to model the motion of nanoparticles through the bloodstream [21], low-density lipoprotein (LDL) particles through the bloodstream [22], otoconia particles in the semicircular canals of the inner ear [23, 24], etc. The simulations involving deformable particles are applied to model the motion of RBCs through the blood vessels of a fish species called zebrafish [25], circulating tumor cells (CTCs) through the microfluidic chips [26], etc.

The results of several simulations are presented in this chapter. These examples include two rather standard examples with simple simulation setups that are used for testing the accuracy of the numerical model and two examples with more complex geometrical domains that are used to illustrate the motion of particles through geometries similar to arteries with stenosis and bifurcation. All simulations are performed using specifically designed software based on the presented numerical model. In the simulations, the motion of RBCs and spherical particles is considered and the behavior of rigid and deformable objects is compared. The obtained results are also compared with the results published in literature and a good agreement of results is obtained that confirms that the presented numerical model can be successfully used to simulate the motion of both rigid and deformable objects in fluid flow.

The chapter is organized as follows: the numerical model is presented in Section 2.2. Section 2.3 contains the results of the numerical simulations. This includes four mentioned examples, with comparisons of results with other experimental and numerical results presented in literature. A short conclusion is given in Section 2.4.

2.2 Numerical Model

In this section, the numerical model is explained in detail. As it was already mentioned, three numerical methods are used and they are linked together into a single numerical model. The LB method [27] was used for fluid flow simulation. The standard equations of motion are used for the modeling of the behavior of a rigid particle. Various types of deformations were considered for the modeling of the behavior of a deformable particle. Discrete formulae are used for determining the reaction force caused by the change of volume, surface area, and bending of the membrane of the deformable particle, according to the approach proposed in literature [28]. Finite element method (FEM) was used to model the membrane surface strain [29–31], and a special material model that is proposed by Skalak [32] was implemented to model the membrane of the red blood cell (RBC). The IBM approach, which was first presented by Peskin [9], was used to model the solid–fluid interaction. Also, a specific approach is used for modeling fluid flow when different values of viscosity of the internal and external fluid are defined.

2.2.1 Modeling Blood Flow

In this chapter, blood flow is simulated using the LB method. The LB method is a discrete method that belongs to the class of problems that are called cellular automata (CA). In this type of simulation methods, the physical system is observed in an idealized manner, such that space and time are discretized and the whole domain is made of a large number of identical cells [33]. In case of the LB method, fluid is observed as a set of particles. For the type of problems that is considered in this chapter, it is important to analyze the changes in the domain at the macroscale level, while it is not significant to analyze each particle separately. This is why the LB method can be applied for this type of simulations. The basic quantity that is introduced in the LB method is called the distribution function f and it is used to describe the behavior of particles in space. Distribution function is defined such that $f(\mathbf{x}, \xi, t)$ represents the probability of a particle to be located within $dxd\xi$ around point $(\mathbf{x}, \xi)$, in moment in time t, where $\mathbf{x}$ and ξ denote the particle position vector and the particle velocity vector, respectively. The basic equation of the LB method that is used to derive all other equations is the Boltzmann equation. This is a partial differential equation that is valid for continuum. It actually represents the equation of balance of the distribution function. If external force is taken into consideration (represented by a force vector $\mathbf{g}$), the Boltzmann equation is given by:

$$\frac{\partial f}{\partial t} + \xi \cdot \frac{\partial f}{\partial \mathbf{x}} + \frac{\mathbf{g}}{m} \cdot \frac{\partial f}{\partial \xi} = \Omega \tag{2.1}$$

In the above equation, Ω represents the collision operator that is used to introduce the change of distribution function due to particle collisions. This operator Ω can be defined in various ways, including simplified and complex functions. One simplified model is proposed by Bhatnagar, Gross, and Krook [34], where it is assumed that the effect of particle collisions is to bring the fluid "closer" to the equilibrium state. This model is known as the single relaxation time (SRT) approximation or the Bhatnagar–Gross–Krook (BGK) model. Operator Ω is in this case defined as:

$$\Omega = -\frac{1}{\tau}\left(f - f^{(0)}\right) \tag{2.2}$$

where τ represents the average time between two collisions, which is also called the relaxation time in the LB method, and $f^{(0)}$ is the equilibrium distribution function, the so-called Maxwell–Boltzmann distribution function [35, 36]. This probability distribution function is taken from statistical mechanics, and the final expression for the mentioned equilibrium distribution function is given by:

$$f^{(0)}(\mathbf{x}, \boldsymbol{\xi}, t) = \frac{\rho(\mathbf{x}, t)}{(2\pi\theta(\mathbf{x}, t))^{D/2}} \exp\left(-\frac{(\mathbf{u}(\mathbf{x}, t) - \boldsymbol{\xi})^2}{2\theta(\mathbf{x}, t)}\right) \tag{2.3}$$

In the above equation, D is the number of physical dimensions ($D = 2$ for two-dimensional space, $D = 3$ for three-dimensional space), the constant θ is equal to $\theta = k_B T/m$, where $k_B = 1,38 \cdot 10^{-23} \frac{J}{K}$ is the Boltzmann constant, T is the absolute temperature in Kelvins (K), and m is the particle mass (in the LB method, it is considered that $m = 1$).

If Eqs. (2.2) and (2.3) are replaced back into Eq. (2.1), the BGK model of the Boltzmann equation is obtained:

$$\frac{\partial f}{\partial t} + \boldsymbol{\xi} \cdot \frac{\partial f}{\partial \mathbf{x}} + \mathbf{g} \cdot \frac{\partial f}{\partial \boldsymbol{\xi}} = -\frac{1}{\tau}\left(f - f^{(0)}\right) \tag{2.4}$$

BGK Boltzmann equation defined in (2.4) is valid for continuum and it is related to a continuous velocity field. In order to simulate this equation, it is necessary to perform appropriate discretization. In the LB method, discretization is performed in two steps. In the first step, the velocity space is discretized and then in the second step, the space and time are discretized. During this process, certain procedures have to be applied, in order to ensure that the system of equations remains valid for fluid dynamics and can be used to simulate fluid flow. In other words, all the assumptions introduced have to ensure that the final equations can be transformed into traditional equations used for CFD (Navier–Stokes equations and the continuity equation). Details of the transformation procedure as well as the discretization procedure will be omitted here, but are explained in detail in literature [37–41].

During the discretization of the velocity space, the Gauss–Hermite quadrature [42, 43] is used to calculate all the integrals in equations. This procedure tends to calculate the value of integral weighted sums and as accurately as possible, by defining the optimal set of abscissae $\boldsymbol{\xi}_i$, $i = 1, 2, ..., q$ and weight coefficients ω_i, $i = 1, 2, ..., q$. The abscissae are practically used to determine the directions along which the fictional particles are moving in the observed domain. The parameters $\boldsymbol{\xi}_i$

and ω_i are related to the lattice structure that is used in numerical simulations. Lattice structure denoted by $DdQq$ contains parameters that define q velocity directions in d-dimensional space.

After mentioning discretization steps, the equation that represents the entire numerical scheme of the LB method is given by:

$$\bar{f}_i(\mathbf{x} + \bar{\xi}_i, t + 1) - \bar{f}_i(\mathbf{x}, t) = -\frac{1}{\bar{\tau}}\left(\bar{f}_i(\mathbf{x}, t) - f_i^{(0)}(\rho, \mathbf{u})\right) + \left(1 - \frac{1}{2\bar{\tau}}\right)F_i(\rho, \mathbf{u})$$

(2.5)

In the numerical implementations of the LB method, Eq. (2.5) is most commonly separated into two steps – collision step and propagation step.

Collision step is described as follows:

$$\bar{f}_i^{out}(\mathbf{x}, t) = \bar{f}_i^{in}(\mathbf{x}, t) - \frac{1}{\bar{\tau}}\left(\bar{f}_i^{in}(\mathbf{x}, t) - f_i^{(0)}(\rho, \mathbf{u})\right) + \left(1 - \frac{1}{2\bar{\tau}}\right)F_i$$

(2.6)

Propagation step is described as follows:

$$\bar{f}_i^{in}(\mathbf{x} + \bar{\xi}_i, t + 1) = \bar{f}_i^{out}(\mathbf{x}, t)$$

(2.7)

In Eqs. (2.6) and (2.7), two values of distribution function are introduced - $\bar{f}_i^{in}$ and $\bar{f}_i^{out}$ that represent values of the discretized distribution function before and after collision. In Eq. (2.6), F_i denotes the external force term that is given by:

$$\mathbf{F}_i = \omega_i \rho \left(\frac{\bar{\xi}_i - \mathbf{u}}{c_s^2} + \frac{(\bar{\xi}_i \cdot \mathbf{u})\bar{\xi}_i}{c_s^4}\right) \cdot \mathbf{g}(\mathbf{x}, t)$$

(2.8)

Also, in Eqs. (2.6) and (2.7), the quantities $\bar{\tau}$ and $\bar{\xi}_i$ are modified version of already defined quantities that are modified during the space and time discretization.

Modified relaxation time $\bar{\tau}$ is given by:

$$\bar{\tau} = \tau + \frac{1}{2}$$

(2.9)

Modified abscissae $\bar{\xi}_i$ are given by:

$$\bar{\xi}_i = c_s \xi_i$$

(2.10)

where c_s is the constant named speed of sound in lattice units. This is actually a scale factor that depends on the chosen lattice structure. For the lattice structures used in this chapter, the value of this constant is equal to:

$$c_s^2 = \frac{1}{3}$$

(2.11)

The weight coefficients ω_i and vectors defining the abscissae $\bar{\xi}_i$ for two lattice structures used in this chapter are listed in Table 2.1.

Table 2.1 Lattice structures – weight coefficients and vectors defining abscissae.

Structure	Weight coefficients	Vectors defining abscissae
D2Q9	$\omega_i = \begin{cases} 4/9 & i = 0 \\ 1/9 & i = 1,2,3,4 \\ 1/36 & i = 5,6,7,8 \end{cases}$	$\boldsymbol{\xi}_i = \begin{cases} (0,0) & i = 0 \\ (\pm 1, 0); (0, \pm 1) & i = 1,2,3,4 \\ (\pm 1 \pm 1) & i = 5,6,7,8 \end{cases}$
D3Q27	$\omega_i = \begin{cases} 8/27 & i = 0 \\ 2/27 & i = 1,\dots,6 \\ 1/216 & i = 7,\dots,14 \\ 1/54 & i = 15,\dots,26 \end{cases}$	$\boldsymbol{\xi}_i = \begin{cases} (0,0,0) & i = 0 \\ (\pm 1, 0, 0); (0, \pm 1, 0); (0, 0, \pm 1) & i = 1,\dots,6 \\ (\pm 1, \pm 1, \pm 1) & i = 7,\dots,14 \\ (\pm 1, \pm 1, 0); (0, \pm 1, \pm 1); (\pm 1, 0, \pm 1) & i = 15,\dots,26 \end{cases}$

In numerical simulation, the domain of interest is divided in a lattice mesh, containing a defined number of nodes in two or three dimensions. During the simulation, the collision and propagation steps described in Eqs. (2.6) and (2.7) are repeated in a series of iterations, until the defined criterion for the simulation end is reached. In these steps, the values of components of distribution function are calculated. These calculated values are then used to calculate the macroscopic quantities, such as pressure and velocity.

The expressions defining these quantities are already listed in this chapter, but in the sequel, final expressions that are used in simulations will be listed. Due to the introduction of modified relaxation time (which is defined in Eq. (2.9)), it is necessary to modify the expressions for calculation of macroscopic quantities, in order to ensure the consistency of the model.

Pressure in a particular node of the lattice mesh is calculated as:

$$p = c_s^2 \rho \theta \tag{2.12}$$

where the value of density in the considered node of the lattice mesh is given by:

$$\rho = \sum_i \bar{f}_i \tag{2.13}$$

The value of velocity in a particular node of the lattice mesh is given by:

$$\mathbf{u} = \frac{1}{\rho} \sum_i \bar{\xi}_i \bar{f}_i + \frac{\mathbf{g}}{2} \tag{2.14}$$

2.2.2 Modeling Solid–Fluid Interaction

In this chapter, the motion of particles through the fluid flow is simulated. This type of simulation requires the modeling of the interaction between these two entities – fluid domain and solid domain (particle). This is achieved by using a solid–fluid interaction approach. In this chapter, the so-called strong coupling approach is used, where solid and fluid form a coupled mechanical system. The equations for solid and fluid domains are solved simultaneously, in each iteration of the simulation, and this coupling is achieved by introducing an additional set of equations that handle the interaction. In this chapter, the IBM approach is used that was developed and presented by Peskin [9]. This method is chosen due to its good compatibility with the LB method. Also, it enables the use of meshes with different density for discretization of solid and fluid domains, since the coupling is performed using interpolation around each node of both domains. The IBM uses a fixed Cartesian mesh to represent the fluid domain and the fluid mesh is composed of Eulerian points. This description is in accordance with the definition of fluid domain that is used in the LB method. On the other hand, the IBM treats the solid as an immersed object, or more specifically, as an isolated part of fluid. The boundary between the two domains is represented by a set of Lagrangian points. This boundary can be treated as rigid or deformable, and this enables the simulation of motion of both rigid and deformable particles within the fluid domain. In the IBM, the solid–fluid interaction is simulated as follows: the fluid is acting on the solid, i.e. on the boundary surface, via the extracted external forces and this causes the motion (and deformation) of solid and vice versa; the solid is opposing the influence of the fluid, by acting on the fluid over the boundary surface with a reaction force that alters the fluid flow.

There are some differences in applying this method for motion of rigid and deformable particles. The common part of the IBM that is valid in both cases is the interpolation scheme. Namely, all quantities necessary for the modeling of solid–fluid interaction are obtained by interpolating the quantities of interest over the neighboring nodes, as it was already mentioned. This way the problem of interpolation over different discretization meshes is resolved, since for each solid boundary point, the influence of a larger number of points from the fixed fluid mesh is considered. Interpolation is performed using the Dirac delta function that is approximated as follows:

$$\delta(\mathbf{x} - \mathbf{X}_B(t)) = D_{ijk}\left(\mathbf{x}_{ijk} - \mathbf{X}_B^l\right) = \delta\left(x_{ijk} - X_B^l\right)\delta\left(y_{ijk} - Y_B^l\right)\delta\left(z_{ijk} - Z_B^l\right)$$

$$(2.15)$$

where $\mathbf{X}_B^l(t)$, $l = 1, 2, ..., G$ represents the coordinates of the l-th Lagrangian boundary point, G is the overall number of points on the boundary, and $\mathbf{x}_{ijk}$ is the position of the considered node in the fluid domain (lattice mesh). This equation is given in its three-dimensional form, and of course, omitting the third dimension, its two-dimensional form can be obtained.

The value of function $\delta(r)$ used in Eq. (2.15) is defined in literature [9]:

$$\delta(r) = \begin{cases} \dfrac{1}{4h}\left(1 + \cos\left(\dfrac{\pi r}{2}\right)\right), & |r| \leq 2 \\ 0, & |r| > 2 \end{cases}$$

$$(2.16)$$

where h denotes the distance between two points of the fixed fluid mesh (in this chapter, since the LB method is considered, $h = 1$).

The Dirac delta function defined this way ensures that the interpolated value is maximal in nodes that coincide absolutely in both meshes and the influence of interpolated quantity is smaller as the distance between considered points increases, as can be seen in Figure 2.1.

Details about the equations that are used specifically for simulation of motion of rigid and deformable particles will be discussed in the sequel of this section.

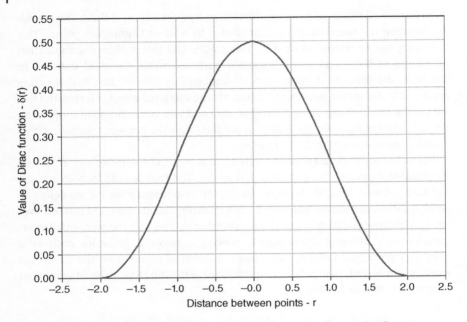

Figure 2.1 Diagram of variation of Dirac delta function depending on the distance between nodes.

2.2.2.1 Modeling the Motion of Rigid Particle

As already mentioned, the force exerted from the surrounding fluid tends to deform the boundary between the fluid and the particle. When the motion of rigid particle is simulated, this force should not cause any deformation of the particle and this possible deformation actually yields to a force that tends to restore the boundary to its original shape. For a rigid particle, the forces exerted from the fluid and solid have to be in equilibrium. There are several ways to determine this force, including direct forcing term [44–46], penalty formulation [10], momentum exchange method [47], etc. In this chapter, the approach proposed by Wu and Shu [48] is used, since this approach ensures that this mentioned force is calculated such that the nonslip boundary condition is satisfied on the immersed object.

In order to ensure that the velocity on the boundary satisfies the nonslip boundary condition, a fluid velocity correction $\delta \mathbf{u}$ is introduced as follows:

$$\delta \mathbf{u} = \frac{1}{2\rho} \mathbf{g}(\mathbf{x}, t) \tag{2.17}$$

If the intermediate fluid velocity is denoted by $\mathbf{u}^*$:

$$\mathbf{u}^* = \frac{1}{\rho} \sum_i \overline{\xi_i \bar{f}_i} \tag{2.18}$$

then the physical velocity $\mathbf{u}(\mathbf{x}, t)$ that is calculated for fluid domain using Eq. (2.14) can be rewritten as:

$$\mathbf{u} = \mathbf{u}^* + \delta\mathbf{u} \tag{2.19}$$

The introduced velocity correction for fluid nodes is an unknown quantity and can be interpolated based on the velocity correction of solid nodes at the boundary $\delta\mathbf{u}_B$:

$$\delta\mathbf{u}(\mathbf{x}, t) = \int_\Gamma \delta\mathbf{u}_B(\mathbf{X}_B, t)\delta(\mathbf{x} - \mathbf{X}_B)ds \tag{2.20}$$

Using the already defined Dirac delta function in Eqs. (2.15) and (2.16), the velocity correction for a two-dimensional problem can be calculated as follows:

$$\delta\mathbf{u}(\mathbf{x}_{ij}, t) = \sum_l \delta\mathbf{u}_B^l(\mathbf{X}_B^l, t)D_{ij}(\mathbf{x}_{ij} - \mathbf{X}_B^l)\Delta s_l \tag{2.21}$$

where Δs_l is the arc length of the boundary element (between two boundary points).

In order to satisfy the nonslip boundary condition, the fluid velocity at every boundary point must be equal to the velocity of the solid body in that boundary point. The velocity in Lagrangian points can be interpolated from the surrounding Eulerian points in fluid:

$$\frac{\partial \mathbf{X}_B^l}{\partial t} = \mathbf{u}_B^l(\mathbf{X}_B^l, t) = \int_\Omega \mathbf{u}(\mathbf{x}, t)\delta(\mathbf{x} - \mathbf{X}_B^l)d\mathbf{x} \tag{2.22}$$

When the already introduced interpolation scheme is applied on Eq. (2.22), the solid velocity can be calculated as follows:

$$\mathbf{u}_B^l(\mathbf{X}_B^l, t) = \sum_{i,j} \mathbf{u}(\mathbf{x}_{ij}, t)D_{ij}(\mathbf{x}_{ij} - \mathbf{X}_B^l)\Delta x\Delta y \tag{2.23}$$

Using Eqs. (2.19) and (2.21), Eq. (2.23) can be transformed:

$$\begin{aligned} \mathbf{u}_B^l(\mathbf{X}_B^l, t) = &\sum_{i,j} \mathbf{u}^*(\mathbf{x}_{ij}, t)D_{ij}(\mathbf{x}_{ij} - \mathbf{X}_B^l)\Delta x\Delta y \\ &+ \sum_{i,j}\sum_l \delta\mathbf{u}_B^l(\mathbf{X}_B^l, t)D_{ij}(\mathbf{x}_{ij} - \mathbf{X}_B^l)\Delta s_l D_{ij}(\mathbf{x}_{ij} - \mathbf{X}_B^l)\Delta x\Delta y \end{aligned} \tag{2.24}$$

If the following changes of variables are introduced (to simplify the expressions):

$$\begin{aligned} \delta_{ij}^B &= D_{ij}(\mathbf{x}_{ij} - \mathbf{X}_B^l)\Delta s_l \\ \delta_{ij} &= D_{ij}(\mathbf{x}_{ij} - \mathbf{X}_B^l)\Delta x\Delta y \end{aligned} \tag{2.25}$$

then Eq. (2.24) can be rewritten in matrix form:

$$\mathbf{A}\cdot\mathbf{X} = \mathbf{B} \tag{2.26}$$

where the matrices are given by:

$$\mathbf{X} = \left\{\boldsymbol{\delta}\mathbf{u}_B^1, \boldsymbol{\delta}\mathbf{u}_B^2, ..., \boldsymbol{\delta}\mathbf{u}_B^m\right\}^T \tag{2.27}$$

$$\mathbf{A} = \begin{bmatrix} \delta_{11} & \delta_{12} & \cdots & \delta_{1n} \\ \delta_{21} & \delta_{22} & \cdots & \delta_{2n} \\ \vdots & \vdots & \ddots & \vdots \\ \delta_{m1} & \delta_{m2} & \cdots & \delta_{mn} \end{bmatrix} \cdot \begin{bmatrix} \delta_{11}^B & \delta_{12}^B & \cdots & \delta_{1m}^B \\ \delta_{21}^B & \delta_{22}^B & \cdots & \delta_{2m}^B \\ \vdots & \vdots & \ddots & \vdots \\ \delta_{n1}^B & \delta_{n2}^B & \cdots & \delta_{nm}^B \end{bmatrix} \tag{2.28}$$

$$\mathbf{B} = \begin{pmatrix} \mathbf{u}_B^1 \\ \mathbf{u}_B^2 \\ \vdots \\ \mathbf{u}_B^m \end{pmatrix} - \begin{bmatrix} \delta_{11} & \delta_{12} & \cdots & \delta_{1n} \\ \delta_{21} & \delta_{22} & \cdots & \delta_{2n} \\ \vdots & \vdots & \ddots & \vdots \\ \delta_{m1} & \delta_{m2} & \cdots & \delta_{mn} \end{bmatrix} \begin{pmatrix} \mathbf{u}_1^* \\ \mathbf{u}_2^* \\ \vdots \\ \mathbf{u}_n^* \end{pmatrix} \tag{2.29}$$

In these expressions, m is the number of Lagrangian (boundary) points and n is the number of surrounding Eulerian points that are used in the Dirac delta interpolation function.

The unknown solid velocity corrections in boundary points $\boldsymbol{\delta}\mathbf{u}_B^l$ can be calculated by solving the system of equations (2.26). Using these velocity corrections, it is straightforward to calculate the forces exerted from the fluid, acting on the solid and vice versa. Using Eq. (2.17), the external force field acting on the fluid is given by:

$$\mathbf{g}(\mathbf{x}, t) = 2\rho\boldsymbol{\delta}\mathbf{u}_B^l \tag{2.30}$$

This evaluated external force (for every Eulerian point that is surrounding the solid) is included first in Eq. (2.8) and then in Eq. (2.6), and this way the immersed solid particle is influencing the fluid flow. At the same time, by using the Newton's third law of action and reaction, this same force is acting on the particle in particular boundary Lagrangian point. The overall influence of the fluid on the particle is expressed through a force and torque that are given by:

$$\mathbf{F}_R = -\sum_l \mathbf{F}\left(\mathbf{X}_B^l\right)\Delta s_l = -\sum_l 2\rho\boldsymbol{\delta}\mathbf{u}_B^l\Delta s_l \tag{2.31}$$

$$\mathbf{M}_R = -\sum_l \left(\mathbf{X}_B^l - \mathbf{X}_R\right) \times \mathbf{F}\left(\mathbf{X}_B^l\right)\Delta s_l = -\sum_l \left(\mathbf{X}_B^l - \mathbf{X}_R\right) \times \left(2\rho\boldsymbol{\delta}\mathbf{u}_B^l\right)\Delta s_l \tag{2.32}$$

where $\mathbf{X}_R$ is the center of mass of the particle. The minus sign in both equations is the consequence of the mentioned Newton's third law.

After the force and torque acting on the particle are calculated, the following equations of motion can be used to determine the position and orientation of the particle:

$$m\frac{d\mathbf{v}_C}{dt} = \mathbf{F}_R \tag{2.33}$$

$$I_0\frac{d\boldsymbol{\omega}}{dt} = \mathbf{M}_R \tag{2.34}$$

where m and I_0 are the particle mass and moment of inertia, respectively, and $\mathbf{v}_C$ and $\boldsymbol{\omega}$ are the velocity and angular velocity of the particle, respectively.

Using these newly calculated data, the position of the particle inside fluid domain can be updated. Afterward, the immersed particle again causes a change in fluid flow and the entire procedure is repeated in iterations until the defined criterion for the simulation end is reached.

2.2.2.2 Modeling the Motion of Deformable Particle

When modeling the motion of deformable particle, the effect of the surrounding fluid on the immersed particle is reflected through a deformation of the boundary between two domains. Due to this deformation, the internal reaction force appears in the particle and this force determines the effect of the particle on the surrounding fluid. This effect is again introduced as an external force field in the fluid domain, in order to simulate the effect of immersed solid body. This external force is given by:

$$\mathbf{g}(\mathbf{x}, t) = \int_\Gamma \mathbf{F}(t)\boldsymbol{\delta}(\mathbf{x} - \mathbf{X}_B(t))ds \tag{2.35}$$

where $\mathbf{F}(t)$ is the force with which solid opposes the effect of the fluid. Calculation of this force in explained in detail in Section 2.2.3 of this chapter.

If the already introduced interpolation scheme using Dirac delta function is performed (defined in Eqs. (2.15) and (2.16)), the force can be calculated as:

$$\mathbf{g}(\mathbf{x}_{ij}, t) = \sum_{l=1}^{m} \mathbf{F}_l(t)D_{ij}\left(\mathbf{x}_{ij} - \mathbf{X}_B^l(t)\right) \tag{2.36}$$

Using a similar approach that was already used for rigid particle, the velocity of solid nodes can be interpolated in terms over the surrounding fluid nodes, and this can be represented as follows:

$$\frac{\partial \mathbf{X}_B(t)}{\partial t} = \mathbf{u}(\mathbf{X}_B(t), t) = \int_S \mathbf{u}(\mathbf{x}, t)\boldsymbol{\delta}(\mathbf{x} - \mathbf{X}_B(t))d\mathbf{x} \tag{2.37}$$

Again, the Dirac delta function is applied, and the solid velocity is given by:

$$\mathbf{u}_B^l(\mathbf{X}_B^l, t) = \sum_{i,j} \mathbf{u}(\mathbf{x}_{ij}, t) D_{ij}(\mathbf{x}_{ij} - \mathbf{X}_B^l) \tag{2.38}$$

Applying the Euler Forward method on Eq. (2.38), the new position of solid boundary points can be calculated:

$$^{t + \Delta t}\mathbf{X}_B^l = {}^t\mathbf{X}_B^l + \mathbf{u}_B^l \Delta t \tag{2.39}$$

Using the obtained position vectors of the Lagrangian points, the deformation of the solid can be calculated and then the procedure explained in Section 2.2.3 is performed to calculate the reaction forces that this deformation has caused. These forces are used to introduce the effect of the immersed particle to the fluid flow that then again causes new deformation of the solid. This process is repeated in iterations until the defined criterion for the simulation end is reached.

2.2.3 Modeling Deformation of the Particle

In this chapter, two types of deformable bodies are considered: the first type considers a particle that is made of a linear elastic material and the second considers a particle that is made of a hyperelastic material. The linear elastic material is used to model the behavior of particles though fluid domain, when large deformations are not involved. The hyperelastic material is used to model the behavior of particles when large deformations are involved. This second model is mainly related to the behavior of RBCs that are highly deformable. In both cases, the outside surface of the particle is discretized using a triangular mesh. If only two-dimensional domain is simulated, then the outer domain of the particle is divided in linear segments. Actually, it is considered that the deformable particle has a surrounding membrane that has negligible thickness and that this membrane opposes the deformation. This definition is again adapted to the structure of RBCs that consists of a highly deformable membrane and an inner structure that behaves like a fluid [49, 50]. In the mentioned mesh, it is possible to calculate the reaction force for every element and for every node of the mesh, with which this particular entity opposes the deformation of the surrounding fluid. The total reaction force of the deformable particle caused by the defined deformation in a particular node is given by:

$$\mathbf{F}_i(t) = \mathbf{F}_i^S + \mathbf{F}_i^B + \mathbf{F}_i^A + \mathbf{F}_i^V \tag{2.40}$$

The four forces in Eq. (2.40) represent the reactions to four types of deformation that are considered in this chapter, and these are: surface strain of the membrane (represented by $\mathbf{F}_i^S$), bending of the membrane (represented by $\mathbf{F}_i^B$), surface area of the membrane (represented by $\mathbf{F}_i^A$), and volume within membrane (represented by

$\mathbf{F}_i^V$). It should be noted that for two-dimensional simulations, only the first two types of deformation are considered. In the sequel of this chapter, the approaches to calculate these four forces will be discussed.

2.2.3.1 Force Caused by the Surface Strain of Membrane

There are two ways to model the surface strain. One is to treat the membrane as a set of nodes that are interconnected with springs and then the spring force is calculated according to the defined deformation. This approach was used by Dupin et al. [28]. This model is used in this chapter to simulate the behavior of the linear elastic material. The force in this case is given by:

$$\mathbf{F}_{j_1}^S = -\mathbf{F}_{j_2}^S = K^S \frac{L_{12} - {}^0L_{12}}{{}^0L_{12}} \mathbf{l}_{12} \tag{2.41}$$

where K^S represents the parameter of resistance to membrane surface strain that is defined for a particular type of deformable body, L_{12} and ${}^0L_{12}$ denote the current distance between nodes j_1 and j_2 and the initial distance between considered nodes, respectively, and $\mathbf{l}_{12}$ denotes the unit vector that connects nodes j_1 and j_2.

The second approach to model the surface strain is used to simulate the behavior of a hyperelastic material. In this case, the relationship between stress and strain can be determined using the strain energy density function. Several hyperelastic material models have been developed. To model the relationship between applied stress and strain in case of RBCs, Mooney–Rivlin and neo-Hookean material models are used [17, 19, 51]. However, Skalak et al. [32] analyzed mentioned models and came to the conclusion that they are not able to simulate the behavior of RBCs accurately enough. They proposed a new material model for the membrane of RBC and their model is used in this chapter.

Strain energy density function in the Skalak material model of deformable membrane is defined in terms of two invariants of Cauchy–Green deformation tensor:

$$W^S = \frac{k_s}{12} \left(I_1'^2 + 2I_1' - 2I_2' \right) + \frac{k_a}{12} I_2'^2 \tag{2.42}$$

where I_1' and I_2' represent the modified invariants, like it is proposed in [32]. Surface elastic shear modulus and area dilation modulus are denoted by k_s and k_a, respectively. These two parameters are defined for a particular type of deformable body.

Before defining the invariants and the modified invariants introduced in Eq. (2.42), two additional quantities have to be defined, which are then used to define the invariants. These quantities are gradient deformation tensor and left Cauchy–Green deformation tensor.

When a deformable body is considered, in the initial moment in time, this body is in its initial configuration that is denoted by ^0B and the coordinates of a particular material point P are defined by $^0\mathbf{x}$. In the current moment in time t, the body is in its current configuration ^tB and the coordinates of the particular material point are defined by $^t\mathbf{x}$. The relationship between coordinates in the current moment in time and the initial moment in time is represented by the deformation gradient tensor, which is defined by:

$$_0^t F_{kj} = \frac{\partial^t x_k}{\partial^0 x_j} \tag{2.43}$$

If the displacement of a material point is defined as:

$$\mathbf{u} = {}^t\mathbf{x} - {}^0\mathbf{x} \tag{2.44}$$

and if this expression is substituted into Eq. (2.43), it is obtained:

$$_0^t\mathbf{F} = \mathbf{I} + \frac{\partial \mathbf{u}}{\partial^0 \mathbf{x}} \tag{2.45}$$

where $\mathbf{I}$ is the unit tensor.

Similarly, the deformation gradient that is calculated for deformed configuration ^tB with respect to the reference configuration ^tB is defined as (the expression is given in index notation):

$$_t^t F_{kj} = \delta_{kj} + \frac{\partial u_i}{\partial^t x_j} \tag{2.46}$$

where δ_{kj} is the Kronecker delta symbol.

Left Cauchy–Green deformation tensor can now be defined as:

$$\mathbf{B} = \mathbf{F}\mathbf{F}^T \tag{2.47}$$

or in index notation:

$$B_{ij} = F_{im}F_{jm} \tag{2.48}$$

The tensor invariants do not depend on the coordinate transformations and can be expressed in terms of the Cauchy–Green deformation tensor $\mathbf{B}$ [52]:

$$I_1 = \text{tr}\mathbf{B} = B_{11} + B_{22} + B_{33} = B_{nn} \tag{2.49}$$

$$I_2 = \frac{1}{2}\left((\text{tr}\mathbf{B})^2 - \text{tr}\mathbf{B}^2\right) \tag{2.50}$$

Also, invariants can be expressed in terms of principal stretches λ_1, λ_2, and λ_3:

$$I_1 = \lambda_1^2 + \lambda_2^2 + \lambda_3^2 \tag{2.51}$$

$$I_2 = \lambda_1^2\lambda_2^2 + \lambda_2^2\lambda_3^2 + \lambda_3^2\lambda_1^2 \tag{2.52}$$

If a thin deformable membrane is considered, in a local coordinate system it can be assumed that the strain is only in one plane and thus the stretch along the z axis can be neglected - $\lambda_3 = 0$.

Modified invariants, as proposed by Skalak et al. [32] and introduced in Eq. (2.42), are given by:

$$I_1' = \lambda_1^2 + \lambda_2^2 - 2 = I_1 - 2 \tag{2.53}$$

$$I_2' = \lambda_1^2\lambda_2^2 - 1 = I_2 - 1 \tag{2.54}$$

Using the definition of modified invariants, it is possible to determine the relationship between invariants and the strain energy density function in the Skalak material model:

$$W^S = \frac{k_s}{12}\left(I_1^2 - 2I_1 - 2I_2 + 2\right) + \frac{k_\alpha}{12}\left(I_2^2 - 2I_2 + 1\right) \tag{2.55}$$

Now, the relationship between stress and strain has to be defined, using the equation for the strain energy density (2.55). Cauchy stress tensor σ can be expressed as [53]:

$$\sigma_{ij} = \frac{1}{J}\overline{T}_{im}F_{jm} \tag{2.56}$$

where $\overline{T}_{im}$ are the components of the first Piola–Kirchhoff stress tensor, and J is the determinant of the deformation gradient:

$$J = \det(\mathbf{F}) \tag{2.57}$$

The first Piola–Kirchhoff stress tensor can be defined as a partial derivative of strain energy density with respect to the deformation gradient:

$$\overline{T}_{im} = \frac{\partial W^S}{\partial F_{im}} \tag{2.58}$$

Substituting Eq. (2.58) into Eq. (2.56), it is obtained:

$$\sigma_{ij} = \frac{1}{J}\frac{\partial W^S}{\partial F_{im}}F_{jm} \tag{2.59}$$

Partial derivative in Eq. (2.59) can be transformed using the chain rule:

$$\frac{\partial W^S}{\partial F_{im}} = \frac{\partial W^S}{\partial I_1}\frac{\partial I_1}{\partial F_{im}} + \frac{\partial W^S}{\partial I_2}\frac{\partial I_2}{\partial F_{im}} + \frac{\partial W^S}{\partial J}\frac{\partial J}{\partial F_{im}} \tag{2.60}$$

Partial derivative of determinant with respect to the deformation gradient is equal to:

$$\frac{\partial J}{\partial F_{im}} = JF_{mi}^{-1} \tag{2.61}$$

Partial derivative of the first invariant with respect to the deformation gradient is equal to:

$$\frac{\partial I_1}{\partial F_{im}} = \frac{\partial \left(F_{np}F_{np}\right)}{\partial F_{im}} = 2F_{np}\frac{\partial F_{np}}{\partial F_{im}} = 2F_{im} \tag{2.62}$$

Partial derivative of the second invariant with respect to the deformation gradient is equal to:

$$\frac{\partial I_2}{\partial F_{im}} = \frac{\partial \left[\frac{1}{2}\left((\mathrm{tr}\mathbf{B})^2 - \mathrm{tr}\mathbf{B}^2\right)\right]}{\partial F_{im}} = \frac{1}{2}\left[2\mathrm{tr}\mathbf{B}\frac{\partial(\mathrm{tr}\mathbf{B})}{\partial F_{im}} - \frac{\partial\left(B_{np}B_{pn}\right)}{\partial F_{im}}\right] = \tag{2.63}$$

$$= \frac{1}{2}\left[2I_1\frac{\partial I_1}{\partial F_{im}} - B_{np}\frac{\partial B_{pn}}{\partial F_{im}} - \frac{\partial B_{np}}{\partial F_{im}}B_{pn}\right] = 2I_1 F_{im} - 2B_{ik}F_{km}$$

Substituting expressions (2.61)–(2.63) into expression (2.60), and then into Eq. (2.59), it is obtained:

$$\sigma_{ij} = \frac{1}{J}\left(\frac{\partial W^S}{\partial I_1}2F_{im} + \frac{\partial W^S}{\partial I_2}\left(2I_1 F_{im} - 2B_{ik}F_{km}\right) + \frac{\partial W^S}{\partial J}JF_{mi}^{-1}\right)F_{jm} \tag{2.64}$$

If Eq. (2.48) is taken into account and also equation

$$\delta_{ik} = F_{ij}F_{jk}^{-1} \tag{2.65}$$

then the equation for Cauchy stress tensor is obtained as follows:

$$\sigma_{ij} = \frac{1}{J}\left(\frac{\partial W^S}{\partial I_1}2B_{ij} + \frac{\partial W^S}{\partial I_2}\left(2I_1 B_{ij} - 2B_{ik}B_{kj}\right) + \frac{\partial W^S}{\partial J}J\delta_{ij}\right) = \tag{2.66}$$

$$= \frac{2}{J}\left(\frac{\partial W^S}{\partial I_1} + I_1\frac{\partial W^S}{\partial I_2}\right)B_{ij} + \frac{\partial W^S}{\partial J}\delta_{ij} - \frac{2}{J}\frac{\partial W^S}{\partial I_2}B_{ik}B_{kj}$$

Based on Eq. (2.55), it is possible to calculate the partial derivatives of strain energy density with respect to the determinant and invariants:

$$\frac{\partial W^S}{\partial I_1} = \frac{k_s}{12}(2I_1 - 2); \frac{\partial W^S}{\partial I_2} = -\frac{k_s}{6} + \frac{k_a}{12}(2I_2 - 2); \frac{\partial W^S}{\partial I_2} = -\frac{k_s}{6} + \frac{k_a}{12}(2I_2 - 2) \tag{2.67}$$

The final expression that defines the stress–strain relationship is given by:

$$\sigma_{ij} = \frac{1}{J}\left[\left(-\frac{k_s}{3} + \frac{k_a}{3}I_1(I_2 - 1)\right)B_{ij} - \left(-\frac{k_s}{3} + \frac{k_a}{3}(I_2 - 1)\right)B_{ik}B_{kj}\right] \tag{2.68}$$

This derived expression is valid for continuum. In order to use this expression in numerical simulations, it is necessary to apply the FEM. Namely, the entire domain, i.e. the membrane of the deformable body is discretized – divided in subdomains and they are called FEs. This discretization enables the numerical simulation of reaction of continuum domain to the external influences, in this case the reaction of membrane of deformable body to the deformation caused by the forces that are exerted from the external and internal fluid. In this chapter, all the internal forces that are caused by membrane deformation are calculated using the FEM in discrete form, in all points of the discretized mesh.

2.2.3.2 Force Caused by the Bending of the Membrane

Bending is observed as a change of angle between two triangular elements (for three-dimensional case) or between two line segments (for two-dimensional case). This approach was proposed by Dupin et al. [28]. The change of angle causes a force in all nodes of both considered entities, i.e. in all four nodes of the two neighboring triangular elements or in all three nodes of the two neighboring line segments. The force in the node that is not common to both entities is given by:

$$\mathbf{F}^B_{j_3} = K^B \frac{\varphi_{i_1 i_2} - {}^0\varphi_{i_1 i_2}}{{}^0\varphi_{i_1 i_2}} \mathbf{n}_{i_1} \tag{2.69}$$

while the force in common node(s) is given by:

$$\mathbf{F}^B_{j_{12}} = \mathbf{F}^B_{j_{21}} = -\frac{1}{2}\mathbf{F}^B_{j_3} \tag{2.70}$$

In Eq. (2.69), K^B represents the parameter of resistance to the bending that is defined for a particular type of deformable body, $\varphi_{i_1 i_2}$ and ${}^0\varphi_{i_1 i_2}$ represent the current and initial angle between elements i_1 and i_2, and $\mathbf{n}_{i_1}$ denotes the unit vector normals of the considered element i_1.

2.2.3.3 Force Caused by the Change of Surface area of the Membrane

The approach for calculation of this force was also proposed by Dupin et al. [28]. This force is calculated for each element of the triangular mesh, and the local change of area of this particular element causes a force in each node that is calculated as:

$$\mathbf{F}^A_{ij} = -K^{Al} \frac{A_i - {}^0A_i}{A_i} \mathbf{w}_j \tag{2.71}$$

where i denotes the index of the element, j is the local index of the node within the particular element ($j = 1, 2, 3$), K^{Al} is the parameter of resistance to the local change of surface area, that is defined for appropriate type of deformable body, and A_i and

^0A_i denote the current surface area of i-th element of the membrane after defined deformation and the initial surface area of i-th element of the membrane, respectively. The unit vector whose direction coincides with the line that connects the center of gravity of the particular element with the considered node is denoted by $\mathbf{w}_j$.

The change of overall surface area of the membrane on each element independently also influences the reaction to deformation. In fact, for the modeling of behavior of RBC, the change of overall surface area has a much greater influence on the reaction force than the local change of surface area [28]. Hence, an additional term is included in Eq. (2.71) in order to calculate more precisely the force caused by the change of surface area:

$$\mathbf{F}_{ij}^A = -\left(K^{Al}\frac{A_i - {}^0A_i}{A_i} + K^{At}\frac{A - {}^0A}{A}\right)\mathbf{w}_j \tag{2.72}$$

where K^{At} is the parameter of resistance to the overall (total) change of surface area, and A and ^0A represent the current overall surface area of the membrane and the initial overall surface area, respectively.

2.2.3.4 Force Caused by the Change of Volume

The change of volume causes a force in each element of the mesh, as it is discussed in [28], and this force is given by:

$$\mathbf{F}_i^V = -K^V\frac{V - {}^0V}{{}^0V}A_i\mathbf{n}_i \tag{2.73}$$

where K^V represents the parameter of resistance to change of volume that is defined for appropriate type of deformable body, and V and ^0V denote the current volume of a deformable body after defined deformation and the initial volume, respectively.

In this chapter, the discretization on triangular elements is applied and hence the calculated force for each element (defined in Eq. (2.73)) is equally distributed on all three nodes of the particular element:

$$\mathbf{F}_{i1}^V = \mathbf{F}_{i2}^V = \mathbf{F}_{i3}^V = \frac{1}{3}\mathbf{F}_i^V \tag{2.74}$$

2.2.4 Modeling the Flow of Two Fluids with Different Viscosity that are Separated by the Membrane of the Solid

Some of the examples in this chapter model the behavior of RBCs, as an example of deformable object immersed in fluid. As it was already mentioned, the structure of the RBC is such that it has a thin membrane and interior that is filled with incompressible liquid, cytoplasm [49]. The density of cytoplasm inside cells and blood plasma in blood vessel is similar, and this difference can be neglected for small Reynolds number. But, in normal RBCs, the viscosity of cytoplasm is around five times greater than the viscosity of blood plasma in which cells are immersed [54, 55]. This should be considered in numerical simulations. Since the only difference between these two fluids is its viscosity, it is possible to use same equations for both fluids, and it is not even necessary to simulate two fluid domains independently, it is enough to assign different values of viscosity to a particular set of particles in the considered domain. The viscosity in the LB method is related to the relaxation time τ using the following equation:

$$\mu = c_s^2 \rho \left(\bar{\tau} - \frac{1}{2} \right) \tag{2.75}$$

Hence, it is sufficient to assign a different value of relaxation time to particular nodes that are inside the RBC and this will automatically ensure that in these nodes, fluid will have different viscosity. Of course, during the motion of RBC, it is necessary to update the viscosity values in nodes that are inside and close to the cell membrane.

There are several ways to perform this update of viscosity values in nodes. An index field can be created, in order to determine the relative position of each node in the fluid domain with respect to the cell membrane. Tryggvasson et al. [56] solved the Poisson equation in each iteration to calculate the mentioned index field. Francois et al. [57] have proposed to update the characteristics of the fluid with respect to the normal distance to the membrane. In this dissertation, the second approach is used, since it is computationally more efficient.

The vector that connects a node in fluid domain (from Eulerian mesh) with its closest node on the solid membrane (from Lagrangian mesh) is denoted by $\mathbf{r}$. The important point to be noted is that along with the intensity of this vector, a sign is attached that indicates whether the fluid node is within the solid membrane (it is taken that $\mathbf{r}$ is positive in this case) or outside the solid membrane (it is taken that $\mathbf{r}$ is negative in this case).

The Heaviside function is defined as:

$$H(\mathbf{r}) = \begin{cases} 0, & \mathbf{r} < -2h \\ \prod_{i=1}^{d} \frac{1}{2} \left(1 + \frac{r_i}{2h} + \frac{1}{\pi} \sin \frac{\pi r_i}{2h} \right), & -2h \le |\mathbf{r}| \le 2h \\ 1, & \mathbf{r} > 2h \end{cases} \tag{2.76}$$

The variable fluid characteristic, in this case the viscosity, can be related to the distance between fluid node and solid membrane, using the Heaviside function as follows:

$$\mu(\mathbf{x}) = \mu_{out} - (\mu_{out} - \mu_{in}) H(\mathbf{r}(\mathbf{x})) \tag{2.77}$$

where the upper indices *in* and *out* are related to values of viscosity inside and outside the solid membrane, respectively. From Eq. (2.76), it is evident that during the motion of the RBC, only the characteristics of fluid nodes whose distance is less than 2 should be updated. This is an additional simplification during implementation and provides additional speedup of the program execution, since it is not necessary to update the characteristics of all fluid nodes, but only of those nodes that are close to the solid.

2.3 Results

In this section, the results of the simulations using the numerical model described in Section 2.2 will be presented. The obtained results are compared with results from literature, in order to evaluate the accuracy of the proposed model. Simulations are performed in two-dimensional and three-dimensional domains, depending on the models that are used for comparison and that were presented in literature.

Three of these examples are also available in the Software on the web, where it is possible to change the initial position of the particle within the domain and then observe the motion of rigid or deformable particle in selected two-dimensional domain, in real time.

In this section, two types of particles are considered. A spherical particle and a RBC. RBC has a biconcave shape, with diameter from 6,7 to 7,7 μm, average thickness from 1,7 to 2,5 μm and volume from 76 to 96 fL. The equation that describes the shape of RBC was empirically defined by Evans and Fung [58]:

$$x^2 + y^2 = r^2$$

$$z = \pm \frac{1}{2} R_0 \left[1 - \frac{x^2 + y^2}{R_0^2}\right]^{0.5} \cdot \left[C_0 + C_1 \frac{x^2 + y^2}{R_0^2} + C_2 \left(\frac{x^2 + y^2}{R_0^2}\right)^2\right] \quad (2.78)$$

where the values of constants are:

$$R_0 = 3.91 \mu m; \quad C_0 = 0.207; \quad C_1 = 2.0; \quad C_2 = -1.123 \quad (2.79)$$

and *r* represents the radius of RBC.

Values of parameters of resistance to the four types of deformation that are discussed in Section 2.2.3 are defined according to the values proposed in literature [28, 59, 60]. Parameter of resistance to change of volume is equal to $K^V = 50$ $pN\mu m^{-1}$. Parameter of resistance to local change of membrane area is equal to $K^{Al} = 1, 67 \cdot 10^{-1} pN\mu m^{-1}$, while the parameter of resistance to global change of membrane area is equal to $K^{At} = 1.67 pN\,\mu m^{-1}$. The parameters that were used to define the Skalak material model of cellular membrane are equal to

$k_s = 0.75 \cdot 10^3 pN\mu m^{-1}$ and $k_\alpha = 75 \cdot 10^3 pN\,\mu m^{-1}$, according to literature [32]. Parameter of resistance to bending of the membrane is equal to $K^B = 10^{-1} pN\mu m^{-1}$.

2.3.1 Modeling the Behavior of Particles in Poiseuille Flow

In this example, the behavior of particles in a straight channel is considered. If a circular particle is fixed in the domain, then the drag force exerted from the surrounding fluid can be calculated. The drag force over the circular particle in a laminar flow has been calculated analytically [61].

The value of the drag force depends on the radius of the particle. This is analyzed in literature [21], where the results obtained using the LB method are compared with the mentioned analytical solution, as well as with results obtained using the DPD method [62].

The motion of a deformable particle through a laminar flow is interesting from the modeling point of view, since it is important to predict the shape of the particle during this motion. In this case, three-dimensional simulation of fluid flow through a straight channel with a square cross-section is performed. The hyperelastic material model, presented in Section 2.2.3.1, was used for the modeling of surface strain. The boundary conditions are defined such that the velocity on the walls is equal to zero, and a pressure difference is set at the inlet and outlet that causes the fluid flow. Geometry of the fluid domain is shown in Figure 2.2. The dimensions of the domain in all three directions are defined such that the following relations are valid: $L_y = L_z$ and $L_x = 2L_y$. The radius of spherical particle is equal to one-fourth of the width of the domain, i.e. $r = \frac{1}{4}L_y$. The particle is initially placed in the center of the channel and hence the coordinates of the center of gravity are given by: $x_C = \frac{L_y}{2}$; $y_C = \frac{L_y}{2}$; $z_C = \frac{L_y}{2}$.

Other parameters regarding the fluid flow, necessary for the simulation using the LB method, are defined to be consistent with the parameters in the example given in literature with which the results were compared [14]. Figure 2.3 shows

Figure 2.2 Example 1 – Geometry of the fluid domain.

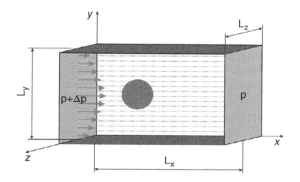

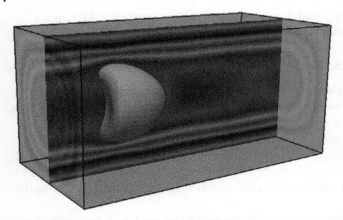

Figure 2.3 Example 1 – Fluid velocity field and current position of the spherical particle.

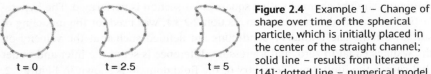

t = 0   t = 2.5   t = 5

Figure 2.4 Example 1 – Change of shape over time of the spherical particle, which is initially placed in the center of the straight channel; solid line – results from literature [14]; dotted line – numerical model.

the velocity distribution and the position of the particle in one moment in time during the simulation, while Figure 2.4 shows the change of shape of spherical particle over time. The change of shape was compared with results presented in literature [14].

Murayama et al. [14] also analyzed the change of shape of spherical particle when the particle is shifted for a defined small amount from the center of the channel. The relative position of the center of gravity of the particle is defined as:

$$y_r = \frac{\left(y_C - \left(L_y/2\right)\right)}{\left(L_y/2\right)} \tag{2.80}$$

Of course, in the case that was considered at the beginning of this section, the spherical particle is placed in the center of the channel and thus the relative position is equal to $y_r = 0$.

The spherical particle is moved for $y_r = 0.063$ and $y_r = 0.13$ in literature [14]. The simulations with shifted particle are performed using the proposed numerical model and a good agreement with results from literature is obtained, as it can be seen in Figures 2.5 and 2.6.

Figure 2.5 Example 1 – Change of shape over time of the spherical particle, which is initially moved for y_r = 0.063; solid line – results from literature [14]; dotted line – numerical model.

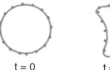

t = 0   t = 2.5   t = 5

Figure 2.6 Example 1 - Change of shape over time of the spherical particle, which is initially moved for y_r = 0.13; solid line – results from literature [14]; dotted line – numerical model.

t = 0   t = 2.5   t = 5

Motion of RBC in a straight channel was also modeled in literature [63]. The boundary conditions are identical to the conditions that are used for the fluid domain in case of the motion of spherical particle. In this case, the viscosity of internal fluid is set to be five times greater than the viscosity of external fluid, which is the case with real RBCs. In the example presented in literature [63], the diameter of the channel was set to be $10\mu m$ and maximum fluid velocity at the inlet is set to be 0.5 mm/s, same conditions were defined in numerical simulation as well. Simulations were performed for two different initial positions of the RBC. The first position is when the RBC is placed vertically, i.e. such that the angle between the largest semiaxis of the cell and x axis is equal to 90°. In the second case, this angle is equal to 45°. Figure 2.7 shows the fluid velocity distribution and the position of RBC during simulation, for the first considered position, and Figure 2.8 shows the fluid velocity field and the position of RBC during simulation, for the second considered position. The change of shape of the RBC over time that is obtained using the presented numerical model for both cases is compared with results from literature and a good agreement of results is obtained, as it can be seen in Figures 2.9 and 2.10.

2.3.2 Modeling the Behavior of Particles in Shear Flow

In this example, the behavior of particles in double shear flow is analyzed.

First, the three-dimensional simulation was performed, with the hyperelastic material model, presented in Section 2.2.3.1, that was used for the modeling of surface strain. The fluid domain is modeled such that the velocities of upper and lower walls are prescribed and these velocities have opposite directions, as it is illustrated in Figure 2.11. Dimensions of the domain in all three directions are equal - $L_x = L_y = L_z$. Along x and z axis, the periodical boundary condition is defined on the boundaries of the domain. First, the behavior of a deformable spherical

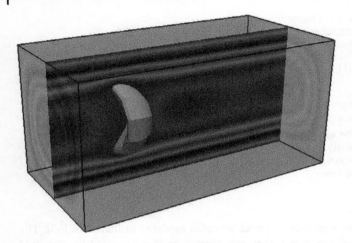

Figure 2.7 Example 1 – Fluid velocity field and current position of RBC – first considered case.

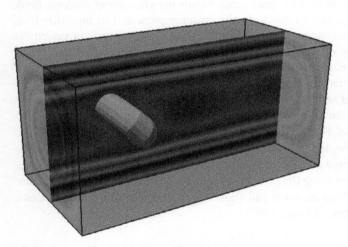

Figure 2.8 Example 1 – Fluid velocity field and current position of RBC – second considered case.

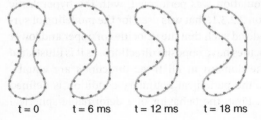

t = 0 t = 6 ms t = 12 ms t = 18 ms

Figure 2.9 Example 1 – Change of shape of RBC over time – first considered case; solid line – results from literature [63]; dotted line – numerical model.

Figure 2.10 Example 1 – Change of shape of RBC over time – second considered case; solid line – results from literature [63]; dotted line – numerical model.

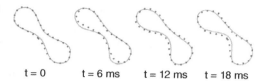

t = 0   t = 6 ms   t = 12 ms   t = 18 ms

Figure 2.11 Example 2 – Geometry of the fluid domain.

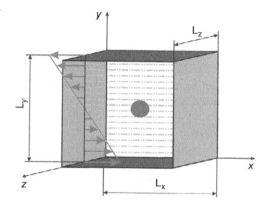

particle that is fixed in the domain is simulated. Diameter of the spherical particle is equal to one-tenth of the dimension of fluid domain - $r = \frac{1}{10} L_x$. Particle is placed in the center of fluid domain, hence the coordinates of the center of gravity of particle are equal to: $x_C = \frac{L_x}{2}$; $y_C = \frac{L_x}{2}$; $z_C = \frac{L_x}{2}$.

If values of velocity of upper and lower wall are denoted by u_w and if the height of the domain is denoted by L_z like in Figure 2.11, then the shear rate can be defined as:

$$\dot{\gamma} = \frac{2u_w}{L_z} \tag{2.81}$$

It is first assumed that the densities of fluid inside and outside the spherical particle are equal ($\rho_{in} = \rho_{out} = \rho$), and the same is assumed for kinematic viscosities. In the first considered case, the ratio between the viscosity of internal fluid and the viscosity of external fluid is equal to 1:

$$\lambda = \frac{\nu_{in}}{\nu_{out}} = 1 \tag{2.82}$$

and in the sequel, other values of coefficient λ will also be considered.

Reynolds number can be defined as:

$$\text{Re} = \frac{\rho \dot{\gamma} r^2}{\nu} \tag{2.83}$$

where r is the radius of spherical particle. Reynolds number in this example is very small, much smaller than 1 and in numerical simulations performed in this chapter, its value is set to be 0.025, which is in accordance with the same example that is analyzed in literature [13, 14].

Another parameter is defined, and that is the dimensionless shear rate:

$$G = \frac{\dot{\gamma} \nu r}{k_s} \tag{2.84}$$

where k_s is the surface elastic shear modulus that is already defined in Section 2.2.3.1. Depending on the value of this parameter, the deformation of the particle changes under the influence of the fluid.

In order to quantify the deformation of the particle appropriately, the Taylor deformation index is defined as:

$$D_{xy} = \frac{L_S - B_S}{L_S + B_S} \tag{2.85}$$

where L_S and B_S represent the largest and smallest semiaxis of the ellipse that is obtained as a cross-section of the spherical particle along the plane that contains the center of gravity of the particle and that is parallel to the coordinate plane $x0y$. These mentioned semiaxes of the ellipse are shown in Figure 2.12.

The results of simulations obtained using the presented numerical model are compared with the same example presented in literature [13, 14], and thus all other simulation parameters are chosen to be equal to those in literature.

During the simulation, particle is gradually deforming, until the final deformation is reached and then the so-called tank-threading behavior begins. The fluid velocity distribution and the position of the solid when the final deformation is reached are shown in Figure 2.13, while Figure 2.14 shows the velocity streamlines. From Figure 2.14, it is evident that fluid is flowing around the particle, while the particle is rotating around its center of gravity.

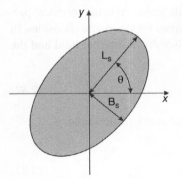

Figure 2.12 Cross-section of a spherical particle during deformation and the semiaxes of the obtained ellipse.

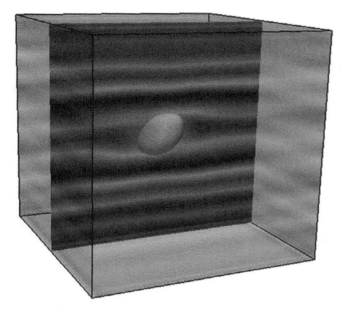

Figure 2.13   Example 2 – Fluid velocity field and the current position of a spherical particle.

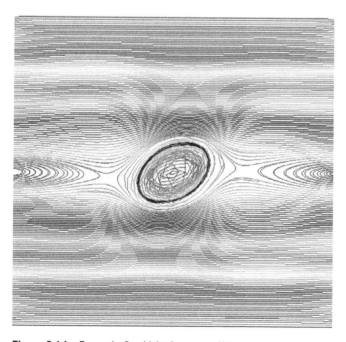

Figure 2.14   Example 2 – Velocity streamlines.

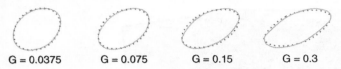

G = 0.0375   G = 0.075   G = 0.15   G = 0.3

Figure 2.15 Example 2 – Comparison of the final shape of spherical particle in simulation using the presented numerical model (dotted line) with results from literature [14] (solid line).

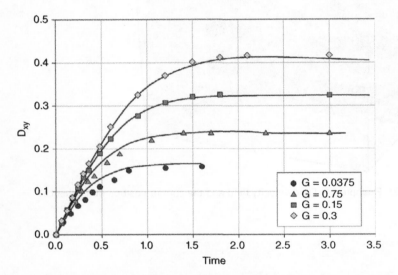

Figure 2.16 Example 2 – Variation of Taylor deformation index over time, for different values of parameter G (solid lines represent results from literature [14], and points represent results of simulation using the presented numerical model).

Varying the dimensionless shear rate G, different values of Taylor deformation index D_{xy} are obtained. Figure 2.15 shows the final deformation of the spherical particle for different values of G. The shape that is shown with dotted line is obtained using the presented numerical model (denoted as LB simulation), while the solid line shows the shape that is obtained in literature [14] and like it can be observed, a good agreement of results is obtained.

Figure 2.16 shows the variation of Taylor deformation index over time, for 4 different values of parameter G. Dots represent values that are obtained using the numerical model presented in this chapter, and lines represent results obtained using other methods in literature [14]. Time is shown in dimensionless form on the diagram and it is calculated as:

$$t_d = iter \cdot \dot{\gamma} \qquad (2.86)$$

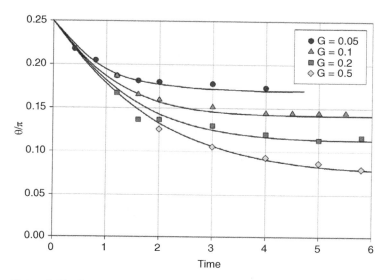

Figure 2.17 Example 2 – Variation of inclination angle over time, for different values of parameter G (solid lines represent results from literature [17], and points represent results of simulation using the presented numerical model).

where *iter* is the current iteration in the simulation software.

Also, the angle between the largest semiaxis of the ellipse and x axis (shown in Figure 2.12) varies over time and with particle deformation. Values of this angle also depend on parameter G and this dependence is analyzed by Ramanujan and Pozrikidis [17]. Figure 2.17 shows the comparison of results that these authors obtained with results obtained using the presented numerical model.

For comparison purposes, a simulation was performed for a rigid elliptical particle. The behavior of elliptical particle was already simulated in literature [21], where the particle was fixed in the single shear flow (the lower wall velocity was set to zero, and the upper wall velocity was equal to the predefined value). The ratio of the major and minor semiaxes of the ellipse was set to be 0.5 and the initial angle was equal to $\pi/2$. The analytical solution for this problem was determined by Jeffery [64], where equations of variation of angular velocity ω, and angle of rotation θ over time are defined. This analytical solution is compared with the results of numerical simulations obtained using the LB method in literature [21] and a good agreement of results is achieved. In this chapter, the similar setup to the one in cited literature is defined, only the elliptical particle rotates in a double shear flow (under the same conditions as the deformable particle), in order to compare the behavior of a rigid particle with the already analyzed behavior of a deformable particle. The change of inclination angle for the rigid elliptical particle

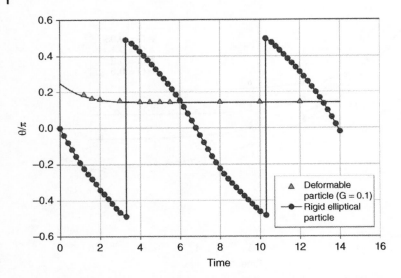

Figure 2.18 Example 2 – Variation of inclination angle over time, for a rigid elliptical particle and a deformable spherical particle - parameter G=0.1 (solid line represents results from literature [17], and points represent results of simulation using the presented numerical model).

and the deformable particle with parameter G equal to 0.1 is shown in Figure 2.18. The inclination angle for the deformable particle (like it was already shown in Figure 2.17) reaches a certain value and then remains fixed, when the particle starts rotating around its axis. In the case of rigid particle, the particle rotates constantly. The elliptical particle rotates following the similar pattern that was described analytically and numerically [21], with the difference in the rotation direction, due to the different fluid flow.

In the already cited paper, Ramanujan and Pozrikidis [17] considered the influence of viscosity of internal fluid on the deformation of the particle. Namely, the ratio between viscosity of fluid inside and viscosity of fluid outside the particle, defined in Eq. (2.82) varies from 5 to 10 for a RBC immersed in blood plasma and varies between 0.1 and 1 for a RBC immersed in the standard medium in experiments. For different values of parameter λ, the deformation of particle is also different. Figure 2.19 shows the variation of Taylor deformation index over time for 4 different values of parameter G, for the case of $\lambda = 0.2$, i.e. when the viscosity of internal fluid is five times smaller than the viscosity of external fluid. The results obtained using the presented numerical model are compared with results from literature [17]. The same variation is shown in Figure 2.20, but here the parameter is set to be $\lambda = 5$, i.e. the viscosity of internal fluid is five times greater than the

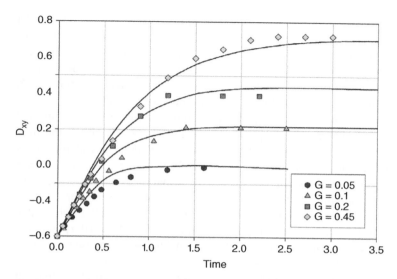

Figure 2.19 Example 2 – Variation of Taylor deformation index over time, for different values of parameter G, λ = 0.2 (solid lines represent results from literature [17], and points represent results of simulation using the presented numerical model).

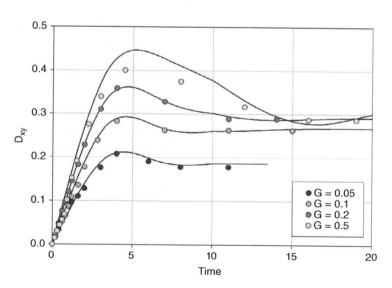

Figure 2.20 Example 2 – Variation of Taylor deformation index over time, for different values of parameter G, λ = 5 (solid lines represent results from literature [17], and points represent results of simulation using the presented numerical model).

viscosity of external fluid. Results obtained using the presented numerical model are again compared with results from literature [17]. It should be noted that in this case the results obtained using the presented numerical model differ from the results obtained in literature. This is consequence of the fact that in this chapter, four deformation parameters are considered, as it is explained in Section 2.2.3, while in literature only the membrane surface strain was considered.

In order to illustrate more appropriately the influence of viscosity of internal fluid on the deformation of the spherical particle, Figure 2.21 shows the final deformation of the particle, for the case of $G = 0.1$, for all three considered values of parameter λ. It is obvious that the largest deformation is reached for $\lambda = 5$ (shown with the solid line), and the smallest deformation is reached for $\lambda = 0.2$ (shown with dashed line), while the dotted line shows deformation for $\lambda = 1$.

Another important aspect that has to be considered when modeling deformable particles that are hyperelastic and that behave similar to the RBCs is the possibility to restore its original shape after the effect of external forces that caused the deformation is excluded. In literature [14], the authors tested this behavior on a spherical particle that is exposed to shear flow. After a certain period of time (in dimensionless units, this time is equal to $t_d = 1$), the upper and lower walls of the domain stop moving and the behavior of particle is analyzed. Again, the variation of Taylor deformation index is observed over time. In this simulation, the value of parameter G is set to be 0.075, and the Reynolds number is equal to 0.25. Figure 2.22 shows this variation, where the results from literature [14] are plotted with solid line, and results obtained using the presented numerical model are plotted with red dots. As it can be seen, a good agreement of results is obtained, which demonstrates that the developed software can accurately model this type of behavior of deformable particles. After a certain period of time (in dimensionless

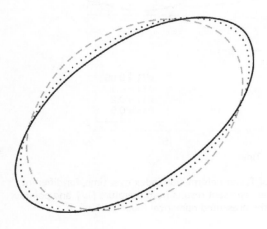

Figure 2.21 Example 2 – Deformation of the particle for G=0.1 for different values of the coefficient λ: Solid line - λ = 5; dotted line - λ = 1; dashed line - λ = 0.2.

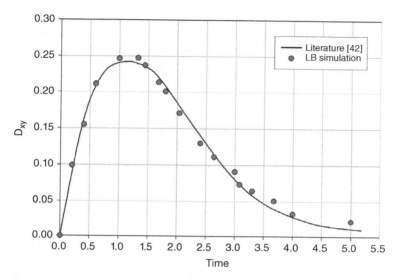

Figure 2.22 Example 2 – Variation of Taylor deformation index over time, for the case of restoration of initial shape.

Figure 2.23 Example 2 – Deformation of the particle during restoration of initial shape for G=0.075; solid circular line - t_d = 0; solid elliptical line - t_d = 1; dashed line - t_d = 5.

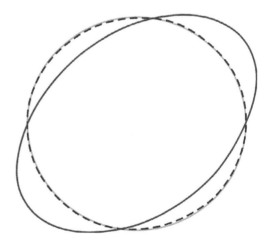

units, this time is equal to $t_d = 5$), the particle restores the shape from the beginning of the simulation, which is illustrated in Figure 2.23.

The behavior of RBC in the double shear flow was also analyzed. The fluid domain is defined in exactly the same way and values of all parameters necessary for LB simulation are the same as in the case of previously presented simulations with spherical particle. The dimensionless shear rate G was set to 0.2 in this simulation. The variation of Taylor deformation index over time was analyzed and the obtained results were compared with results published by Ramanujan and Pozrikidis [17]. Good agreement of results was obtained for the ratio between viscosity of internal fluid and viscosity of external fluid $\lambda = 1$ (which is shown in Figure 2.24), and for ratio $\lambda = 5$ (which is shown in Figure 2.25).

Apart from changing its shape and deforming over time, RBC also performs a rotation around its axis. Depending on the initial angle between the largest "semi-axis" of the RBC and the x axis, the change of angle of rotation is different. In all simulations in this case, the ratio between viscosity of internal fluid and viscosity of external fluid is set to be $\lambda = 5$, like it is the case of real RBCs in human blood. Results obtained using the presented numerical model are compared with results from literature [17] and a good agreement of results is obtained, as it can be seen in Figure 2.26.

During the motion of RBC in double shear flow, there is another phenomenon that can be observed. RBC can either rotate around its axis without changing its shape drastically (this is so-called tumbling behavior) or it can be practically fixed at a particular angle, while the membrane rotates around the center of gravity (this is the so-called tank-threading behavior). The behavior of RBC depends on the value of the capillary number [65]. Capillary number Ca is defined as the ratio

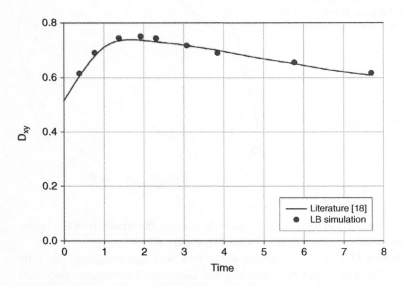

Figure 2.24 Example 2 – Variation of Taylor deformation index over time, $\lambda = 1$.

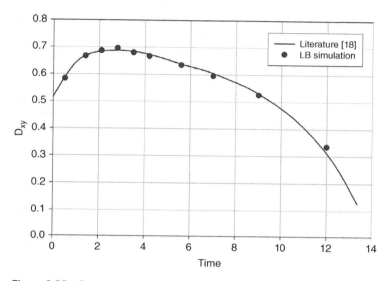

Figure 2.25 Example 2 – Variation of Taylor deformation index over time, $\lambda = 5$.

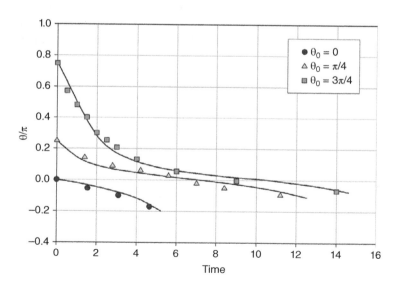

Figure 2.26 Example 2 – Variation of inclination angle over time, for different values of initial inclination angle (solid lines represent results from literature [17], and points represent results of simulation using the presented numerical model).

between fluid stress and characteristic membrane stress, where the characteristic membrane stress is related to the shear elasticity of the membrane.

$$Ca = \frac{\nu_{out} \, \dot{\gamma} \, r}{k_s} \tag{2.87}$$

Capillary number is actually equal to the dimensionless shear rate G that is defined in Eq. (2.84). Krüger et al. [65] concluded that if the value of this parameter is below the critical value, then the RBC will perform a tumbling motion, without larger deformation. On the other hand, if the value of this parameter is above the critical value, then the RBC will perform a tank-threading motion. They also calculated that this critical value is between 0.2 and 0.3. The change of shape of RBC over time when the capillary number is set to be $Ca = 0.1$ is shown in Figure 2.27, while Figure 2.28 shows the change of shape of RBC over time when the capillary number is set to $Ca = 0.5$. In both figures, the solid line represents results obtained by Krüger et al. [65], while the dotted line represents results obtained using the presented numerical model.

The streamlines for both considered cases are shown in Figure 2.29 (for $Ca = 0.1$) and in Figure 2.30 (for $Ca = 0.5$).

It is also interesting to analyze the motion of particle in double shear flow. Since the results will be compared with results published in literature [11], the simulation setup was defined accordingly. In this case, a two-dimensional simulation was performed, with the linear elastic material model, presented in Section 2.2.3.1, that was used for the modeling of surface strain. The main goal of this example is to compare the behavior of both particles under the same conditions. A circular

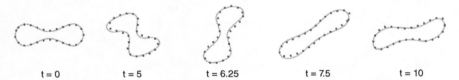

| t = 0 | t = 5 | t = 6.25 | t = 7.5 | t = 10 |

Figure 2.27 Example 2 – Change of shape of RBC over time, for $Ca = 0.1$; solid line – results from literature [65]; dotted line – simulation using the presented numerical model.

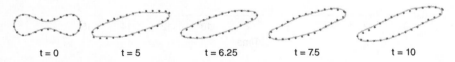

| t = 0 | t = 5 | t = 6.25 | t = 7.5 | t = 10 |

Figure 2.28 Example 2 – Change of shape of RBC over time, for $Ca = 0.5$; solid line – results from literature [65]; dotted line – simulation using the presented numerical model.

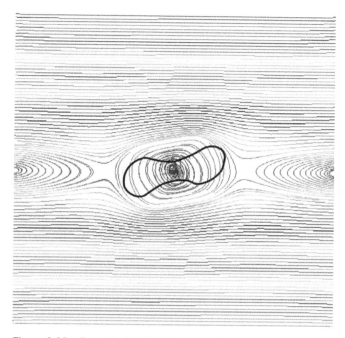

Figure 2.29 Example 2 – Velocity streamlines for $Ca = 0.1$.

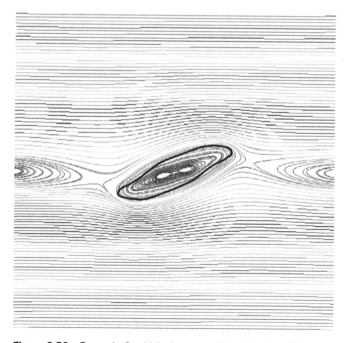

Figure 2.30 Example 2 – Velocity streamlines for $Ca = 0.5$.

particle is placed in fluid domain and is free to move. The boundary conditions are the same as before - the velocities of upper and lower walls are prescribed and these velocities have opposite directions (as it is illustrated in Figure 2.11). Dimensions of the domain are L_y and $L_x = 10L_y$. Along x axis, the periodical boundary condition is defined. Diameter of the spherical particle is set to $r = \frac{1}{8}L_y$. Particle is placed such that the coordinates of the center of gravity of the particle are equal to: $x_C = \frac{5L_y}{8}$; $y_C = \frac{L_y}{4}$.

This type of simulation was already performed in literature [66] and it was concluded that a particle placed anywhere in this domain tends to migrate to the centerline of the channel. In this chapter, the motion of both rigid and deformable particles was simulated and the obtained trajectories are shown in Figure 2.31. As it can be observed, both particles tend to move toward the centerline, and for the rigid particle, this tendency is lower than the tendency for the rigid particle. This is due to the deformation that occurs for the deformable particle. The shapes that this particle has during the motion in several moments in time are shown below the graph in Figure 2.31, while the positions of both particles in these specific moments in time are shown with dots on the graph. It can be observed that the deformable particle deforms to a certain shape and remains in this shape

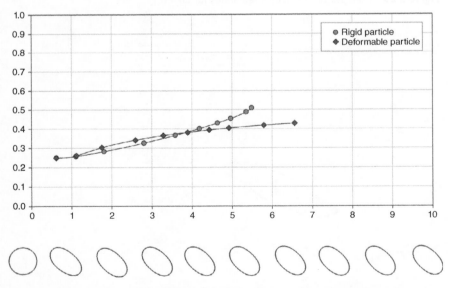

Figure 2.31 Example 2 – Motion of rigid and deformable particle through the domain during double shear flow; solid lines below the graph represent the shapes of deformable particle during motion in specific moments in time (denoted by diamonds).

throughout the simulation. This behavior is in accordance with the behavior of a fixed particle in double shear flow that was previously discussed in this example.

 This example has been simulated by others as well, using different techniques for solid–fluid interaction. In this chapter, the results obtained using the presented numerical model are compared with the results presented in literature [10, 11, 47]. Figures 2.32 and 2.33 show the change of x and y components of the dimensionless velocity during simulation. On both graphs, the line represents results from literature [10, 11, 47] which are obtained for a rigid particle, and velocity changes for both rigid and deformable particles obtained using the presented numerical model are plotted with dots, in order to compare the values for two types of particles. The change of x component for both particles is similar, only the velocity of the deformable particle remains higher, which can be also observed in Figure 2.31, since the deformable particle has a longer trajectory along the x axis than the rigid one. On the other hand, the value of y component at the beginning is greater for the deformable particle, since it first moves closer to the centerline (shown in Figure 2.31), but later both particles have the same tendency to move toward the centerline and the velocities remain at the similar value.

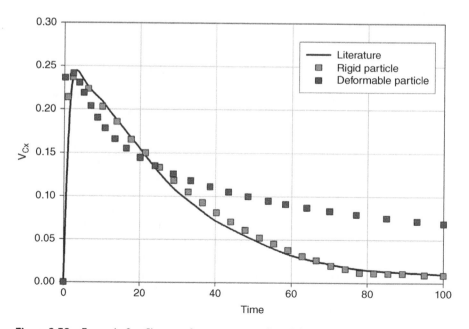

Figure 2.32 Example 2 – Change of x component of particle velocity during simulation of double shear flow.

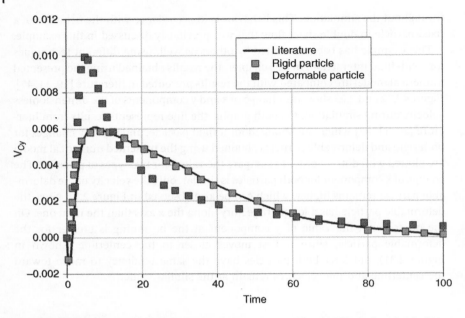

Figure 2.33 Example 2 – Change of y component of particle velocity during simulation.

2.3.3 Modeling Behavior of Particles in Stenotic Artery

In this example, the behavior of particles in fluid domain that represents a stenotic artery is analyzed. Since the results will again be compared with results published in literature [11], the simulation setup was defined accordingly. In this case, a two-dimensional simulation was performed, with the linear elastic material model, presented in Section 2.2.3.1, that was used for the modeling of surface strain. In this example, the behavior of rigid and deformable particles under same conditions will be compared. A circular particle is placed in fluid domain and is free to move. Dimensions of the domain are L_y and $L_x = 4L_y$. The domain is defined such that the stenosis is modeled as two parts (halves) of a sphere that are placed at the middle of the domain along x axis, and the shortest distance between the spheres along y axis is defined to be equal to $d = \frac{1.75}{8}L_y$, according to the simulation setup defined in cited literature [11]. The stenosis is simulated by applying the bounce-back boundary condition in the LB simulation. The shape and position of the stenosis is illustrated in Figure 2.34. Along the x axis, the periodical boundary condition is defined. The pressure difference on the inlet and outlet is prescribed as the initial condition and the velocity of the upper and lower wall is set to be equal to zero.

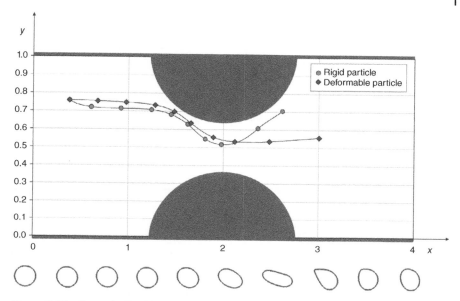

Figure 2.34 Example 3 – Motion of rigid and deformable particle through the stenotic artery; solid lines below the graph represent the shapes of deformable particle during motion in specific moments in time (denoted by diamonds).

Diameter of the spherical particle is set to $r = \frac{1}{16}L_y$. Particle is placed such that the coordinates of the center of gravity of the particle are equal to: $x_C = L_y$; $y_C = \frac{3}{4}L_y$.

The obtained trajectories of motion of both rigid and deformable particle are shown in Figure 2.34. As it can be observed, the motion through the stenosis is similar in both cases, while the deformable particle later continues to move faster and closer to the centerline of the domain, while the rigid one moves toward the initial position on the other side of the domain. This difference is again due to the deformation that occurs for the deformable particle. The shapes that this particle has during the motion in several moments in time are shown below the graph in Figure 2.34, while the positions of both particles in these specific moments in time are shown with dots on the graph. It can be observed that the deformable particle changes its shape and deforms significantly while passing through the stenosis, and that it starts returning to its initial shape afterward.

Figures 2.35 and 2.36 show the change of x and y components of the dimensionless velocity during simulation. On both graphs, the line represents results from literature [11] which are obtained for a rigid particle, and velocity changes for both rigid and deformable particles obtained using the presented numerical model are plotted with dots, in order to compare the values for two types of particles. The

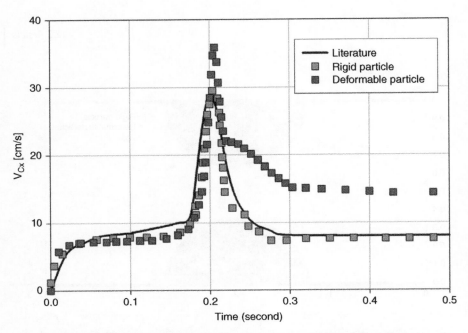

Figure 2.35 Example 3 – Change of x component of particle velocity during simulation of fluid flow through a stenotic artery.

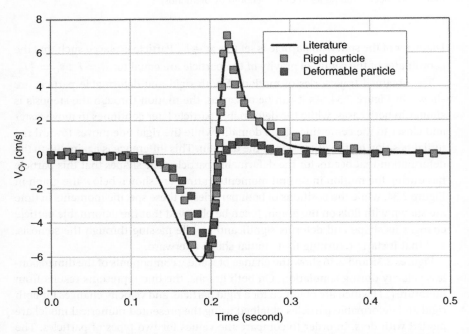

Figure 2.36 Example 3 – Change of y component of particle velocity during simulation of fluid flow through a stenotic artery.

deformable particle moves faster after passing through the stenosis, but generally, the change of the x component of velocity has a similar shape for both particles. On the other hand, since the deformable particle continues to move closer to the centerline of the domain (shown in Figure 2.34), the change of the y component of velocity is different and there is no growth of this component that is observed for the rigid particle.

A similar example was simulated using the presented numerical model in a three-dimensional domain in literature [67]. The obtained results are compared with experimental results published in literature [68]. Also, similar simulations were performed in literature [26, 69], where the fluid domain with complex boundaries actually represents a microfluidic chip for separation of circulating tumor cells. The results obtained in numerical simulations for the number of captured cells are compared with experimental results. In both cited papers, a good agreement of results is obtained and this demonstrates that this method can be used to accurately model the behavior of particles in complex domains with constrictions, like microfluidic chips.

2.3.4 Modeling Behavior of Particles in Artery with Bifurcation

In this example, the behavior of particles in fluid domain that represents an artery with bifurcation is analyzed. In this case, a two-dimensional simulation was performed, with the linear elastic material model, presented in Section 2.2.3.1, that was used for the modeling of surface strain. Again, the behavior of rigid and deformable particles under same conditions will be compared. A circular particle is placed in fluid domain and is free to move. Dimensions of the domain are L_y and $L_x = 3L_y$. The domain is defined such that the inlet branch of the bifurcation and two outlet branches have the same diameter, equal to $\frac{L_y}{3}$. The outlet branches are forming an angle of 30°. The boundaries of the bifurcation are simulated by applying the bounce-back boundary condition in the LB simulation. The shape of the artery is illustrated in Figure 2.37. The velocity is prescribed as a Poiseuille velocity profile at the inlet and the outflow condition is defined on the outlet branches. Diameter of the spherical particle is set to $r = \frac{1}{20}L_y$. Particle is placed such that the coordinates of the center of gravity of the particle are equal to: $x_C = L_y$; $y_C = \frac{L_y}{2}$.

The obtained trajectories of motion of both rigid and deformable particle are shown in Figure 2.37. The motion of rigid particles was already simulated in literature [21], using the multi-scale FE-DPD method. In this chapter, the same simulation setup was used and the obtained results are similar for the rigid particle. When the particle gets close to the bifurcation, a strong repulsive force is acting on the particle, and due to this force, the particle rebounds from the wall and

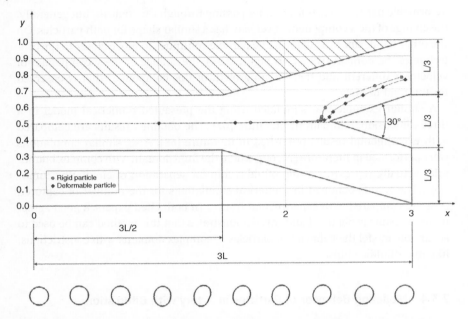

Figure 2.37 Example 4 – Motion of rigid and deformable particle through the artery with bifurcation; solid lines below the graph represent the shapes of deformable particle during motion in specific moments in time (denoted by diamonds).

continues to move through the upper branch. As it can be observed from Figure 2.37, the deformable particle moves slightly closer to the bifurcation wall and also comes closer to the bifurcation part of the artery during collision. This is due to the deformation of this type of particle, whose shapes during the motion in several moments in time are shown below the graph in Figure 2.37. The positions of both particles in these specific moments in time are shown with dots on the graph. In this example, the deformable particle changes its shape slightly during the collision with the bifurcation, and later returns to its initial shape.

This example was also simulated in three-dimensional domain, to compare the results with the experiment conducted by Secomb et al. [68]. This experiment was performed ex-vivo on a capillary bifurcation in the rat mesentery. The mesentery was removed from the abdominal cavity of the rat and then it was kept in appropriate laboratory conditions. Catheters were inserted into the artery to allow blood flow, the flow was then monitored and microscopic images and videos were generated. Details of the entire experimental procedure are described in literature [68, 70]. Special attention is dedicated to the artery with bifurcation.

The numerical model presented in this chapter is used to simulate the motion of RBC through a geometry similar to the artery with bifurcation. The geometry of the fluid domain is schematically shown in Figure 2.38. Diameter of the inlet

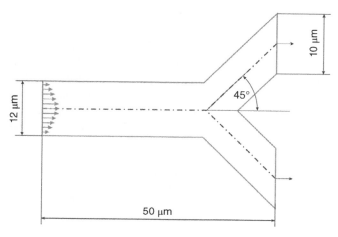

Figure 2.38 Example 4 – Geometry of the three-dimensional artery with bifurcation.

Figure 2.39 Example 4 – Fluid pressure field and initial position of RBC.

artery is set to be 12 μm, diameter of outlet arteries is set to be 10 μm. Velocity at the inlet is set to be 1 mm/s, and the velocity on the walls of the artery is equal to zero. In simulations, the viscosity of internal fluid inside the RBC is set to be five times greater than the viscosity of external fluid.

Figure 2.39 shows the position of RBC, as well as the pressure distribution at the beginning of the simulation, when the cell starts moving through the artery. The simulated motion of the RBC through the artery with bifurcation is shown in Figure 2.40. Change of shape of the RBC is illustrated in Figure 2.41.

A similar example was simulated using the presented numerical model in literature [25], where the three-dimensional fluid flow and motion of RBC was modeled in order to simulate the blood circulation of a fish species called zebrafish. The obtained results for the shapes of the RBC were compared with results obtained from videos that were taken during in-vivo experiments. The good agreement of these results presented in literature demonstrates that this method can be used

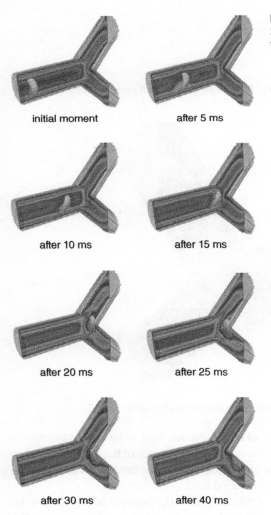

Figure 2.40 Example 4 – Simulation of motion of RBC through an artery with bifurcation.

initial moment

after 5 ms

after 10 ms

after 15 ms

after 20 ms

after 25 ms

after 30 ms

after 40 ms

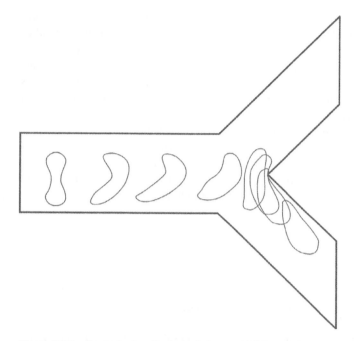

Figure 2.41   Example 4 – Change of shape of RBC over time.

to accurately model the behavior of particles in extremely complex domains, like the caudal vein plexus of zebrafish.

2.4   Conclusion

The presented numerical model and software can help to better understand and solve several real problems in biomedical engineering and improve medical treatments, such as the development of plaque and thrombosis; behavior, motion, and separation of cancer cells within the microfluidic chips; analysis and improvement of intravascular distribution and delivery of stem cells; and drugs and genetic material to target organs. In all these processes, it is difficult and expensive to perform experiments. Thus, numerical simulations are offering a good and efficient alternative.

In this chapter, the numerical model that successfully models both two-dimensional and three-dimensional fluid flow with immersed objects is presented. Movement of both rigid and deformable particles was analyzed in simple and more complex geometries that were used to define the fluid domain. Overall, 4 different examples are presented. Two examples are used to validate the accuracy of the presented numerical model and the other two examples are relevant in the field of biomedicine, because they illustrate the motion of particles through a bifurcation and an artery with stenosis. In all these examples, the behavior of two types of particles (rigid and deformable) is compared. Also, within this numerical model, it is enabled to model the motion of deformable particle, i.e. RBC, that is filled with fluid that has different characteristics than the surrounding plasma, which is absolutely identical to the real conditions in human organism.

Comparing the results obtained using the developed software with results that were published in literature, as well as with experimental results, it is shown that the presented numerical model and the developed software have a great potential for a practical application in modeling real phenomena in biomedical engineering on a microscopic level.

References

1 Hatzikirou, H., Brusch, L., Schaller, C. et al. (2010). Prediction of traveling front behavior in a lattice-gas cellular automaton model for tumor invasion. *Computers and Mathematics with Applications* **59** (7): 2326–2339.
2 Sequeira, A., Artoli, A.M., Silva-Herdade, A.S., and Saldanha, C. (2009). Leukocytes dynamics in microcirculation under shear-thinning blood flow. *Computers and Mathematics with Applications* **58** (5): 1035–1044.
3 Migliorini, C., Qian, Y., Chen, H. et al. (2002). Red blood cells augment leukocyte rolling in a virtual blood vessel. *Biophysical Journal* **83** (4): 1834–1841.
4 Sun, C. and Munn, L.L. (2008). Lattice-Boltzmann simulation of blood flow in digitized vessel networks. *Computers and Mathematics with Applications* **55** (7): 1594–1600.
5 Sun, C., Migliorini, C., and Munn, L.L. (2003). Red blood cells initiate leukocyte rolling in postcapillary expansions: A lattice Boltzmann analysis. *Biophysical Journal* **85** (1): 208–222.
6 Mateescu, D., Mekanik, A., and Paidoussis, M.P. (1996). Analysis of 2-D and 3-D unsteady annular flows with oscillating. *Journal of Fluids and Structures* **10**: 57–77.
7 Li, R., Tang, T., and Zhang, P.W. (2002). A moving mesh finite element algorithm for singular problems in two and three space. *Journal of Computational Physics* **177**: 365–393.
8 Hu, H.H., Patankar, N.A., and Zhu, M.Y. (2001). Direct numerical simulations of fluid–solid systems using arbitrary Lagrangian. *Journal of Computational Physics* **169**: 427–462.
9 Peskin, C.S. (1977). Numerical analysis of blood flow in the heart. *Journal of Computational Physics* **25** (3): 220–252.

10 Feng, Z. and Michaelides, E. (2004). The immersed boundary-lattice Boltzmann method for solving fluid-particles interaction problem. *Journal of Computational Physics* **195** (2): 602–628.

11 Wu, J. and Shu, C. (2010). Particulate flow simulation via a boundary condition-enforced immersed boundary-lattice Boltzmann scheme. *Communication Computational Physics* **7** (4): 793–812.

12 Đukić, T. (2012). *Modeliranje solid-fluid interakcije primenom LB metode.* Kragujevac: Master rad.

13 Krüger, T., Varnik, F., and Raabe, D. (2011). Efficient and accurate simulations of deformable particles immersed in a fluid using a combined immersed boundary lattice Boltzmann finite element method. *Computers and Mathematics with Applications* **61**: 3485–3505.

14 Murayama, T., Yoshino, M., and Hirata, T. (2011). Three-Dimensional Lattice Boltzmann Simulation of Two-Phase Flow Containing a Deformable Body with a Viscoelastic Membrane. *Communication Computational Physics* **9** (5): 1397–1413.

15 Decuzzi, P. and Ferrari, M. (2006). The adhesive strength of non-spherical particles mediated by specific interactions. *Biomaterials* **27** (30): 5307–5314.

16 Pozrikidis, C. (1995). Finite deformation of liquid capsules enclosed by elastic membranes in simple shear flow. *Journal of Fluid Mechanics* **297**: 123–152.

17 Ramanujan, S. and Pozrikidis, C. (1998). Deformation of liquid capsules enclosed by elastic membranes in simple shear flow: Large deformations and the effect of fluid viscosities. *Journal of Fluid Mechanics* **361**: 117–143.

18 Eggleton, C.D. and Popel, A.S. (1998). Large deformation of red blood cell ghosts in a simple shear flow. *Physics of Fluids* **10** (8): 1834–1845.

19 Sui, Y., Chew, Y.T., Roy, P., and Low, H.T. (2008). A hybrid method to study flow-induced deformation of three-dimensional capsules. *Journal of Computational Physics* **227** (12): 6351–6371.

20 Boryczko, K., Dzwinel, W., and Yuen, D.A. (2003). Dynamical clustering of red blood cells in capillary vessels. *Journal of Molecular Modeling* **9**: 16–33.

21 Filipović, N., Isailović, V., Đukić, T. et al. (2012). Multi-scale modeling of circular and elliptical particles in laminar shear flow. *IEEE Transactions on Biomedical Engineering* **59** (1): 50–53.

22 Filipovic, N., Zivic, M., Obradovic, M. et al. (2014). Numerical and experimental LDL transport through arterial wall. *Microfluidics and Nanofluidics* **16** (3): 455–464.

23 Djukic, T. and Filipovic, N. (2017). Numerical modeling of the cupular displacement and motion of otoconia particles in a semicircular canal. *Biomechanics and Modeling in Mechanobiology* **16** (5): 1669–1680.

24 Djukic, T., Saveljic, I., and Filipovic, N. (2019). Numerical modeling of the motion of otoconia particles in the patient-specific semicircular canal. *Computational Particle Mechanics* **6**: 767–780.

25 Djukic, T., Karthik, S., Saveljic, I. et al. (2016). Modeling the behavior of red blood cells within the caudal vein plexus of zebrafish. *Frontiers in Physiology* https://doi. org/10.3389/fphys.2016.00455.

26 Djukic, T., Topalovic, M., and Filipovic, N. (2015). Numerical simulation of isolation of cancer cells in a microfluidic chip. *Journal of Micromechanics and Microengineering* **25** (8): 084012.

27 Chen, S. and Doolen, G. (1998). Lattice Boltzmann method for fluid flows. *Annual Review of Fluid Mechanics* **30** (1): 329–364.

28 Dupin, M., Halliday, I., Care, C. et al. (2007). Modeling the flow of dence suspensions of deformable particles in three dimensions. *Physical Review E, Statistical Nonlinear, and Soft Matter Physics* **6** (75): 066707.

29 Kojic, M., Filipovic, N., Stojanovic, B., and Kojic, N. (2008). *Computer Modeling in Bioengineering: Theoretical Background, Examples and Software*. Chichester, England: John Wiley and Sons.

30 Kojić, M., Slavković, R., Živković, M., and Grujović, N. (1998). *Metod konačnih elemenata 1*. Kragujevac: Mašinski fakultet Kragujevac.

31 Filipović, N. (1999). Numeričko rešavanje spregnutih problema deformabilnog tela i strujanja fluida. PhD. dissertation. Mašinski fakultet Kragujevac.

32 Skalak, R., Tozeren, A., Zarda, R.P., and Chien, S. (1973). Strain energy function of red blood cell membranes. *Biophysical Journal* **3** (13): 245–264.

33 Wolfram, S. (1986). Cellular Automaton fluids 1: Basic theory. *Journal of Statistical Physics* **3/4**: 471–526.

34 Bhatnagar, P.L., Gross, E.P., and Krook, M. (1954). A model for collision processes in gases. i. small amplitude processes in charged and neutral one-component systems. *Physical Review E* **77** (5): 511–525.

35 Carter, A.H. (2001). *Classical and Statistical Thermodynamics*. New Jersey: Prentice–Hall, Inc.

36 Kennard, E.H. (1938). *Kinetic Theory of Gases*. New York: McGraw-Hill.

37 Grad, H. (1949). Note on the N-dimensional Hermite polynomials. *Communications on Pure and Applied Mathematics* **2** (4): 325–330.

38 Grad, H. (1949). On the kinetic theory of rarefied gases. *Communications on Pure and Applied Mathematics* **2** (4): 331–407.

39 Huang, K. (1987). *Statistical Mechanics*. New York: John Wiley.

40 Malaspinas, O.P. (2009). Lattice Boltzmann method for the simulation of viscoelastic fluid flows. PhD. dissertation. Switzerland.

41 Đukić, T. (2015). Modeling motion of deformable body inside fluid and its application in biomedical engineering. PhD. dissertation. Kragujevac: Faculty of Engineering, University of Kragujevac.

42 Krylov, V.I. (1962). *Approximate Calculation of Integrals*. New York: Macmillan.

43 Davis, P.J. and Rabinowitz, P. (1984). *Methods of Numerical Integration*. New York.

44 Fadlun, E., Verzicco, R., Orlandi, P., and Mohd-Yusof, J. (2000). Combined immersed-boundary finite-difference methods for three-dimensional complex flow simulations. *Journal of Computational Physics* **161** (1): 35–60.

45 Feng, Z. and Michaelides, E. (2005). Proteus: A direct forcing method in the simulations of particulate flows. *Journal of Computational Physics* **202** (1): 20–51.

46 Uhlmann, M. (2005). An immersed boundary method with direct forcing for the simulation of particulate flows. *Journal of Computational Physics* **209** (2): 448–476.

47 Niu, X.D., Shu, C., Chew, Y.T., and Peng, Y. (2006). A momentum exchange-based immersed boundary-lattice Boltzmann method for simulating incompressible viscous flows. *Physics Letters A* **354** (3): 173–182.

48 Wu, J. and Shu, C. (2009). Implicit velocity correction-based immersed boundary-lattice Boltzmann method and its application. *Journal of Computational Physics* **228** (6): 1963–1979.

49 Evans, E.A. and Skalak, R. (1980). *Mechanics and Thermodynamics of Biomembranes*. Boca Raton FL, USA: CRC Press.

50 Maeda, N. and Shiga, T. (1993). Red cell aggregation, due to interactions with plasma proteins. *Journal of Blood Rheology* **7**: 3–12.

51 Barthès-Biesel, D., Diaz, A., and Dhenin, E. (2002). Effect of constitutive laws for two-dimensional membranes on flow-induced capsule deformation. *Journal of Fluid Mechanics* **460**: 211–222.

52 Živković, M. (2006). *Nelinearna analiza konstrukcija*. Kragujevac, Srbija: Mašinski fakultet, Kragujevac.

53 Živkovic, M. (2011). *Mehanika kontinuuma za analizu metodom konačnih elemenata*. Kragujevac, Srbija: WUS Austria.

54 Skalak, R. and Chien, S. (1987). *Handbook of Bioengineering*. New York, USA: McGraw-Hill.

55 Waugh, R.E. and Hochmuth, R.M. (2006). Mechanics and deformability of hematocytes. In: *Biomedical Engineering Fundamentals* (eds. J.D. Bronzino and D. R. Peterson). Boca Raton, Florida, USA: CRC Press.

56 Tryggvason, G., Bunner, B., Esmaeeli, A. et al. (2001). A front-tracking method for the computations of multiphase flow. *Journal of Computational Physics* **169** (2): 708–759.

57 Francois, M., Uzgoren, E., Jackson, J., and Shyy, W. (2003). Multigrid computations with the immersed boundary technique for multiphase flows. *International Journal of Numerical Methods for Heat and Fluid Flow* **14** (1): 98–115.

58 Evans, E.A. and Fung, Y.C. (1972). Improved measurements of the erythrocyte geometry. *Microvascular Research* **4** (4): 335–347.

59 Bagchi, P., Johnson, P.C., and Popel, A.S. (2005). Computational fluid dynamic simulation of aggregation of deformable cells in a shear flow. *Journal of Biomechanical Engineering* **127** (7): 1070–1080.

60 Dao, M., Lim, C.T., and Suresh, S. (2003). Mechanics of human red blood cell deformed by optical tweezers. *Journal of the Mechanics and Physics of Solids* **51**: 2259–2280.

61 Chwang, A.T. and Wu, T.Y. (1976). Hydromechanics of low-Reynolds-number flow, Part 4, Translation of spheroids. *Journal of Fluid Mechanics* **75** (4): 677–689.

62 Filipović, N., Kojić, M., Decuzzi, P., and Ferrari, M. (2010). Dissipative particle dynamics simulation of circular and elliptical particles motion in 2D laminar shear flow. *Microfluidics and Nanofluidics* **10** (5): 1127–1134.

63 Xu, Y.-Q. and Tian, F.-B. (2013). An efficient red blood cell model in the frame of IB-LBM and its application. *International Journal of Biomathematics* **6** (1): 1250061–1250083.

64 Jeffery, G.B. (1922). The motion of ellipsoidal particles immersed in a viscous fluid. *Proceedings of the Royal Society of London A* **102** (715): 161–180.

65 Krüger, T., Gross, M., Raabe, D., and Varnik, F. (2013). Crossover from tumbling to tank-treading-like motion in dense simulated suspensions of red blood cells. *Soft Matter* **9** (37): 9008–9015.

66 Feng, J., Hu, H.H., and Joseph, D.D. (1994). Direct simulation of initial value problems for the motion of solid bodies in a Newtonian fluid. Part 2, Couette and Poiseuille flows. *Journal of Fluid Mechanics* **277**: 271–301.

67 Ðukić, T. and Filipović, N. (2015). Numerical simulation of behavior of red blood cells and cancer cells in complex geometrical domains, *IEEE 15TH International Conference On Bioinformatics And Bioengineering (BIBE)*, Belgrade, Serbia.

68 Pries, A.R. and Secomb, T.W. (2008). Blood flow in microvascular networks. In: *Microcirculation*, 2ee (eds. R.F. Tuma, W.N. Durán and K. Ley), 3–36. Academic Press, ISBN 9780123745309, https://doi.org/10.1016/B978-0-12-374530-9.00001-2.

69 Djukic, T. (2020). Numerical modeling of cell separation in microfluidic chips. In: *Computational Modeling in Bioengineering and Bioinformatics* (ed. N. Filipovic). Academic Press, ISBN 9780128195833, https://doi.org/10.1016/B978-0-12-819583-3.00010-2.

70 Secomb, T., Styp-Rekowska, B., and Pries, A.R. (2007). Two-Dimensional simulation of red blood cell deformation and lateral migration in microvessels. *Annals of Biomedical Engineering* **35** (5): 755–765.

3

Application of Computational Methods in Dentistry

Ksenija Zelic Mihajlovic[1,2], Arso M. Vukicevic[3], and Nenad D. Filipovic[3,4]

[1] *Laboratory For Anthropology, Institute of Anatomy, Faculty of Medicine,*
University of Belgrade, Serbia
[2] *School of Dental Medicine, University of Belgrade, Serbia*
[3] *Faculty of Engineering, University of Kragujevac, Kragujevac, Serbia*
[4] *Bioengineering Research and Development Center Kragujevac,*
Kragujevac, Serbia

3.1 Introduction

Stomathognatic system with its anatomy and function represents a very complex mechanism. In order to simulate its basic function, it is necessary to have a very precise and accurate model which assumes development of model with accurate geometry, material prosperities, and occlusal loads that acts on it. Furthermore, in dentistry, a huge part of research is focused on artificial materials. Some of them are used to replace destroyed part of a tooth – for example dental filings, crowns, and dental posts, others are replacements of the whole tooth or teeth – dental implants, and fixed or removable prosthesis. All artificial materials that are mentioned above are required to function in oral environment and are subjected to mechanical occlusal forces (derived by muscles during biting). Thus, research related to the quality and survival of these materials, but also living tissues functioning together with these materials, must consider their biomechanical behavior.

Among various computational methods, finite element method (FEM) has become widely accepted and very often used in dental research. The number of published articles that has been using FEM as a research tool in dentistry has significantly increased during the last decade.

Computational Modeling and Simulation Examples in Bioengineering, First Edition. Edited by
Nenad D. Filipovic. © 2022 The Institute of Electrical and Electronics Engineers, Inc.
Published 2022 by John Wiley & Sons, Inc.

In the following chapter, the following will be discussed:

- The advantages of using FEM in research related to dentistry
- The basic parts of a stomathognatic system that need to be model for finite element analysis (FEA)
- Development of the use of FEM in dental research
- Application of the FEM in major dental topics
- Examples of using the FEM in dental research

3.2 Finite Element Method in Dental Research

It was already stated that dental materials must fulfill the minimum need of stability and resistance to occlusal force (biting force) in wet and warm oral environment. After testing mechanical properties of a new material in the laboratory, there is a need to test its survival in a more realistic artificial model of jaws. In order to perform this kind of testing, the model must be as accurate as possible. However, it is very difficult to build even a simplified model because, in this case, there are a large number of different tissues, with different material properties involved. Also, the function we are attempting to recreate is very complex and difficult to simulate [1]. Furthermore, stress analyses performed on artificial models are destructive and lead to significant deterioration of the model, especially when critical force is applied [2–4]. But most importantly, experimental method usually does not provide enough information. For example, the only information we can get is if the tooth, implant, bone, or other structure is damaged or broken. Still, these methods give insufficient information regarding stress level required to induce fracture and stress and strain distribution throughout the structures.

Computational methods have brought many opportunities in scientific dental research in both basic and clinically oriented investigations [5]. We can point out the FEA as the most appropriate research tool for biomechanical analysis. It is a nondestructive and noninvasive method which enables creating very complex organic systems and visualization of superimposed structures [6]. One of the most important advantages of using FEA simulation is lack of deterioration of the model after simulation. In FEA simulation, force direction, its location and magnitude can be easily changed if needed. Material properties and boundary conditions are also changeable. It is relatively easy to generate a series of FE models that differ only in one property or condition. This allows focusing on the influence of a specific factor of our choice. Most importantly, FEA simulation gives information about stress and strain distribution throughout the whole model, and in all structures. Therefore, critical breaking points can be visualized and stress magnitude can be calculated.

Most important advantages of the use of FEM as a research tool in dentistry are as follows:

- Possibility to model all necessary anatomical structures
- Possibility to simulate jaw movements in different functions
- Possibility to analyze mechanical stress and strain and get its numerical values through complete model
- Possibility to change only one key feature in each model or simulation in order to stress out the influencing factor
- Possibility to test fatigue
- Possibility to repeat measurements and simulations without changing anything else (mechanical models are damaged after testing and the simulations cannot be repeated in completely the same situation)
- The models and simulations are reusable and can be used again in different research studies
- It enables visualization of all structures and stress distribution throughout them
- It is a noninvasive technique

However, the complexity of a stomathognatic system (as well as any other system in a living organism) is still impossible to fully and precisely duplicate. But the results obtained by FEA are only as accurate as the model is realistic.

Main disadvantages are as follows:

- It is almost impossible to create a completely accurate model.
- Any mistakes in input data – morphology, mechanical properties, boundary and loading conditions – will give totally incorrect result.
- It is difficult to perform validation of the simulation. The best way is to perform in vivo validation, but usually this is impossible or ethically inappropriate.

During the last few decades, a significant progress has been achieved in order to obtain better and more truthful models. The following parts of this chapter refer to some important concerns that need to be addressed in performing FEA of jaws and teeth.

3.2.1 Development of FEM in Dental Research

First models and FE simulations had various simplifications regarding the modeling of geometry (2D models, very large finite elements, neglecting of various important structures...) [6, 7]. Also, assumptions of loads and boundary conditions and material properties were sometimes applied [8].

In an early study of Barbenel from 1972 [9], biomechanical behavior of temporomandibular joint was investigated in two dimensions and all anatomical components were assumed to be rigid. However, since oral tissues such as teeth and bone have irregular and nonuniform morphology but also anisotropic material properties, 3D models generally would provide more accurate results than 2D models. One of the first detailed 3D models published by Koolstra and coworkers in 1988 [10] and validated in 1992 by the same authors [11] had detailed geometric information regarding skull size, muscle attachments, and cross-sectional areas obtained from cadaveric studies.

In some FEA studies, jaw bone material was specified as homogeneous and isotropic [12, 13]. However, it has been experimentally confirmed that the cortical bone (outside layer of bone) shows anisotropy [14, 15] meaning that material properties such as elastic moduli differ depending on the direction of the applied force. For example, in the case of mandibula (lower jaw), Schwartz-Dabney and Deschow in 2003 [16] found orthotropic material properties of mandibular cortical bone and emphasized that the maximal rigidity considerably varies along the cortical bone. They separated mandibular cortex into 31 different zones on both facial and lingual side. Apicella et al. [17] also emphasized the importance of including the cortical bone orthotropicity. In early FE studies, materials were specified as isotropic as it was very difficult to build a model with anisotropic properties. Also, this consumed substantial amount of time and software. But, later, with the development of the more sophisticated software, it has become more common to employ anisotropicity on the FE model [18].

3.2.1.1 Morphology and Dimensions of the Structures – Application of Digital Imaging Systems

As mentioned before, in early studies, morphology and dimensions of structures were determined based on cadaveric studies [19]. With the advances of digital imaging systems, (computerized tomography – CT, magnetic resonance imaging – MRI, and cone beam computerized tomography – CBCT) it is now possible to obtain very precise information about the geometry of bone and teeth structures [20–23]. Even more, with the use of micro-computerized tomography (MicroCT) [24], the visualization of very small structures can be achieved, and it is possible to extrapolate specific data to an FEA model.

In literature, stress on FE models under load is often expressed using Von Mises stress criterion [25, 26]. However, it has been suggested that the analysis of the principal stress is more informative concerning the fracture mechanisms in teeth, given that tensile and compressive stresses cause different outcomes on different tooth tissues (enamel and dentin) [27]. Namely, tooth tissues have different properties and different tensile and compressive strengths, which are addressed in the analysis of maximum principal stress. This is also true for other materials (natural or artificial) in the FE models of jaws and should be taken into account when FE simulations are performed and analyzed.

3.2.1.2 FE Model – Required/Composing Structures

A realistic FEA model of the jaws has to include several structures (Figure 3.1):

- Bone
  - Compact bone (cortical bone, cortex) – thin outside layer
  - Spongious bone (trabecular bone, cancellous bone) – inner part of the bone
- Teeth
  - Enamel (thin outside layer of the tooth crown)
  - Dentine (main tooth tissue)
  - Pulp cavity with pulp tissue (hollow inside of the tooth filed with pulp tissue)

(a)

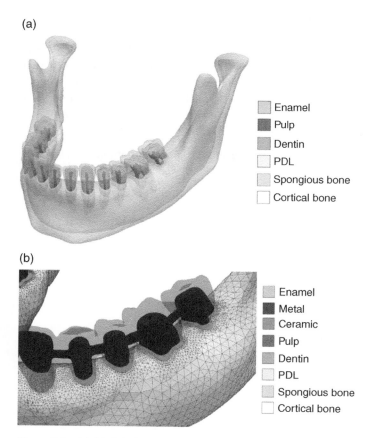

(b)

Figure 3.1 (a). Model of complete lower jaw with all needed anatomical and histological structures (b). Detail of a model with artificial materials (metalo-ceramic bridge).

- Periodontal ligament (PDL)
- Simplified temporomandibular joints (blocs of bone, which will be constrained in all directions and separated with soft tissue with the aim to amortize the stress distribution and prevent over-constrain)
- Artificial material if needed

3.2.1.3 Simulating Occlusal Load

Occlusal loading refers to the stress generated on teeth during biting and clenching. This stress is further distributed on the surrounding bone and soft tissue. Occlusal load (or bite force) can be simulated in different ways on an FE model (Figure 3.2).

1) The simplest way is to apply force on a single node or element, or on a small area of the occlusal surface of the tooth (occlusal surface is the one that is in contact with the antagonistic teeth during clenching).
2) Another way is very similar; however, the force is not applied through a single point but through a sphere (representing the cusp of the antagonistic tooth) that is transferring pressure to the occlusal surface and making contact with the tooth in two, three or four points; in other words, modeling occlusal load as nonlinear contact between tooth occlusal surface and cusp of its antagonistic tooth.

(a)   (b)

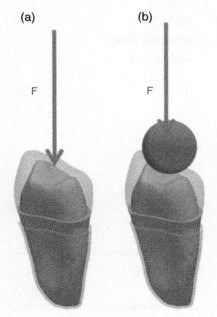

Figure 3.2 (a). Linear static occlusal load on tooth surface (b). Modeling occlusal load as nonlinear tooth-to-tooth contact.

F   F

Although these two solutions seem similar, they are quite different. The first one is the conventional linear elastic FE model. The second one uses nonlinear contact analysis that gives results much closer to reality [28].

Comparison of these two approaches revealed significant differences in favor of the later one. Namely, it has been shown that with the first method maximum voluntary force of 618N (experimentally measured in a 34-year old man) would fracture the enamel of the crown, which is inconsistent with the empirical experience. The second method gave much more realistic levels of stresses [5, 29].

However, the best way of simulating the occlusal load, e.g. bite force is modeling of muscle force [30, 31]. In other words, developing a realistic model of the jaws morphology and function. In order to drive biomechanical musculoskeletal model, information about muscle activations is required. Nowadays, muscular activation is obtained through the use of electromyography (EMG) [5]. This diagnostic tool gives information about the muscle force. However, how to simulate muscle forces in a realistic way is probably the most challenging part of an FEA dental research. The masticatory muscles originate on the skull and are inserted into the mandible (lower jaw), allowing jaw movements through contraction. There are four main masticatory muscles. Three of them are in charge of abduction of the mandible and one assists in its abduction. Each muscle fiber is attached in a specific place on lower jaw and corresponding place on skull. When the force is simulated on the FE model, they should be spread over the areas of the model that match the attachment areas of the particular muscle (Figure 3.3). It has been strongly suggested that because of the complexity of muscle force directions, creating skeletal muscles as

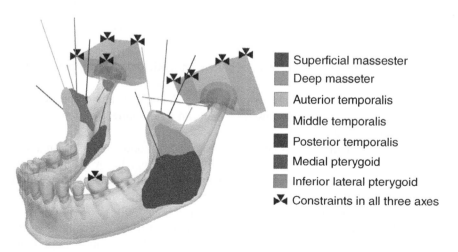

Figure 3.3 Muscles attachment areas, direction of forces, and constrains.

one-dimensional lines of action might cause major errors [32, 33]. The locations of those attachment sites and the orientations of muscle lines in contraction can be obtained from cadaveric dissections but also novel-motion capturing techniques are introduced [33].

The goal of modeling muscle attachment sites and simulating their activation is producing realistic bite force. By combining muscular activation with biomechanical models of maxillo-mandibular system, a bite force can be computed [30]. Bite force is crucial in all FEA research related to dentistry. The survival and morbidity rate of any artificial structure or material in the mouth depends mostly on the level of force applied on it, but also on its mechanical properties. Another phenomenon that significantly influenced the structures in the mouth is fatigue of the material [34]. However, fatigue is also the consequence of the cycling bite force.

The question of muscle activity was addressed by Groning et al. in a study from 2012 [30]. They have reported that great care should be taken when determining the orientation of muscle forces. To do this properly, we need to model complete masticatory system with lower jaw and skull, which is very time and software consuming. However, they stated that if only lower jaw is modeled, the attachment sites on the skull are needed but can be obtained by adding a simplified model of the skull just in order to obtain proper direction of the muscle force. The same is applicable if only upper jaw is modeled-lower jaw can be added as a simplified model. If we do not have all needed data for the same individual, muscle directions could be estimated based on the morphology of a skull with similar size and morphology [30].

3.2.1.4 Boundary Conditions

Overall, models related to dental research can be divided in two main groups:

1) Models of isolated parts of bone (usually with teeth or implants)
2) Models of the masticatory system

In the first group, only a small part of bone with tooth and all necessary tissues and materials is modeled.

In the second group, the whole system is modeled, but if the aim of the study is referred to the lower jaw, upper jaw, and skull can be simplified (as stated in the previous section) and vice versa.

To ensure that FE model does not undergo rigid body motion under loading, the model has to be sufficiently constrained, but should not be over-constrained because this can cause stress and strain artifacts. Models of bone parts are usually constrained in all three directions. This is performed by applying boundary conditions on the outer surface of the modeled part of bone [30].

FE models of complete mandibles are constrained at three points: at the bite point and at the two joints (Figure 3.3). This is used when the most important function of masticatory system is simulated – food processing (chewing). Namely, during chewing, food is moved through mouth by cheeks and tongue, but, during one bite, pressure is located only on one spot at the time. This is repeated many times but on different spots (teeth) [30]. Thus, there should be only one constrained point on the occlusal surface of the teeth for one simulation [30].

Another important question refers to constrain at the joint level. Temporomandibular joint (TMJ) is consisted out of three main structures – mandibular condyle (segment of the mandible), articular surface of the temporal bone and articular cartilage disc. There are two joints-left and right. The articular cartilage disc has buffering effect and prevents transferring high occlusal stress to the cranium.

The condyles can be constrained directly in all three directions [35, 36]. But, in order to simulate buffering effect of the articular cartilage, indirect constrain of condyles is applied. This is performed by creating simplified models of TMJ. Each TMJ is made of a block of temporal bone and a disk representing the TMJ cartilage (Figure 3.3). Each block of temporal bone is constrained in all three directions [37–39]. Groning found that the inclusion of a simplified TMJ model has a major effect on the results [30]. It is a way to avoid over-constraining of the FE mandible models, which occurs if the condyles are constrained directly.

3.2.1.5 Importance of Periodontal Ligament, Spongious, and Cortical Bone
In early studies some important structures (line periodontal ligament or alveoli socket) have been neglected because they are very small [40, 41]. It was assumed that their influence is minor [42]. However, it was found that the importance of these structures if crucial. For example, the modeling of periodontal ligament (PDL) was ignored [43]. Periodontal ligament is a tissue connecting the tooth roots with the alveolar sockets and has an important role in load transfer from the teeth to the alveolar bone (bone adjacent to the tooth).

In 2011, Panagiotopoulou et al. performed a sensitivity study and found that including isotropic, linear elastic, and homogeneous PDL affects strains only in the region of the alveolar bone near to the loaded tooth, but not other remote areas of the mandible [44]. Their results suggested that PDL can be excluded from FEA studies if the area of interest is not close to the alveolar region. On the other hand, the results of Groning et al. in 2011 indicate that including the PDL significantly reduces the overall stiffness of the mandibular corpus [38]. This means that models without any interface between teeth and bone are expected to be too stiff, and authors suggested that this can be explained by simple beam mechanics. Their results also show that strain magnitudes increase unevenly across the whole mandible while strain distribution changes when PDL is included. Based on these findings, it can be concluded without doubt that PDL is an important structure in the dentoalveolar mechanism because it buffers the stress distribution through teeth to the bone. Therefore, its inclusion in the FE models is absolutely necessary.

Nevertheless, not only inclusion of PDL is necessary but also its adequate modeling. Most of FE studies incorporate PDL as a liner elastic material. However, this tissue shows nonlinear, anisotropic and viscoelastic behavior. Although time and software consuming, there have been many articles attempting to generate more accurate approach for modeling PDL. They are summarized in a review of Fill and coworkers [45]. Four types of constitutive modeling PDL approaches were distinguished: linear-elastic, hyperelastic, viscoelastic, and multiphase models. For further details about the variations of modeling PDL, readers can refer to [45] and [46].

The presence of the PDL is one of the main differences between natural teeth and dental implants. Implants are connected directly to the bone and the buffering effect is, therefore, absent.

The jaw bones (as well as other bones in the human body) consist of two parts. Inner part is spongious or trabecular bone. It is a spongy-like network consisting of very small interconnected bony trabeculae. The outside dense layer is the cortical bone or cortex. It is still not possible to model trabecular architecture in FE models. Thus, spongious bone tissue is usually modeled as a homogeneous bulk material. Groning et al. found that mechanical properties of cortical bone tissue should be determined very accurately, since it has been shown that this heavily affects the FEA results [30]. But, changing Young's modulus of the cancellous bone material has only a minor effect on the results. It was stated that this is the consequence of the large difference between Young's modulus of cortical and spongious bone tissue.

3.2.2 Overview of FEM in Dental Research – Most Important Topics in the Period 2010–2020

Many topics have been addressed in the field of dentistry by means of FEM. In some of them, this research tool has been used more often than in others. Basically, FEM deals with stress and strain distribution throughout the materials and complex structures while looking for weak points. Thus, most of the studies are related to artificial materials or structures replacing teeth or parts of teeth and their biomechanical behavior inside oral environment. In another group, there are studies analyzing normal conditions in dentoalveolar complex and behavior of certain structures, distributions of stress, and especially development of pathological conditions. These studies are very important as many of them gave important information about basic physiological processes. Furthermore, significant improvement of FE models, simulations, and analysis of the results has been obtained, thanks to these basic studies. Special group of studies is the one dealing with orthodontic questions. Namely, in orthodontics, teeth are moved by means of orthodontic appliances with the aim to place them correctly in the jaws. Finally, in some articles main topic is the influence of an outside force – trauma, such as risk of bone or tooth fracture in contact sports. In the following part of this chapter, the application of FEM in these research topics will be addressed.

3.2.2.1 FEM in the Research Related to Implants, Restorative Dentistry, and Prosthodontics

Most of the articles in this part of research were focusing on dental implants. This is probably because this part of dentistry is in expansion, growing, and developing very fast in last few decades. In the literature search (PubMed-indexed journals), number of dentistry-related research, which used FEM as a method, was almost 3000 in all times (searched on 24 December 2019). More than 30% of the articles published in this field were studies related to dental implants. During the last decade, more than a half of all FEM studies in dentistry were published – 1676 papers in total. And to confirm the previous statement, FEM studies of dental implants accounted for 37% (625). Even with all this research already performed and published, there are still a lot of open questions in this field, many controversial topics and a huge need for further scientific investigations. We must not forget that this is a part of dental treatments that is very expensive and because of that attracts a lot of attention. From the manufacturer's point of view, dental implantology brings a huge income, but on the other hand, failure is a bigger problem than in other dental branches. This is why a lot of money is invested in implant-related research overall. Coming back to our main topic, FEM is developing into a perfect tool for studying biomechanical behavior of implants and attached tissues. This is mainly because clinical or animal trials that might give enough information about mechanical behavior, stress distribution, or behavior of an implant inside a functioning living organism are impossible and ethically unacceptable. Thus, models of masticatory complex are needed in order to understand biomechanics. It has already been mentioned that completely realistic FE model is impossible to build, but, on the other hand, these models are much more realistic than artificial jaw models. Also, FEM allows focusing on one key morphological or material property as it is possible to build series of many FEM models that are different in only this one particular property. FEM is also repeatable and, last but not the least important, nondestructive.

Since the introduction of dental implants by Brånemark [47], they have become an unavoidable part of dental treatment [48]. Their everyday implantation is increasing ever since. As stated in the paragraph above, a great number of studies have been carried out in order to address problems that have arisen in this field along the way. Our overall knowledge of the behavior of dental implants inside the functioning masticatory system is increasing all the time and a huge part of this knowledge is owed to FEM.

Nowadays, implant survival is about 95% [49]. However, this refers only to the early complications – the loss of implant because osseointegration failed [49]. This usually occurs before prosthetic restoration (within approximately three months after surgical implantation). Namely, after implantation, bone tissue is expected to grow in between screws of implant and to be fully adherent to implant material. This is called osseointegration. However, other complications or even loss of implant can occur after the osseointegration and prosthetic restoration and these are named – late complications [49]. Severe loss of bone adjusted to implant is the main cause of the implant failure in the period after osseointegration. This pathological state starts as inflammation of peri-implant soft tissue (mucositis) and finally develops into peri-implantitis. Factors causing this bone loss are, first of all, biological. Namely, bacteria and their products continually cause inflammation, which destroys bone tissue. Some systematic factors are also important to be mentioned – diabetes, smoking, severe illnesses, and others. But also, very significant role in the mechanism of peri-implantitis development has the mechanical factors. First, it should be noted that stress distribution in dental implants is entirely different from that of a natural tooth [50]. The reason for this is complete absence of the periodontal ligament around implants. The intact tooth root is covered by periodontium, which gives the possibility of 25–100 µm micro-movements of tooth on axial direction. In contrast, dental implants, after osseointegration, can move just about 3–5 µm at the same axis. Tooth mobility is the consequence of the mechanism of periodontal ligaments, while implant mobility is only possible because of restricted deformation of the bone. As stated before, periodontal ligament has the role of a damper and reduces the effect of the stress on nearby bone. Moreover, periodontium is able to control the level of occlusal load because of proprioceptive sensors. The role of these sensors is protective – they stope the muscle contraction when the occlusal load is too high. As stated in the mechanostat theory by Frost [51], mechanical forces are strongly affecting both bone growth and resorption. Consequently, the osseointegrated implant can be overloaded and cannot survive due to this excessive loading [52]. Overloading can happen in natural teeth as well. But, thanks to proprioceptive sensors and other properties of living tissue, (especially PDL) usually this does not lead to the loss of a tooth. On the other hand, if the occlusal forces surrounding dental implant are spread homogenously, then the bone remains healthy and the implant-supported occlusal rehabilitation can be claimed for successful. If we want to prevent bone loss and implant failure, it is essential to be aware of the importance of biomechanical behavior of dental implants and bony tissues around them, to study and analyze it. The results of these studies and analyses, along with conclusions and guidelines for practitioners, could increase the survival rate of implant-supported restorations. Hence, as listed before, there has been a dramatic increase in the number of biomechanical studies related to the field of implant dentistry. Different methods have been used for the purpose of investigating stress distribution inside the peri-implant bone: photoelastic model, strain gauge analysis, and three-dimensional (3D) FEA. Because of many advantages listed in the beginning of this chapter, finite element method (FEM) was found to be the most commonly used method [49].

Based on the previous overview of the implant-related issues, it is obvious that the central question in this field of research was, does overload of implant-supported restorations can be a risk predictor for peri-implant bone loss and finally, implant failure? Several FEM studies gave positive answer to this question. Occlusal overload has been predicted to be a risk factor for peri-implant bone loss on animal models as well [49]. Further questions were focusing on the specific factors for this outcome. Which morphological feature causes overload? Which material property is inadequate? It has been stressed out that pattern of stress distribution in peri-implant bone is affected by the number of implants, size, material properties of implant elements and also by the quality and quantity of bone, which the implant was placed in [49]. It can be noticed that the biomechanics of implant-bone complex depends on many factors. In the previous sentence, only some of them were mentioned. And furthermore, many are still unknown. Most of FE studies related to dental implants are focused on these factors and their results are crucial for further clinical guidelines.

As the time goes on, some novel technical solutions are developing in implantology – for example, short implants [53, 54], complete dentures carried on four of six implants ("All on 4," "All on 6") [55, 56], and new materials. After the basic in vitro testing, biomechanical analyses of the implant-bone complex are required. This is why the number of FEM studies in dental implant research is not decreasing, but on the contrary, rapidly increasing.

The field of prosthodontics in dentistry basically deals with missing teeth replacement with fixed or removable dental prosthesis and also restoration of severally destroyed teeth with artificial dental crowns. There are several crucial topics in this field – biomechanics of fixed and removable dental prosthesis and the problem of posts and core of the tooth. The last one is particularly interesting, especially for FE analysis. It is clinically evident that a tooth that has undergone and procedure of a root canal treatment would break more easily, than others. Root canal treatment is one of the most commonly used dental procedures, during which pulp tissue from the root canal is removed, canal is substantially spread and finally filled with sealer and gutta-percha. Afterward, tooth crown is restored. Nevertheless, this removal of large amount of dental hard tissue makes tooth weaker, and this cannot be re-established with crown restoration alone. Posts are small sticks that are inserted in the root canal of the tooth in order to give him more strength. If tooth is heavily damaged, posts together with the core (part of the post–core system that is located in the coronal part of the tooth) are used to carry the artificial dental crown or tooth filing. However, they are made of different materials (gold, metal alloy, fiberglass, fiberglass reinforced with composite material...) and in different designs. From a clinician point of view, it is extremely important to know which post to use in which situation. This information can be obtained by FE analysis [57–59]. In the literature search (PubMed-indexed journals), there were 357 published articles investigating biomechanics of post and core system in all times and 228 of them were published in the last 10 years.

To name just some other interesting topics in the prosthodontic field, there is a huge interest in all-ceramic crowns and bridges (fixed dental prosthesis) [60]. Besides excellent aesthetics (which is probably the main cause of their invention), all-ceramic prosthesis show high biocompatibility, little plaque accumulation, and low thermal conductivity. Nevertheless, their load-bearing capacity is the main topic in focus of biomechanical research [61]. Zirconia (yttria-stabilized tetragonal zirconia polycrystal (Y-TZP)) was found to be stable, even in the molar region, where occlusal forces are on their higher level, but many questions are still open. In a review of Anusavice and coworkers, it is stated that there is no single in vitro test that will give enough information for clinical performance of these aesthetic prosthesis and that use of FEA can be much more informative [61].

Mixing together research in the field of implants and studies of aesthetically superior materials, there are numerous studies of implant-supported zirconia prosthesis [62]. To make this part of dentistry even more interesting, recently, zirconia implants, instead of golden standard-titanium implants, have been introduced and applied in dental practice [63]. But, since the failure of the first aesthetic implants (alumina implants), which led to withdrawal of this product from market, scientists and clinicians are being much more careful [64]. The cause of their failure was inadequate mechanical stability during function in oral environment.

In the restorative dentistry FEM has also found its place. Restorations of tooth crown or tooth filing replace the part of tooth crown that is lost usually due to dental caries or fracture. Tooth crown contains two very different hard tissues. Mostly, tooth is build out of dentin – tissue with similar strength as bone. Outside layer is enamel. It is a very hard but brittle material. When restoration material is applied, it replaces at the same time both tissues – enamel and dentin. But its mechanical properties are different from both tooth tissues. This is why failure of dental restoration is, even more than for implants, caused by inadequate stress distribution. FEM is, nowadays, used as a standard tool to study the localization and magnitude of stresses in restorations and in teeth [65, 66]. In these FE analyses, it is crucial to use principal stress analysis because of a great difference in resistance to compression and tensile stress between dentine, enamel, and restoration material [27]. The drawbacks in this field are the same as in others – it is still not possible to fully replicate the whole complex; in this case, tooth restoration complex. Research of the field in question is focused on behavior of new materials, new techniques of tooth preparation, and methods for material application. Also, very interesting property of composite restoration materials is shrinkage during polymerization. Namely, composite materials in dental practice are polymerized (hardened) very fast with the use of blue light. However, during this process, the material is shrieked a little bit. This phenomenon attracted attention of many researches [66, 67]. It has been discovered that the shrinkage of the composite material inside the tooth crown cause shrinkage stress on the border of the tooth cavity [68]. This is because bonding of the material to the tooth is very strong. The presence of this stress can further cause damage of the tooth or failure of restoration during function. Simulation of shrinkage is possible in finite element methodology [27, 69]. Some options are showed in the examples at the end of this chapter.

3.2.2.2 FEM in Analysis of Biomechanical Behavior of Structures in Masticatory Complex

Many studies have been performed that analyze separately biomechanical behavior of some tissues in normal conditions and also in some pathological states. Furthermore, through FEM, the development of several pathological lesions was investigated.

The research related to occlusal load distribution has been performed for a long time now. Traditionally, accepted theory of buttresses has been developed quite a while ago [70, 71]. This theory claims that occlusal load is distributed to the skull through specific osseous trajectories known as buttresses. The buttresses were anatomically illustrated as regions in the mid-facial bones that have appearance of vertical and horizontal pillars (Figure 3.1) composed of a thick cortical bone [70]. Ten buttresses were proposed, seven of them were assumed to be vertical and have the role of transferring most of the occlusal load, and three of them were horizontal with the aim to stabilize the vertical ones [70, 71]. Early application of FEM appeared the most promising for clarifying the pattern of occlusal load distribution through the mid-facial bones [72]. However, few published FE studies addressing this problem provided conflicting results that, in some cases, differed significantly from the theory of buttresses [20, 21]. Despite these inconsistencies, it is shown that FE can be used in addressing this question.

The sophisticated connection between teeth and alveolar bone (part of the jaws that contains teeth) has been known as periodontal ligament or PDL. It has also been named bone-periodontal ligament-tooth complex or fibrous joint. In the previous text, the importance of this structure and its presence in FE model has already been emphasized. However, some studies addressed biomechanical behavior of this complex alone [46, 73]. This topic is important because of its role in the masticatory system, and, according to biomechanical research, it was found to be crucial.

Further research is related to the effect of the occlusal trauma on PDL and alveolar bone. There is an empirical opinion that occlusal trauma (extensive occlusal force) will lead to faster and more severe loss of bone surrounding tooth for the duration of periodontitis [74]. During periodontitis, bone support of teeth is lost slowly, but ultimately, and without an appropriate treatment, the affected tooth is finally lost. This serious disease is caused by bacteria and their products [75–77]. However, in the clinical experience, it was noticed that teeth with inadequate position (with nonphysiological load distribution) or teeth under too high occlusal pressure undergo greater and more severe bone loss [74, 78]. The consensus about the precise role of the occlusion on developing periodontal disease has not been obtained so far [74]. But, it is worth noting that several FE analyses have been performed in order to solve exactly this particular problem [78–80].

Another interesting question is the development of non-caries cervical lesions. These lesions are developed on the junction of crown and root of the tooth, which is known as the cervical area. The term non- caries cervical lesion (NCCL) refers to cervical tooth wear not associated with caries. These lesions are usually attributed to abrasion from toothbrush or erosion by acid. However, Grippo in 1991 [81] reported that these lesions are associated to extensive non-axial loading forces resulting in high stress concentration exactly in this cervical area of tooth. He called it abfraction. A number of FEA studies supported the abfraction theory. However, this question still remains open. In a systematic review of Duanghtip and coworkers, out of 69 studies, 56 studies (81%) showed some evidence supporting the role of stress in the development of NCCL and only 13 (19%) did not [82]. Most of these studies used, as expected, the FE method [83].

It can be concluded that these basic studies are important as they give guidelines for creating proper FE models for other research.

3.2.2.3 FEM in Orthodontic Research

In orthodontics, the goal is to move malpositioned teeth with the aim of placing them into proper alignment. As a result, occlusal stress will be distributed adequately, food impaction between teeth will be avoided, and good conditions for maintaining oral hygiene will be achieved. Treatment, during which teeth are slowly moved, is performed with the application of dental braces or other appliances. Also, if needed, treatment can modify facial growth. Orthodontic tooth movement occurs when forces of the orthodontic appliance are brought to the teeth, resulting in several forms of displacement in the periodontal ligament (PDL) [44]. The stress in the PDL sets off the cellular reaction, which results in resorption (on one side of the tooth) and apposition (on the other side of the tooth) of alveolar bone and leads to tooth dislocation. Several topics are interesting here – the behavior and resistance of materials used for orthodontic appliances, the strength of their bonding to teeth, distribution of stress through the system, etc.

[84]. But more important is the analysis of the stress developed inside teeth, bone ant TMJ, its intensity, distribution and finally, damage to the living tissue that can be caused [23, 85]. In the literature research (PubMed), 605 papers, which used FEM in orthodontics research, were published. More than a half of them (371) were published during the last decade.

3.2.2.4 FEM in Studies of Trauma in the Dentoalveolar Region

Mechanical trauma in the region of face and teeth can cause significant injuries with large consequences for the patient's health and physical appearance. The prediction of the extent of a possible injury can be beneficial in contact sports or other situations that are expected to be dangerous. In other words, if it is known how big effect a blow in a certain area of the head will have, protection measures could be undertaken. One example is the study of Antic and coworkers [18, 22] in which the authors addressed the question of the presence of impacted lower third molar on the strength of the mandible. When lower jaw fractures, the location of this fracture is of crucial importance. The question is, do some situations can be beneficial, or in other words, lead to more favorable fracture location or magnitude? On the other hand, can the fracture be avoided if some prevention measures were undertaken? The FE analysis gave insight into the stress distribution and high stress concentration in the lower jaw during frontal and lateral blow in the face. Both the presence and the position of the lower third molar were found to be influencing the fracture pattern of the mandibular angle and condylar regions (these are the regions in close contact with the lower third molar). Other studies similar to this one are conducted and interesting results were obtained [13, 86]. FEM was proven to be very good method of choice for this king of research.

3.3 Examples of FEA in Clinical Research in Dentistry

In all mentioned fields of dentistry, simulations of different circumstances can be made. Two of them are particularly interesting for clinicians:

1) Simulation of critical breaking load (the case when, during chewing, a small, very hard object, for example a cherry or a fish bone, is unwillingly bitten);
2) Simulation of cyclic loading (this represents normal everyday use of the teeth) and investigation of the material fatigue.

Both of these simulations are described in the following section. These examples are taken from our studies published in two articles in 2015 [27, 69]

3.3.1 Example 1– Assessment of Critical Breaking Force and Failure Index

3.3.1.1 Background

In research related to restorative and endodontic dentistry, one of the main questions raised by many authors over years is, is the endodontically treated tooth weaker than healthy tooth and if yes, why and how much [87, 88]? In everyday dental practice, it is noticed that endodontically treated teeth are fractured more often than those with intact pulp tissue and even those with large tooth fillings. During endodontic (root canal) treatment, the root canal and pulp chamber in tooth is widened and a lot of tooth tissue (dentine) is removed from the inner side of the tooth. Also, a large part of dentine and enamel is removed from the coronal part. After the filing of the root with sealer and gutta-percha, tooth crown is restored with dental filing materials. However, it is difficult to discover is the main reason for tooth fracture. Is it directly linked to the lack of dental tissue or are there some other reasons involved [89]? And if the removal of the dental tissue is the principle reason for toot fracture, which part of tooth is in danger the most? In other words, which part of the tooth should we attempt to keep intact in order to prevent its fracture?

In order to answer these questions, we present how to perform a numerical FEM analysis and experimental validation of the influence of:

1) two-surface Class II cavity preparation restored with composite resin (in this situation, two walls of the tooth crown are replaced with composite material)
2) two steps of major dentine removal in the process of root canal treatment
   a) access cavity preparation
   b) root canal enlargement

Conducting of this study demanded designing of a framework, which included all necessary elements of the finite element methodology and gives reliable results. Furthermore, validation of the simulation is desirable.

It is obvious that unrestored tooth will fracture more easily than the restored one. Thus, after each of these steps (cavity preparation, access cavity preparation and root canal enlargement), the tooth crown was restored and FEA was performed only on restored teeth models.

To estimate the influence on tooth fracture resistance, Failure Indices based on maximum principal stress criterion (MPSC) were determined.

Validation of the FE analysis was performed in vitro on the same teeth that were used for developing FE models.

3.3.1.2 Materials and Methods

The graphical illustration of the workflow is shown in Figure 3.4. The figure describes the sequential steps that should be performed in order to develop finite

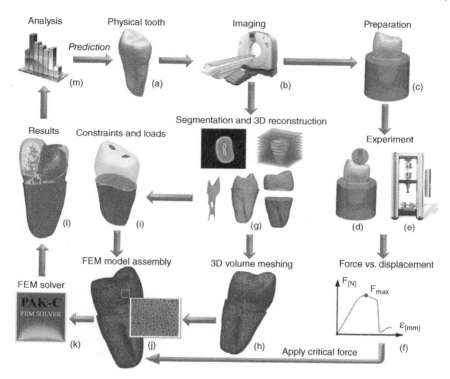

Figure 3.4 Schematic overview of the sequential steps performed in this study. (a, b) refer to preparation of a physical tooth and its scanning (CT or CBCT). (c–e) shows the preparation of the tooth for mechanical testing. (c, g, h, and j) show the phases of FE model preparation as well (i) which represents the point in this process where the loads and constrains are being set up. The FE analysis (k) gives the results (l) and then the results of the mechanical testing (f) and FEA are analyzed (m).

element model and perform analyses using a specific data/model. In the rest of this section, we describe key steps of our study from 2014 [27], which include preparation of tooth, imaging and three-dimensional reconstruction, and FEA.

3.3.1.2.1 Tooth Preparation

In this study, after obtaining the approval of Ethical Committee of The School of Dental Medicine, University of Belgrade, two intact upper second premolars that were extracted for orthodontic purposes and from the same patient were used. One tooth was left intact, while the other one underwent root canal treatment in vitro.

First step was the preparation of a mesio-occlusal (MO) Class II preparation. This means that two walls of the tooth crown were removed. The second step, as stated above, was performing an endodontic access opening to the root canals. During this step, a large amount of dentine was removed from the entrance to the root canals. Finally, the root canal enlargement was performed as follows: A size-10 file reached the working length (WL) without preparation. Serial step-back technique was used in order to enlarge the canal to the size 30 at full WL, with a 0.1 mm increment every 1 mm coronally. Irrigation with 1.0% NaOCl was used throughout the preparation. The canal was filled with cold gutta-percha and endodontic sealer. The tooth crown was restored with composite material.

3.3.1.2.2 *Imaging of Considered Teeth*

In order to obtain accurate 3D geometry for FEA, both intact and treated teeth were scanned by using CT scanner. In our study, we used the Siemens Somatom Sensation 16 (Figure 3.4b). The acquired scans were stored in the DICOM format for further reconstruction and analysis.

3.3.1.2.3 *Assessment of the Critical Force*

Validation of the FE analysis was performed by subjecting both teeth to the mechanical fracture test. With the aim to mimic physiological conditions, both teeth were fixed in auto-polymerizing acrylic resin (Technovit 4000, Kulzer, Wehrheim, Germany) [88]. Furthermore, periodontal ligament (PDL) was also imitated by covering the teeth root with a 0.1 mm thick layer of polyurethane impression material (Impregum, F, 3M ESPE, St Paul, MN). The teeth were sunken in water for 5 min during polymerization of the acrylic resin to prevent their overheating, which could influence their mechanical properties.

Assessment of the fracture force was performed by using the compression test (Figure 3.4e). Continuous compression of the subject (by forcing displacements) was performed while measuring the change of force in time. The experiment was performed on the professional equipment for the material testing ZWICK ROELL Z 100 Zwick GmbH & Co. KG. In order to better reproduce real mastication conditions (to mimic the cusp of the opposite tooth), the compressive loading was applied using a steel bar with a round tip. The tip was placed in the center of the tooth, so that it was in contact only with buccal and lingual cusps of the teeth. This way, we tried to reproduce conditions when a force is acting parallel to the axial axis of the tooth (physiological conditions) (Figure 3.4d–f). The speed of applied displacements was 5 mm/min. The accuracy of the system that measures the force was 1N, and the accuracy of the compressive strain was approximately 0.001 mm.

3.3.1.2.4 Development of the Finite Element Models

3.3.1.2.4.1   Reconstruction of 3D Models   After the acquisition of teeth CT scans as DICOM files, the segmentation and reconstruction of corresponding surface meshes from CT were performed by means of the Mimics 10.01 (Materialise, Leuven, Belgium). The process includes semi-automatic image thresholding and annotation, so that the accurate segmentations of material are obtained in each DICOM slice. Briefly, user is responsible to annotate material in 2D planes, while software is able to reconstruct-triangulate 3D surface (using the marching cubes algorithm) from a series of 2D slices (volume). At the end, a separate STL surface meshes were generated for each material of interest, as shown in Figure 3.4g.

3.3.1.2.4.2   Post-Processing of Raw 3D Mesh and Generation of "Virtual Models"   In most cases, the quality of a mesh retrieved from the Mimics software is not suitable for performing FEA. This is because user is not able to precisely segment each DICOM slice, and because the algorithm that reconstructs 3D mesh from the segmented 2D slices gives too dense mesh. Particularly, at this stage, the size of one triangle (composing blocks of the mesh) is determined with the resolution of CT scanner, which is commonly much smaller compared to dimensions of a tooth. As a consequence, the mesh contains much more triangles that are necessary to describe the surface of teeth; and moreover, many of triangles are of low quality (they have sharp edges), which negatively affects accuracy of FEA. Accordingly, additional mesh refinement and assembly of different composing parts of the model were done in Geomagic Studio 10 (Geomagic GmbH, Stuttgart, Germany) software. Geomagic provides features for manipulation of triangular surface meshes, which enables us to repair low-quality elements and cut, merge, or smooth out the regions of interest. It is important to understand that each contact surface between two materials needs to have shared nodes/triangles. Otherwise, we will not be able to model a tooth as one body (composed of various materials) but as a series of disconnected bodies. In the same software, we selected specific points (regions of triangles) that will be used by finite element solver to define boundary conditions. For each boundary condition, we selected and saved a separate STL file.

Intact tooth was labeled as Model 1. Structures that were included are as follows (Figure 3.5):

1) Enamel
2) Dentin
3) Pulp chamber with pulp tissue

Dentin   Enamel   Pulp   Gutta-percha   Composite resin   PDL

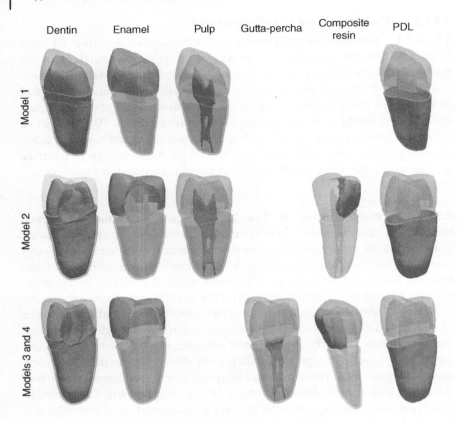

Model 1

Model 2

Models 3 and 4

Figure 3.5 Presentation of different structures in each tooth model.

4) Periodontal ligament (PDL), which was modeled as a 250-µm-thick shell surrounding the root, starting at 2 mm apically to the enamel edge (cemento-enamel junction)

A simplification of the model was applied by avoiding bone modeling. This was resolved by strangulation of the external nodes on the PDL.

In order to generate Models 2–4, we used data obtained from the scans of the contra lateral (treated) tooth.

In addition to the listed parts of the tooth, Model 2 included cavity preparation two-surface MO Class II restored with composite material (Figure 3.5).

Model 3 included the same structures as Model 3 and an access opening to the pulp chamber without root canal enlargement (Figure 3.5). The root canals were filled with cold gutta-percha.

Model 4 was the same as Model 3 but with root canal enlargement. The tin sealer was neglected in the FEA modeling (Figure 3.5).

The Model 4 corresponded to the in vitro treated tooth and was based on its CT scans. Models 2 and 3 were made by editing existing meshes.

3.3.1.2.4.3 Discretization of Three-Dimensional Meshes and Assignment of Material Characteristics
At the moment, each part of considered tooth models is defined with a single STL file, while one tooth is obtained by merging its composing parts. For the purpose of FEA, modeled volumetric domains need to be discretized into elements, in our case hexahedrons. In order to do this, we used the TetGen (Hang Si, WIAS, Berlin, Germany) meshing software, which uses triangular surface mesh and provides tetrahedral volumetric elements. At this point, high-quality four-nodal tetrahedral elements (TET4) mesh was retrieved, as shown in Figure 3.4h. Since our solver uses eight-nodal hexahedral elements, we afterward split the tetrahedral elements into eight-nodal hexahedrals by using the internal Matlab scripts (Figure 3.4j). Additionally, the script was able to identify each composing part of teeth, so that we were able to assign them corresponding material properties following **Table 3.1**.

3.3.1.2.4.4 Definition of Boundary Conditions – Constraints
Constraints applied to the nodes on the outer surface of the PDL are shown with dark gray color. The aim of the constrains was to restrict displacements in all three directions, which mimics fixation of teeth in mandible.

3.3.1.2.4.5 Finite Element Analysis
FEAs were performed by using PAK software [92, 93] (Figure 3.4k). All the analyses were static, linear, and all materials in the simulation were assumed to be homogenous, isotropic, and linear.

Table 3.1 Mechanical properties of dental structures and restorative materials.

Material	Young's modulus [Mpa] (Ref.)	Poisson's ratio (Ref.)
Pulp	6.8 [90]	0.45 [90]
Dentin	$18.6*10^3$ [91]	0.31 [91]
Enamel	$84.10*10^3$ [91]	0.30 [91]
PDL	0.68 [8]	0.45 [91]
Composite resin	$16.6*10^3$ [88]	0.24 [88]
Gutta-percha	70 [8]	0.40 [91]

3.3.1.2.4.6 Shrinkage Stress The volumetric change of the restoration during polymerization shrinkage was simulated in Models 2–4 as the analogy of thermal expansion in a heat transfer analysis following the procedure from literature [94].

3.3.1.2.4.7 Definition of Boundary Conditions – Loads We experimentally determined critical breaking force for the treated tooth, which is represented as Model 4. Then, we applied this force on each model with the aim to analyze stress distribution in each of them under the same conditions.

To prove the validity of the FEA simulation, the results of the mechanical testing of the intact tooth was compared with the FE simulation of Model 1 (intact tooth). The critical breaking force obtained for intact tooth was applied on Model 1.

This way, we were able to compare distribution of stresses from simulation with the experimentally obtained crack propagations.

The load for all models was applied on the buccal and lingual cups (dark spots on the occlusal surface of the tooth crown (Figure 3.4i)) at the same time, in order to develop the resulting force parallel to axial axis of the tooth.

In Models 2–4, which contained composite material, the previously calculated polymerization stress was taken into account.

The performed FEA simulations showed the stress distributions, which enable one to identify areas of stress concentrations. However, additional post-processing of the FEA results need to be performed in order to objectively assesses the risk of failure. Briefly, afterward analysis is necessary because each material has specific properties, which need to be accounted in order to numerically express the possibilities of failure of any given material under particular loads. In this chapter, we briefly explain how to numerically estimate the risk of fracture (Failure Index) for each tooth model (assuming that it is composed of various materials: enamel, dentin, PDL) by using the MPSC. The starting assumption of the MPSC is that tensile stresses are expressed with positive values and compressive stresses are expressed with negative values:

- the peak of tensile stress is defined as the maximum principal stress σ_1;
- while the peak of compressive stress is defined as the minimum principal stress σ_3.

According to the MPSC, the failure of a considered material occurs when its computed maximum principal stresses exceeds it experimentally determined tensile strength σ_{TS} or when the computed minimum principal stress is lower that its experimentally determined compressive strength σ_{CS}.

Therefore, it could be concluded that the safe operation zone for an arbitrary material is defended with the following equation:

$$\sigma_{CS} < \sigma_3 < \sigma_1 < \sigma_{TS}$$

Thus, the failure index was expressed as:

$$FI = \sigma_{PS}/\sigma_{SM}$$

where σ_{PS} is the principal stress and σ_{SM} is defined with the sign of computation of principal stresses.

If:

$\sigma_{PS} > 0$, then for the σ_{SM} is equal to the tensile strength of the analyzed material ($\sigma_{SM} = \sigma_{TS}$),

If:

σ_{PS} is negative (compressive stress), σ_{SM} takes the value of compressive strength ($\sigma_{SM} = \sigma_{CS}$).

Therefore, the failure of a material initiates when:

Failure Index, $FI = 1$

Calculations for Failure Indexes $FI(\sigma_1)$ and $FI(\sigma_3)$ correspond to stress distributions σ_1 and σ_3, respectively.

Beside principal stresses, the FEA analysis included the computation of the most commonly used effective (Von Misses) stress.

Finally, a weakening of each tooth model was computationally estimated using the following formula:

$$\text{Weakening}[\%] = 100 * [\text{Failure Index}(i) - \text{Failure Index}(1)]/\text{Failure Index}(1)$$

where Failure Index(1)is the value of the 1st model and "i" is the number of considered model, respectively.

3.3.1.3 Results and Discussion

This study was conducted with the aim to propose a framework for FEM analysis with experimental validation of the influence of occlusal load on tooth fracture resistance (Figure 3.4). The models were created on the basis of CT data of real teeth. This approach is much more appropriate in comparison with simplified FE models (created on the basis of approximate dimensions of teeth). The present framework also gives an option for the experimental validation of the computational results. Thus, we are performing numerical FEM analysis and verifying the results so the approach can be further used.

Apart from being a validation of the FEA, mechanical testing also gave us the magnitude of the critical breaking force. The results of the experiment showed that the treated tooth fractures under the load of 710N and the intact tooth under the load of 1025N. In the FEA simulation, all teeth models were loaded axially with the lower force intensity (710N). Additional validation of the FEA results was obtained by applying the load of 1025N on the model of the intact tooth. Briefly, if the stress analysis shows failure for the same loading as in the experiment and location of this critical stress is in the location of the tooth fracture line, then we can conclude that FEA

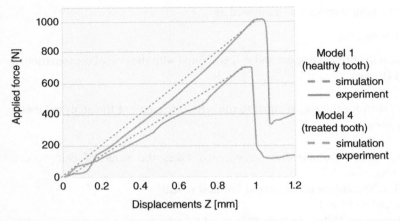

Figure 3.6 Diagram of compressive displacement (strain) dependencies on compressive force in experiment (intact and treated tooth) and in FE simulations which correspond to these teeth (Model 1 and Model 4).

is accurate. The result of the compression test is showed on Figure 3.6. It illustrates the distribution of compressive displacements (strain) under the compressive force. Our findings showed that both teeth fractured in the buccal cervical region. This graph from Figure 3.6 also shows the compressive force dependencies on the compressive displacement (strain) generated by the FE simulation for Model 1 (representing intact tooth) and Model 4 (representing treated tooth). The discrepancy between results obtained in the experiment and FE simulation is less than 10%.

Another important issue in this example is inclusion of the residual polymerization stress. This phenomenon is important in studies that include light – cured composite materials because of their shrinkage. Models 2–4 had previously calculated polymerization stress included because they were restored by composite material. The shrinkage simulation on FE models was performed as the analogy of thermal expansion [94]. There was a sharp transition between composite filing and enamel in the principal stresses distribution (Figure 3.7a–c). The compressive stresses are visible in the filing and they are the outcome of the shrinkage during polymerization (Figure 3.7g–i). Then again, as composite resin is bonded to enamel and dentin, in these tissues, in the area adjacent to the composite filing, the tensile stresses are generated (Figure 3.7d–f). These tensile stresses reached 2.35MPa in the Model 4, while maximal compressive stress in the composite was 9.4MPa (Figure 3.7f, i). In other models, the stress levels were lower (Figure 3.7d, e, g, and h). Failure Indexes were calculated for each material, and it was revealed that tensile stresses are very close to the enamel tensile strength. Thus, micro-cracks in the enamel can be caused by polymerization shrinkage. This was also reported in studies of Chuang and coworkers [94, 100].

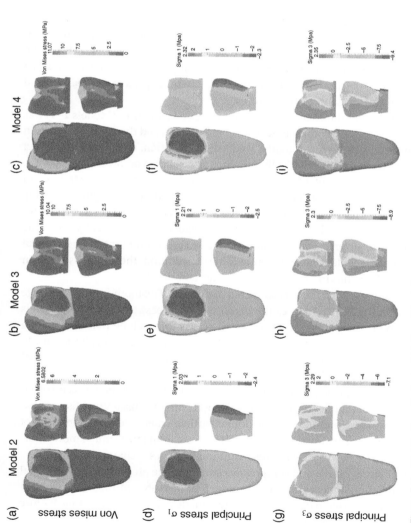

Figure 3.7 Distribution of principal (d–i) and Von Mises stress (a–c) observed during the simulation of composite shrinkage and corresponding Failure Indices.

Table 3.2 Values of the material strength (σ_{SM}) were adopted from literature:

Material	Tensile strength	Compressive strength	Reference
Dentine	$\sigma_{TS} = 105, 5\text{MPa}$	$\sigma_{CS} = -297\text{MPa}$	[95–97]
Enamel	$\sigma_{TS} = 24, 7\text{MPa}$	$\sigma_{CS} = -384\text{MPa}$	[98, 99]
Composite material	$\sigma_{TS} = 77\text{MPa}$	$\sigma_{CS} = -294\text{MPa}$	[94]

The specific aim of this study was to find out what are the individual and joint influences of composite restoration and two phases of root canal treatment on tooth resistance to fracture. We simulated critical breaking force which would, in reality, correspond to unwilling biting of, for example, a cherry, a bone, or some other hard small object. This is a situation that often happens in the real life and also a significant cause of teeth fractures.

To fulfill this aim, the effect of each of these three procedures (Models 2–4) was evaluated by means of biomechanical behavior analysis and the stress distribution was compared to of the intact tooth (Model 1). In addition, to estimate this effect on the tooth weakening numerically, a Failure Index based on MPSC was applied. The proposed formula of MPSC uses the tensile and compressive strength of enamel, dentin, and composite material (Table 3.2). Thus, on the figures showing Failure Indexes for each model, the differences in material properties (compressive and tensile strength) of these structures were considered. Consequently, the same stress can cause diverse Failure Indexes and as result, different biomechanical behavior of enamel, dentine, and dental filing. In addition, the stress distribution through interfaces linking two materials is more noticeable.

The biggest part of the tooth is consisted of dentine. The tooth is broken if large amount of dentine is lost. On the other hand, enamel is easy to replace because it is only covering the tooth crown. Therefore, it is easy to conclude that the loss of a tooth is never caused by enamel fracture [96]. However, we find it important to present biomechanical behavior of enamel and tooth filing under critical occlusal load in this study as well.

In this example, principal stresses were chosen to illustrate the stress distribution in teeth models under loads. However, most commonly used Von Misses stress equivalents was also presented in the original article order to allow comparison of the results with other studies. Principal stresses were found to be much more informative, especially in this case because hard tooth tissues – enamel and dentine – have very different tensile and compressive strength. The presentation of principal stresses and Failure Indexes in Model 1 under load of 710N and 1025N are given in Figure 3.8, while principal stresses on other models and corresponding Failure Indexes are given on Figure 3.9.

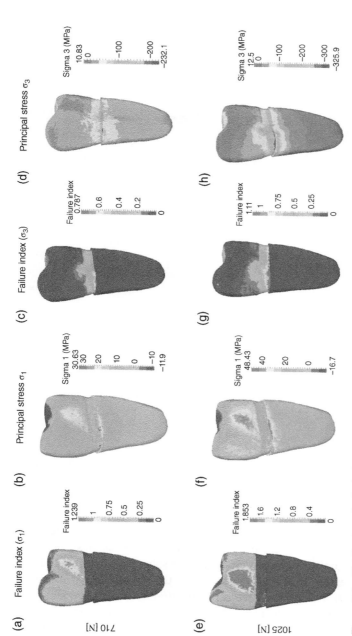

Figure 3.8 Distribution of principal stresses and Failure Indices in Model 1 under loads of 710N (a–d) and 1025N (e–h).

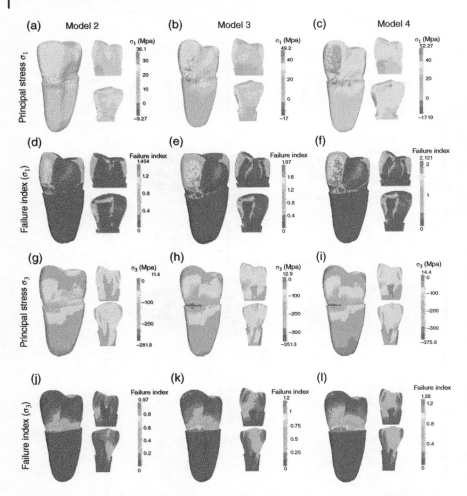

Figure 3.9 Distribution of principal stresses (a–c, g–i,) and corresponding Failure Indices (d–f, j–l) in Models 2–4 under load of 710N.

To be more specific, in all models, compressive stress is generated on the occlusal surface, at the contacts between occlusal surface and the antagonistic tooth (in the model, the antagonistic tooth is represented by a sphere) (Figures 3.8 and 3.9). In addition, compressive stress is also noticeable in the cervical area of all models (Figures 3.8 and 3.9). The cervical area is the junction of tooth crown and tooth root – in this area enamel is ending in knife-edge shape. These findings are important in context of the development of non-caries cervical lesions (NCCL), which is discussed earlier in this chapter. The level of compressive stress (during the loading with 710N) in dentin in this area did not reach dentine compressive strength (FI(σ_3) = 0,787) for intact tooth (Model 1), which means that simulations showed no risk for fracture with 710N applied on intact tooth. When Model 1 was loaded with 1025N (the force that fractured the intact tooth in vitro), risk of failure was obvious FI(σ_3) = 1.11. Tensile stresses generated in the enamel on lateral sides of the tooth (approximal sides) exceeds, under the load of 710N, tensile strength of the enamel FI(σ_1) = 1.239 in a small zone (Figure 3.8). As a result, cracks in the enamel can occur under these conditions (Figure 3.8).

Stress behavior in other three models was analyzed in the same manner as previously explained. The results showed that new areas of stress accumulation arise after removing part of the tooth crown and replacing it with composite material (Figure 3.9). This was expected because of the known differences between material properties. The stress generated because of the shrinkage of the composite was previously calculated and already discussed. However, the presence of this material weakened the tooth to some degree. For all three models, we used a formula to calculate the weakening of the tooth. Tooth with only composite filing was found to be weaker by 23.25% (Table 3.3). Fortunately, this was not enough to fracture the tooth. Since it was shown in the literature that occlusal load is usually under 700N in the posterior area [101] of the mouth (molars and premolars), fracture of under these circumstances is unlikely. This is in line with the everyday experience.

Table 3.3 Maximum effective stress, maximum displacement, maximum tensile and compressive principal stress, Maximum Failure Index in dentin and enamel and estimated weakening of each model of tooth under the load of 710N and 1025N on Model 1.

Model	1		2	3	4
Effective stress [Mpa]	126.32	187.15[b]	155.33	176.1	180.6
Displacement in model [mm]	0.649	0.913[b]	0.781	0.864	0.890
Tensile stress in enamel [Mpa]	30.63	48.43[b]	36.1	49.2	52.27
Failure Index in enamel [−]	1.239	1.853[b]	1.454	1.97	2.12
Compressive Stress in dentin [Mpa]	−232.1	−325.9[b]	−281.8	−351.3	−375.8
Failure Index in dentin [−]	0.787	1.11[b]	0.97	1.2	1.28
Weakening [a][%]	0		23.25	52.5	62.6

[a] Weakening[%] =100* [FailureIndex(i) − FailureIndex(1)]/FailureIndex(1)
[b] values correspond to the load of 1025N applied on Model 1

The situation is dramatically changed after the first part of the root canal treatment is performed (Model 3). The level of stress is increased, new stress accumulation areas appeared, and the tooth is further weakened by 29.25%. The Failure Index revealed reaching of the fracture risk under the load of 710N. Location of the fracture in this case would be the cervical area. We can suppose that the cause of this increased tooth fragility is the modification of the stress conduction.

Novel changes in the biomechanical behavior were not noticed after the fulfillment of the root canal treatment. Only higher stress levels were found. These findings suggest that the enlargement of root canals had small additional influence on tooth weakening (10.1%).

This study was able to notice and recognize the most important macromechanical factors for tooth weakening after some very common dental procedures. In addition, it was able to measure the weakenings through application of Failure Index based on MPSC.

The method was also experimentally validated, and the results were found to be well founded. Furthermore, the results are in agreement with everyday experience. The discrepancy between results obtained in the experiment and FE simulation is less than 10%.

Use of principal stress criterion was also found to be extremely important in this kind of research where several very different materials are in close contact.

Overall, the framework given by this study was proven to be reliable and can be used for further studies.

3.3.2 Example 2 – Assessment of the Dentine Fatigue Failure

In this example, another very interesting FE simulation will be presented. The example is based on our study published in 2015 [69].

3.3.2.1 Background

Namely, in the previous example, all important parts of the framework needed to conduct FE analysis were determined. But the force that was applied was extreme and exceeded the strength of dental tissue. On the other hand, materials usually fail due to cycling loading, which causes fatigue of the material. In the context of dento-alveolar complex, cycling load refers to mastication. The intensity of a single load is much lower, but this load is occurring on everyday basis. Cyclic mastication load is the cause of cyclic stress and, in time, may induce initiation and progression of micro-cracks and, ultimately, tooth fracture or "fatigue failure" [102]. Tooth fracture is related to dentin fracture, since this tissue makes the greatest part of the tooth. However, teeth are not expected to be fractured as a result of fatigue because they are created to withstand such forces. On the other hand, when teeth are weakened as a result of some lesions or required dental treatment, fatigue becomes more important. During these treatments, natural tissues are replaced with materials that are not perfect although being sophisticated and of high quality. These materials can suffer from fatigue at also remaining natural tissues can be more prone to fatigue after these changes. But, to be specific, what is really changed after the dental treatment? According to the previous example, and also numerous of other researches, stress distribution changes are the basic cause of tooth weakening. However, this kind of research may spread light to other unexpected causes.

If we want to test the fatigue of dentine as a material, we can do in vitro experiments with the bulk of dentine [103, 104]. But this will not give enough information because the morphology of the tooth plays an important role in its resistance to the load. Next, we can perform cycling load on teeth in vitro, but there is once again a limitation – this test is not repeatable on the same specimen (tooth), which is often needed. In addition, specific measuring of stress, strain, displacements, temperature, demand very expensive apparatus, and usually is not informative enough for a complex tooth structure. In such situations, FEA or FEM can be a tool of choice.

In the present study, we aimed to propose a procedure for numerical analysis of fatigue behavior of intact and treated teeth. Particularly, the influence of the same dental treatments as in the previous example was tested.

Shrinkage stress, described thoroughly in the previous example, was taken into account, but its level was a variable of interest in the present study as well.

3.3.2.2 Materials and Methods
3.3.2.2.1 Obtaining Teeth Geometry
Finite element models were the same as in the previous example. To summarize briefly, we had two intact teeth (maxillary second premolars) with similar morphology obtained from the same person. One remained intact and the other one underwent in vitro root canal treatment. Two walls of the crown were removed to simulate two surface cavities and, after root canal treatment, the cavity was filled with composite material. Both teeth were scanned by means of computerized tomography (Siemens Somatom Sensation 16, Munich, Germany).

After the acquisition, Mimics 10.01 software (Materialise, Leuven, Belgium) was used for image segmentation and generation of surface meshes for each of the tooth materials. Additional mesh refinement and assembly of different parts of models were performed in Geomagic Studio 10 (Geomagic GmbH, Stuttgart, Germany).

The models were the same as in the first example in this chapter except the elimination of Model 3.

3.3.2.2.2 FEM Procedures

Three-dimensional discretization (meshing) of developed modes was performed by using the TetGen (Hang Si, WIAS, Berlin, Germany) software. Since TetGen results are coming in the form of the four-nodal tetrahedral elements, they were afterward split with the in-house script into the eight-node hexahedrons to fit the solver used for FEA. Parts of teeth that were assigned to corresponding building materials are given in Table 3.4. As well as in the first-use case, all the materials were assumed to be homogenous, isotropic, and linear. The numerical analysis of mechanical behavior of the teeth was performed by means of the PAK software (University of Kragujevac, Republic of Serbia), which has been previously applied and validated on several problems in biomechanics [92, 93].

Following literature [94], the polymerization shrinkage was modeled as thermal expansion in the heat transfer analysis. For these purposes, we adopted the value of Young's modulus-curing depth as:

$$E = E_{Z250} - E_{Z250}e^{0.722d}$$

where E is the Young's modulus after polymerization, and d is the depth from the irradiated resin surface (Figure 3.10d). Two types of the restoration shrinkage stresses were simulated: high – which corresponded to the 0.5% linear shrinkage of the composite material [105]; and low– which corresponded to the 0.05% linear shrinkage of the composite material. Three types of mastication forces were considered, low (100N), medium (150N), and high (200N). For each mastication load, we assumed that the considered period of chewing was four years, which correspond to approximately 1 000 000 cycles [1]. The load was applied on surfaces

Table 3.4 Material properties of each modeled structure.

Material	Young's modulus [MPa]	Poison's ratio	Thermal conductivity [J · m⁻¹ · s⁻¹ · °C⁻¹]	Density [kg · m⁻³]	Specific heat [J · kg⁻¹ · °C⁻¹]	Thermal expansion [(m/m)⁻¹ · °C⁻¹]
Pulp	6.8	0.45	0.67	1000	4200	1.01e-5
Dentin	18.6e+3	0.31	0.59	1960	1600	1.01e-5
Enamel	84.10e+3	0.3	0.93	2800	712	1.15e-5
PDL	0.68	0.45	—	—	—	—
Composite resin	16.6e+3	0.24	1.087	2100	200	2. 70e-5
Gutta-percha	70	0.40	332.4	2700	1042	16. 20e-5

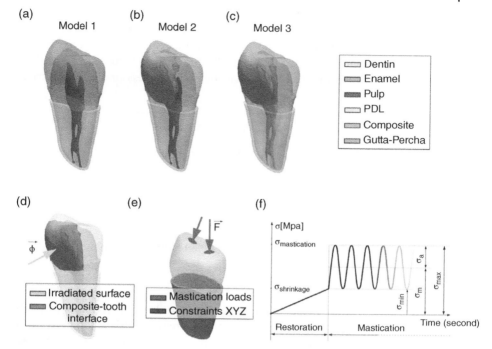

Figure 3.10 FEM procedures: (a–c) Considered models, (d) schematic view of tooth-composite interface and surface that is being irradiated with polymerization light (e) boundary conditions, and (f) sketch of stress changes caused by restoration and mastication.

of the buccal and lingual cup concurrently (the half of the load was applied on each site) in order to get the occlusal load parallel to the long axis of the tooth (Figure 3.10e). Nodes on the outer surface of the PDL were constrained in all three directions (Figure 3.10e).

3.3.2.2.2.1   Dentin Fatigue Assessment and Residual Lifetime Prediction   The fatigue of dentin was preformed following the S-N approach, assuming that there are no preexisting flaws. For these purposes, the Goodman's Fatigue Failure Index (FFI) was calculated following the equation:

$$\sigma_{qa}/\sigma_e + \sigma_{qm}/\sigma_u = \text{FFI}$$

where endurance strength $\sigma_e = 50\text{MPa}$ and ultimate stress $\sigma_u = 160\text{MPa}$ were experimentally determined material properties of dentin that were adopted from the literature [96, 106].

It is important to underline that the stresses caused by shrinkage and mastication are usually multiaxial, with dominant tensile stresses in the dentin [27]. Because of that, we applied the equivalent stress theory (EST) since it has shown to be a reliable approach for the multiaxial fatigue analysis of materials with ductile behavior. Following to the EST, equivalent nominal stress amplitude was computed as:

$$\sigma_{qa} = \sqrt{\left[(\sigma_{a1} - \sigma_{a2})^2 + (\sigma_{a2} - \alpha_{a3})^2 + (\sigma_{a3} - \sigma_{a1})^2 \right]/2}$$

where σ_{ai}, $(i = \overline{1,3})$ are principal alternating nominal stresses $\sigma_{ai} = (\sigma_{maxi} - \sigma_{mini})/2$

Equivalent nominal mean stress was computed as:

$$\sigma_{qm} = \sqrt{\left[(\sigma_{m1} - \sigma_{m2})^2 + (\sigma_{m2} - \alpha_{m3})^2 + (\sigma_{m3} - \sigma_{m1})^2 \right]/2}$$

where σ_{mi}, $(i = \overline{1,3})$ are principal nominal mean stresses $\sigma_{mi} = (\sigma_{maxi} + \sigma_{mini})/2$

As for the ratio:

$$R = \sigma_{min}/\sigma_{max}$$

it is important to notice that σ_{mini} corresponds to shrinkage stress (for intact tooth $\sigma_{mini} \cong 0$) and σ_{maxi} corresponds to the different levels of occlusal load (Figure 3.10f). After the computation of FFI, places with values <1 were assumed to be safe from the fatigue failure (allowable operating zone in Figure 3.11i). Places with FFI values >1 were assumed to be the places of potential crack initiation and propagation (unsafe operating zone in Figure 3.11i).

Three distinctive phases of residual fatigue crack propagation are shown in Figure 3.12. Here, the aim was to estimate the period of stable crack growth (region II) for dentin material, and find the number of mastication cycles before dentin failure, assuming that the failure starts with an unstable crack growth, region III. Following Figure 3.12, the stable crack growth is defined as the period between the crack initiation (region I) and unstable crack growth (region III). In this use-case, we assumed that the period of stable crack propagation should last at least

Fatique safety assesment   **Figure 3.11** Goodman's diagram.

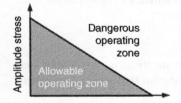

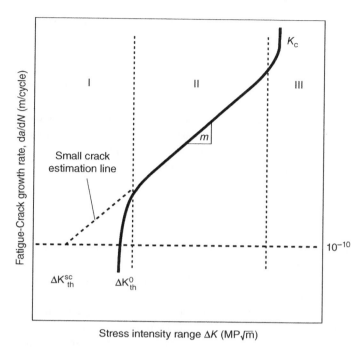

Figure 3.12   Three characteristic phases of fatigue crack growth (region I – crack initiation, region II – stable crack growth, region III – unstable crack growth).

one million mastication cycles (or approximately four years). The numerical analysis stable fatigue crack propagation in dentine was performed by using the Paris law:

$$da/dN = C(\Delta K)^m$$

Parameters are as follows:

da represents the incremental changes in crack length (Δa);
dN represents the number of cycles (ΔN);
C and m are fatigue crack propagation coefficient and exponent, respectively;
ΔK is the stress intensity range.

The stress intensity range is defined as:

$$\Delta K = Y \Delta \sigma (\pi a)^{1/2}$$

where Y is correction factor; $\Delta \sigma$ is far-field stress range ahead of the crack tip, and a is the crack length.

Starting from the Paris power law, the number of remaining loading cycles N (in the phase of stable crack propagation) was estimated as a function of the current crack length a as: $N = \dfrac{2}{(m-2)CY^m(\Delta\sigma)^m\pi^{\frac{m}{2}}}\left(a_0^{1-m/2} - a^{1-m/2}\right)$

where a_0 represents the initial crack size, and $\Delta\sigma = (\sigma_{max} - \sigma_{min})$ is the stress range [107].

The initial crack size was estimated as $a_0 = \pi^{-1}(K_{th}/Y\sigma_e)^2$, and it corresponds to the minimal value of crack size from which Paris power law may be applied. Parameter K_{th} is the fatigue threshold; σ_e is the endurance limit.

For dentin, the fatigue threshold parameter has a value of:

$$K_{th} = 1.06\text{MPA}\sqrt{m}, Y = 1.12$$

for a shallow flaw, fracture toughness is $K_c = 1.8\text{MPa}m$, and scaling constant parameters are $C = 6.24$ and $m = 8.76$ [106]

With the application of the previous equations and parameters, the value of initial crack size was: $a_0 = 107\mu m$

while the critical crack size was calculated from the fracture toughness as: $a_c = \pi^{-1}(K_c/Y\sigma_e)^2 = 312\mu m$

Therefore, unstable crack propagation (which may cause fatigue fracture) will occur when the crack size reaches the critical value a_c, and this moment will be reached after N cycles (which can be calculated from the equation given above).

3.3.2.3 Results and Discussion

Mesh-independency was reached at approximately 300 000 elements for all three models. The results obtained with the FEA are presented in terms of the Von Mises stress (VMS), Fatigue Failure Index (FFI), and stress ratio (R). The cases considered to be representative for the results are given in Figures 3.13–3.16, while overall results are given in Table 3.2. For the models given in Figures 3.13–3.16 with FFI>1, the diagrams for residual lifetime prediction are shown in Figure 3.17.

For the Models 2 and 3 with FFI>1, whose results are given in Figures 3.15 and 3.16, Goodman diagrams were constructed and presented in Figure 3.17a and b. As it was previously mentioned, the triangle on Goodman diagram represents the nodes safe from fatigue failure (FFI<1). The points outside the triangle indicate places where fatigue failure may occur. It may be noted that slightly higher number of such points appear in Model 3 compared to Model 2, which is in accordance with the obtained FFI values in Figures 3.15 and 3.16.

This study presents the workflow for numerical analysis of dentin fatigue in treated teeth. During fatigue of dentine, as for the other natural or artificial materials, deterioration of mechanical properties of this material occurs over a long period of

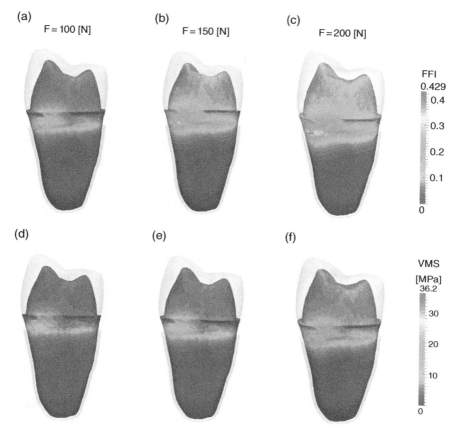

Figure 3.13 FEA results for Model 1 under occlusal load of 100N,150N and 200N (VMS – Von Mises Stress (d–f), FFI – Fatigue Failure Index (a–c)).

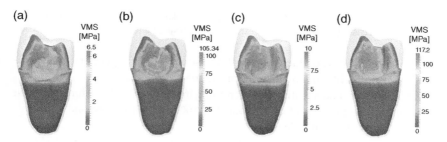

Figure 3.14 Developed shrinkage stresses for the considered restoration cases (a) low shrinkage stress, Model 2, (b) high-shrinkage stress, Model 2, (c) low-shrinkage stress Model 3, (d) high-shrinkage stress Model 3.

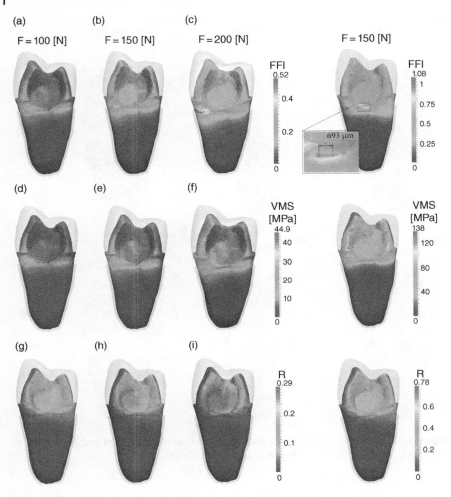

Figure 3.15 FEA results for Model 2; (a) – (i) low-shrinkage stress cases, (j) – (l) high-shrinkage stress cases (VMS – Von Mises Stress, FFI – Fatigue Failure Index, R – stress ratio).

use. The simulation is conducted through the stress-based fatigue total-life evaluation and residual lifetime prediction for fatigue crack expansion.

All models were loaded with three increasing intensity of occlusal force — 100N, 150N, 200N. The number of cycles was 1 000 000, which corresponds approximately to four years of mastication.

The results showed no risk for fatigue failure of the intact tooth (Model 1) even for the worst-case scenario (force of 200N resulted with FFI = 0.42, which is remote from the unsafe zone). These results were expected.

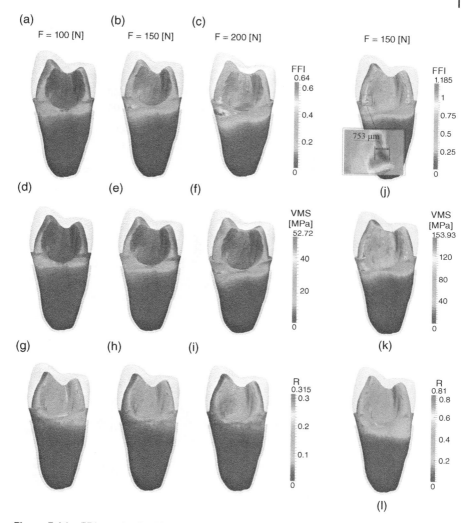

Figure 3.16 FEA results for Model 3; (a) – (i) low-shrinkage stress cases, (j) – (l) high-shrinkage stress cases (VMS – Von Mises Stress, FFI – Fatigue Failure Index, R – stress ratio).

For other models, not only the intensity of the force was changeable but also the level of pre-calculated shrinkage stress (caused by polymerization shrinkage of composite material). We applied two levels of shrinkage stress – low and high. The results pointed out the key factors of dentin fatigue. The crucial factor was found to be the level of shrinkage stress. The mastication stress intensity also plays an important role. In contrast, the role of root canal treatment was minor. To be

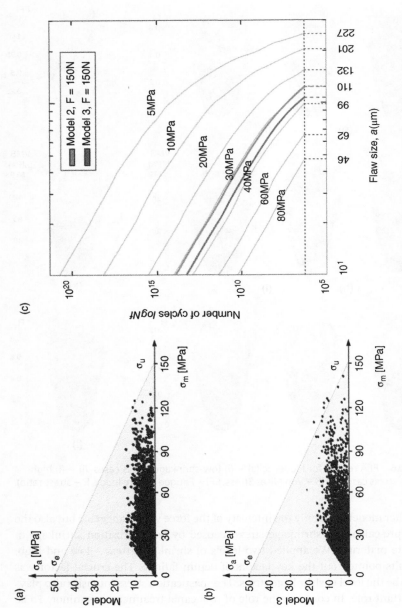

Figure 3.17 Fatigue failure diagrams: (a), (b) Goodman's diagrams for Model 2 and Model 3 (F = 150N and increased shrinkage stress); (c) residual fatigue lifetime vs. crack size for different levels of fatigue loading.

specific, it can be noticed that failure occurs in Models 2 and 3 if we use high level of shrinkage and medium occlusal load (150N). But, when we use low level of shrinkage, both models are away from the unsafe zone even when we apply high occlusal load (200N). The influence of the root canal treatment can be noticed but it was not crucial – when FFI was calculated for Models 2 and 3 with low-shrinkage stress and medium mastication load (150N), it was found that Model 3 (Figure 3.16b) showed increased risk of failure, but, as previously stated, models were significantly away from the dangerous zone. However, when the high shrinkage stress was assumed, and considering the same mastication load of 150N, fatigue failure risk reaches critical levels – FFI = 1.08 for Model 2 (Figure 3.15j) and FFI = 1.18 for Model 3 (Figure 3.16j). One can notice that the tooth with only composite filing (Model 2) is just a bit above the critical value for fracture (FFI = 1.08). Thus, if the mastication forces are a bit lower than 150N, fractures will not occur.

The proposed approach gives also information about the highest acceptable crack size and maximal number or remained cycles for the assumed level of stresses. This is acquired by the analysis of the residual lifetime diagram. Additionally, the size of the crack is measured on virtual models (the red region Figure 3.15j). This way, our approach provides the estimation of highest acceptable crack size for a given case, and measuring the crack propagation during the fatigue simulation.

For models with FFI>1 the residual lifetime prediction was computed based on the equations given in the **Materials and Methods section.** It was assumed that the initial flaw had the stable crack-propagation (Figure 3.17c). For Model 2, in the region with FFI<1 under load of F = 150N, stress range was approximately 30.5MPa and for Model 3 (F = 150N) was 36.7MPa. With the assumption that cracks are present in the regions with FFI>1 (or that such regions are in danger of potential crack initiation and propagation), the remaining number of cycles until fatigue failure could be estimated on the Figure 3.17 based on the length of these regions. Based on the diagram, highest acceptable size of crack (which will not result in fatigue failure for one million cycles) is 110 µm for Model 2 and 96 µm for Model 3 (**Table 3.5**). The residual mastication cycles number for the crack lengths may be acquired from the vertical axis if the stress range is acknowledged. The stress range is specified by the FE analysis. Lifetime diagrams for Models 2 and 3 are shown in Figure 3.17. Damage lengths (regions with FFI>1) were 693 µm and 753 µm for Models 2 and 3, respectively. As the measured damage lengths were larger than the highest acceptable, it can be concluded that maximal allowed shrinkage stress, which (under considered mastication conditions) will not cause dentin fatigue failure, should be less than the high-shrinkage stress adopted in this study (Figure 3.14b, d). Regarding the cases of high-shrinkage stress, which are not presented on figures (Models 2 and 3 with loads of F = 150N and F = 200N), for the developed shrinkage stress above 100MPa variations in mastication loads

Table 3.5 Overall FEA results.

Model	Loads [N]	Shrinkage stress [MPa]	Mastication stress [MPa]	R	FFI	$a_c(10^6$ cycles) [μm]
1	100	0	19.4	0	0.23	148
1	150	0	28.9	0	0.31	118
1	200	0	36.2	0	0.42	97
2	100	6.5	26.6	0.29	0.31	144
2	150	6.5	35.3	0.26	0.40	115
2	200	6.5	44.6	0.17	0.52	94
2	100	105.3	130.4	0.78	0.97	125
2	150	105.3	138.1	0.73	1.08	110
2	200	105.3	149.5	0.64	1.23	79
3	100	10	37.5	0.31	0.35	131
3	150	10	52.7	0.27	0.57	108
3	200	10	66.9	0.19	0.64	86
3	100	117.2	139.3	0.87	1.03	123
3	150	117.2	153.9	0.83	1.18	96
3	200	117.2	165.8	0.78	1.30	62

(100–200N) do not cause major changes in term of fatigue safety, since all cases are in the dangerous zone **(Table 3.5)**. For that reason, only cases of F = 15N were shown as representative examples, while the rest are given in **Table 3.5**.

The given proposal for numerical expression of fatigue and residual lifetime predictions was able to identify risk factors in this specific scenario. The results are in line with empirical knowledge from everyday experience.

At the end, we would like to state a few challengeable issues from the aspect of biomechanics and modeling tooth fatigue fracture that refer to this study. First of all, the results of this study referrers to these particular circumstances. Thus, the results and conclusions can only be taken into consideration and be used as an assumption for what can be expected in other cases of compositely restored teeth with or without root canal treatment. Moreover, the literature gives different values for critical stress definitions (endurance limits and threshold stress intensity ranges) – we acknowledge that more liberal or restrictive values could be found in literature which was not taken under consideration in this study. It should be also emphasized that dentin is an extremely heterogeneous material. There are several varieties of dentin but furthermore, from the root canal to the enamel, there are also fiber-like tubules whose material properties and biomechanical behavior are changing depending on their direction. Dentine is in fact able to resist fatigue crack propagation. But this property changes with age and also with dehydration caused by the diseases of pulp or periodontium [89]. Additionally, some dental treatments have influence on mechanical properties of dentin. Giannini et al. reported that ultimate tensile strength (UTS) of enamel, dentin, and enamel–dentin junction varied according to its nature and location [108]. The same authors also reported that the evaluated UTS was affected by the frequency of loading. The stress-ratio contributes to the fatigue crack growth in dentin increasing the C (fatigue crack growth coefficient) while decreasing m (fatigue crack growth exponent) and ΔK (stress intensity range) [108]. However, the authors considered only stress-ratio in range of -0.5 to 0.5, while our study showed that stress-ratio may be higher than 0.5 in the case of increased shrinkage stress. For that reason, in our study the constant values were adopted for C and m [109], and this is the possible limitation of this study. It means that diagrams in Figure 3.17c should be taken cautiously before performing additional experimental studies.

References

1 DeLong, R. and Douglas, W.H. (1991). An artificial oral environment for testing dental materials. *IEEE Transactions on Bio-Medical Engineering.* 38 (4): 339–345.

2 Sedgley, C.M. and Messer, H.H. (1992). Are endodontically treated teeth more brittle? *Journal of Endodontics.* 18 (7): 332–335.

3 Reeh, E.S., Douglas, W.H., and Messer, H.H. (1989). Stiffness of endodontically-treated teeth related to restoration technique. *Journal of Dental Research.* 68 (11): 1540–1544.

4 Reeh, E.S., Messer, H.H., and Douglas, W.H. (1989). Reduction in tooth stiffness as a result of endodontic and restorative procedures. *Journal of Endodontics.* 15 (11): 512–516.

5 Rohrle, O., Saini, H., and Ackland, D.C. (2018). Occlusal loading during biting from an experimental and simulation point of view. *Dental Materials: Official Publication of the Academy of Dental Materials.* 34 (1): 58–68.

6 Trivedi, S. (2014). Finite element analysis: a boon to dentistry. *Journal of Oral Biology and Craniofacial Research.* 4 (3): 200–203.

7 Geng, J.P., Tan, K.B., and Liu, G.R. (2001). Application of finite element analysis in implant dentistry: a review of the literature. *The Journal of Prosthetic Dentistry.* 85 (6): 585–598.

8 Ruse, N.D. (2008). Propagation of erroneous data for the modulus of elasticity of periodontal ligament and gutta percha in FEM/FEA papers: a story of broken

links. *Dental Materials: Official Publication of the Academy of Dental Materials.* 24 (12): 1717–1719.
9 Barbenel, J.C. (1972). The biomechanics of the temporomandibular joint: a theoretical study. *Journal of Biomechanics.* 5 (3): 251–256.
10 Koolstra, J.H., van Eijden, T.M., Weijs, W.A., and Naeije, M. (1988). A three-dimensional mathematical model of the human masticatory system predicting maximum possible bite forces. *Journal of Biomechanics.* 21 (7): 563–576.
11 Koolstra, J.H. and van Eijden, T.M. (1992). Application and validation of a three-dimensional mathematical model of the human masticatory system in vivo. *Journal of Biomechanics.* 25 (2): 175–187.
12 Tang, Z., Tu, W., Zhang, G. et al. (2012). Dynamic simulation and preliminary finite element analysis of gunshot wounds to the human mandible. *Injury.* 43 (5): 660–665.
13 Lei, T., Xie, L., Tu, W. et al. (2012). Blast injuries to the human mandible: development of a finite element model and a preliminary finite element analysis. *Injury.* 43 (11): 1850–1855.
14 Nomura, T., Gold, E., Powers, M.P. et al. (2003). Micromechanics/structure relationships in the human mandible. *Dental Materials: Official Publication of the Academy of Dental Materials.* 19 (3): 167–173.
15 Bonnet, A.S., Postaire, M., and Lipinski, P. (2009). Biomechanical study of mandible bone supporting a four-implant retained bridge: finite element analysis of the influence of bone anisotropy and foodstuff position. *Medical Engineering & Physics.* 31 (7): 806–815.
16 Schwartz-Dabney, C.L. and Dechow, P.C. (2003). Variations in cortical material properties throughout the human dentate mandible. *American Journal of Physical Anthropology.* 120 (3): 252–277.
17 Apicella, D., Aversa, R., Ferro, F. et al. (2010). The importance of cortical bone orthotropicity, maximum stiffness direction and thickness on the reliability of mandible numerical models. *Journal of Biomedical Materials Research Part B, Applied Biomaterials.* 93 (1): 150–163.
18 Antic, S., Vukicevic, A.M., Milasinovic, M. et al. (2015). Impact of the lower third molar presence and position on the fragility of mandibular angle and condyle: a three-dimensional finite element study. *Journal of Cranio-Maxillo-Facial Surgery: Official Publication of the European Association for Cranio-Maxillo-Facial Surgery.* 43 (6): 870–878.
19 Fanuscu, M.I. and Chang, T.L. (2004). Three-dimensional morphometric analysis of human cadaver bone: microstructural data from maxilla and mandible. *Clinical Oral Implants Research.* 15 (2): 213–218.
20 Janovic, A., Milovanovic, P., Saveljic, I. et al. (2014). Microstructural properties of the mid-facial bones in relation to the distribution of occlusal loading. *Bone.* 68: 108–114.

21 Janovic, A., Saveljic, I., Vukicevic, A. et al. (2015). Occlusal load distribution through the cortical and trabecular bone of the human mid-facial skeleton in natural dentition: a three-dimensional finite element study. *Annals of anatomy Anatomischer Anzeiger: Official Organ of the Anatomische Gesellschaft.* 197: 16–23.

22 Antic, S., Milicic, B., Jelovac, D.B., and Djuric, M. (2016). Impact of the lower third molar and injury mechanism on the risk of mandibular angle and condylar fractures. *Dental Traumatology: Official Publication of International Association for Dental Traumatology.* 32 (4): 286–295.

23 Wu, J., Liu, Y., Wang, D. et al. (2019). Investigation of effective intrusion and extrusion force for maxillary canine using finite element analysis. *Computer Methods in Biomechanics and Biomedical Engineering.* 22 (16): 1294–1302.

24 Chen, G., Fan, W., Mishra, S. et al. (2012). Tooth fracture risk analysis based on a new finite element dental structure models using micro-CT data. *Computers in Biology and Medicine.* 42 (10): 957–963.

25 Cattaneo, P.M., Dalstra, M., and Melsen, B. (2003). The transfer of occlusal forces through the maxillary molars: a finite element study. *American Journal of Orthodontics and Dentofacial Orthopedics: Official Publication of the American Association of Orthodontists, its Constituent Societies, and the American Board of Orthodontics.* 123 (4): 367–373.

26 Gross, M.D., Arbel, G., and Hershkovitz, I. (2001). Three-dimensional finite element analysis of the facial skeleton on simulated occlusal loading. *Journal of Oral Rehabilitation.* 28 (7): 684–694.

27 Zelic, K., Vukicevic, A., Jovicic, G. et al. (2015). Mechanical weakening of devitalized teeth: three-dimensional finite element analysis and prediction of tooth fracture. *International Endodontic Journal.* 48 (9): 850–863.

28 Murakami, N. and Wakabayashi, N. (2014). Finite element contact analysis as a critical technique in dental biomechanics: a review. *Journal of Prosthodontic Research.* 58 (2): 92–101.

29 Staines, M., Robinson, W., and Hood, J. (1981). Spherical indentation of tooth enamel. *Journal of Materials Science* 16 (9): 2551–2556.

30 Groning, F., Fagan, M., and O'Higgins, P. (2012). Modeling the human mandible under masticatory loads: which input variables are important? *Anatomical Record.* 295 (5): 853–863.

31 Groning, F., Jones, M.E., Curtis, N. et al. (2013). The importance of accurate muscle modelling for biomechanical analyses: a case study with a lizard skull. *Journal of the Royal Society, Interface* 10 (84): 20130216.

32 Rohrle, O. and Pullan, A.J. (2007). Three-dimensional finite element modelling of muscle forces during mastication. *Journal of Biomechanics.* 40 (15): 3363–3372.

33 Rohrle, O., Saini, H., Lee, P.V.S., and Ackland, D.C. (2018). A novel computational method to determine subject-specific bite force and occlusal loading during mastication. *Computer Methods in Biomechanics and Biomedical Engineering.* 21 (6): 453–460.

34 Batalha-Silva, S., de Andrada, M.A., Maia, H.P., and Magne, P. (2013). Fatigue resistance and crack propensity of large MOD composite resin restorations: direct versus CAD/CAM inlays. *Dental Materials: Official Publication of the Academy of Dental Materials.* 29 (3): 324–331.

35 Haskell, B., Day, M., and Tetz, J. (1986). Computer-aided modeling in the assessment of the biomechanical determinants of diverse skeletal patterns. *American Journal of Orthodontics.* 89 (5): 363–382.

36 Reina, J.M., Garcia-Aznar, J.M., Dominguez, J., and Doblare, M. (2007). Numerical estimation of bone density and elastic constants distribution in a human mandible. *Journal of Biomechanics.* 40 (4): 828–836.

37 Tanaka, E., Tanne, K., and Sakuda, M. (1994). A three-dimensional finite element model of the mandible including the TMJ and its application to stress analysis in the TMJ during clenching. *Medical Engineering & Physics.* 16 (4): 316–322.

38 Groning, F., Fagan, M.J., and O'Higgins, P. (2011). The effects of the periodontal ligament on mandibular stiffness: a study combining finite element analysis and geometric morphometrics. *Journal of Biomechanics.* 44 (7): 1304–1312.

39 Groning, F., Liu, J., Fagan, M.J., and O'Higgins, P. (2011). Why do humans have chins? Testing the mechanical significance of modern human symphyseal morphology with finite element analysis. *American Journal of Physical Anthropology.* 144 (4): 593–606.

40 Boryor, A., Geiger, M., Hohmann, A. et al. (2008). Stress distribution and displacement analysis during an intermaxillary disjunction--a three-dimensional FEM study of a human skull. *Journal of Biomechanics.* 41 (2): 376–382.

41 Dumont, E.R., Piccirillo, J., and Grosse, I.R. (2005). Finite-element analysis of biting behavior and bone stress in the facial skeletons of bats. *The Anatomical Record Part A, Discoveries in Molecular, Cellular, and Evolutionary Biology.* 283 (2): 319–330.

42 Wood, S.A., Strait, D.S., Dumont, E.R. et al. (2011). The effects of modeling simplifications on craniofacial finite element models: the alveoli (tooth sockets) and periodontal ligaments. *Journal of Biomechanics.* 44 (10): 1831–1838.

43 Wroe, S., Moreno, K., Clausen, P. et al. (2007). High-resolution three-dimensional computer simulation of hominid cranial mechanics. *Anatomical Record.* 290 (10): 1248–1255.

44 Panagiotopoulou, O. and Cobb, S.N. (2011). The mechanical significance of morphological variation in the macaque mandibular symphysis during mastication. *American Journal of Physical Anthropology.* 146 (2): 253–261.

45 Fill, T.S., Toogood, R.W., Major, P.W., and Carey, J.P. (2012). Analytically determined mechanical properties of, and models for the periodontal ligament: critical review of literature. *Journal of Biomechanics.* 45 (1): 9–16.

46 Nikolaus, A., Currey, J.D., Lindtner, T. et al. (2017). Importance of the variable periodontal ligament geometry for whole tooth mechanical function: a validated numerical study. *Journal of the Mechanical Behavior of Biomedical Materials.* 67: 61–73.

47 Branemark, P.I., Adell, R., Breine, U. et al. (1969). Intra-osseous anchorage of dental prostheses. I. Experimental studies. *Scandinavian Journal of Plastic and Reconstructive Surgery.* 3 (2): 81–100.

48 Reddy, M.S., Sundram, R., and Eid Abdemagyd, H.A. (2019). Application of finite element model in implant dentistry: a systematic review. *Journal of Pharmacy & Bioallied Sciences.* 11 (Suppl 2): S85–S91.

49 Maminskas, J., Puisys, A., Kuoppala, R. et al. (2016). The prosthetic influence and biomechanics on peri-implant strain: a systematic literature review of finite element studies. *Journal of Oral & Maxillofacial Research.* 7 (3): e4.

50 Schenk, R.K. and Buser, D. (1998). Osseointegration: a reality. *Periodontol 2000* 17: 22–35.

51 Frost, H.M. (2003). Bone's mechanostat: a 2003 update. *Anatomical Record Part A: Discoveries in Molecular, Cellular, and Evolutionary Biology* 275 (2): 1081–1101.

52 Sakka, S., Baroudi, K., and Nassani, M.Z. (2012). Factors associated with early and late failure of dental implants. *Journal of Investigative and Clinical Dentistry* 3 (4): 258–261.

53 Sotto-Maior, B.S., Senna, P.M., da Silva, W.J. et al. (2012). Influence of crown-to-implant ratio, retention system, restorative material, and occlusal loading on stress concentrations in single short implants. *The International Journal of Oral & Maxillofacial Implants.* 27 (3): e13–e18.

54 Sotto-Maior, B.S., Senna, P.M., da Silva-Neto, J.P. et al. (2015). Influence of crown-to-implant ratio on stress around single short-wide implants: a photoelastic stress analysis. *Journal of Prosthodontics: Official Journal of the American College of Prosthodontists.* 24 (1): 52–56.

55 Soto-Penaloza, D., Zaragozi-Alonso, R., Penarrocha-Diago, M., and Penarrocha-Diago, M. (2017). The all-on-four treatment concept: systematic review. *Journal of Clinical and Experimental Dentistry.* 9 (3): e474–e488.

56 Bhering, C.L., Mesquita, M.F., Kemmoku, D.T. et al. (2016). Comparison between all-on-four and all-on-six treatment concepts and framework material on stress distribution in atrophic maxilla: a prototyping guided 3D-FEA study. *Materials Science & Engineering C, Materials for Biological Applications.* 69: 715–725.

57 Shaikh, S.Y., Mulani, S., and Shaikh, S.S. (2018). Stress distribution on root dentin analogous to natural teeth with various retentive channels design on the face of the root with minimal or no coronal tooth structure: a finite element analysis. *Contemporary Clinical Dentistry.* 9 (4): 630–636.

58 Al-Omiri, M.K., Mahmoud, A.A., Rayyan, M.R., and Abu-Hammad, O. (2010). Fracture resistance of teeth restored with post-retained restorations: an overview. *Journal of Endodontics.* 36 (9): 1439–1449.

59 Mamoun, J. (2017). Post and core build-ups in crown and bridge abutments: biomechanical advantages and disadvantages. *The Journal of Advanced Prosthodontics.* 9 (3): 232–237.

60 Dogan, S. and Raigrodski, A.J. (2019). Cementation of zirconia-based toothborne restorations: a clinical review. *Compendium of Continuing Education in Dentistry.* 40 (8): 536–540.

61 Anusavice, K.J., Kakar, K., and Ferree, N. (2007). Which mechanical and physical testing methods are relevant for predicting the clinical performance of ceramic-based dental prostheses? *Clinical Oral Implants Research.* 18 (Suppl 3): 218–231.

62 Sotto-Maior, B.S., Carneiro, R.C., Francischone, C.E. et al. (2019). Fatigue behavior of different CAD/CAM materials for monolithic, implant-supported molar crowns. *Journal of Prosthodontics: Official Journal of the American College of Prosthodontists.* 28 (2): e548–e551.

63 Nishihara, H., Haro Adanez, M., and Att, W. (2019). Current status of zirconia implants in dentistry: preclinical tests. *Journal of Prosthodontic Research.* 63 (1): 1–14.

64 Kohal, R.J., Att, W., Bachle, M., and Butz, F. (2008). Ceramic abutments and ceramic oral implants. An update. *Periodontology 2000* 47: 224–243.

65 Ausiello, P.P., Ciaramella, S., Lanzotti, A. et al. (2019). Mechanical behavior of class I cavities restored by different material combinations under loading and polymerization shrinkage stress. a 3D-FEA study. *American Journal of Dentistry.* 32 (2): 55–60.

66 da Rocha, D.M., Tribst, J.P.M., Ausiello, P. et al. (2019). Effect of the restorative technique on load-bearing capacity, cusp deflection, and stress distribution of endodontically-treated premolars with MOD restoration. *Restorative Dentistry & Endodontics.* 44 (3): e33.

67 Ausiello, P., Ciaramella, S., Di Rienzo, A. et al. (2019). Adhesive class I restorations in sound molar teeth incorporating combined resin-composite and glass ionomer materials: CAD-FE modeling and analysis. *Dental Materials: Official Publication of the Academy of Dental Materials.* 35 (10): 1514–1522.

68 Ghavami-Lahiji, M. and Hooshmand, T. (2017). Analytical methods for the measurement of polymerization kinetics and stresses of dental resin-based composites: a review. *Dental Research Journal.* 14 (4): 225–240.

69 Vukicevic, A.M., Zelic, K., Jovicic, G. et al. (2015). Influence of dental restorations and mastication loadings on dentine fatigue behaviour: image-based modelling approach. *Journal of Dentistry.* 43 (5): 556–567.

70 Cryer, M.H. (1916). *The Internal Anatomy of the Face.* Philadelphia: Lea & Febiger.

71 Rowe, N.L. and Williams, J.L. (1985). *Maxillofacial Injuries*, vol. I. Edinburgh: Churchill Livingstone.

72 Tanne, K., Miyasaka, J., Yamagata, Y. et al. (1988). Three-dimensional model of the human craniofacial skeleton: method and preliminary results using finite element analysis. *Journal of Biomedical Engineering.* 10 (3): 246–252.

73 Favino, M., Gross, C., Drolshagen, M. et al. (2013). Validation of a heterogeneous elastic-biphasic model for the numerical simulation of the PDL. *Computer Methods in Biomechanics and Biomedical Engineering.* 16 (5): 544–553.

74 Jepsen, S., Caton, J.G., Albandar, J.M. et al. (2018). Periodontal manifestations of systemic diseases and developmental and acquired conditions: consensus report of workgroup 3 of the 2017 world workshop on the classification of periodontal and peri-implant diseases and conditions. *Journal of Periodontology.* 89 (Suppl 1): S237–S248.

75 Papapanou, P.N., Sanz, M., Buduneli, N. et al. (2018). Periodontitis: consensus report of workgroup 2 of the 2017 world workshop on the classification of periodontal and peri-implant diseases and conditions. *Journal of Periodontology.* 89 (Suppl 1): S173–S182.

76 Lang, N. and Lindhe, J. (eds.) (2015). *Clinical Periodontology and Implant Dentistry.* London: John Wiley and Sons.

77 Zelic, O. (2007). *Atlas of Periodontology,* 3e. Službeni glasnik: Belgrade.

78 Reddy, R.T. and Vandana, K.L. (2018). Effect of hyperfunctional occlusal loads on periodontium: a three-dimensional finite element analysis. *Journal of Indian Society of Periodontology.* 22 (5): 395–400.

79 Zhang, H., Cui, J.W., Lu, X.L., and Wang, M.Q. (2017). Finite element analysis on tooth and periodontal stress under simulated occlusal loads. *Journal of Oral Rehabilitation.* 44 (7): 526–536.

80 Poiate, I.A., de Vasconcellos, A.B., de Santana, R.B., and Poiate, E. (2009). Three-dimensional stress distribution in the human periodontal ligament in masticatory, parafunctional, and trauma loads: finite element analysis. *Journal of Periodontology.* 80 (11): 1859–1867.

81 Grippo, J.O. (1991). Abfractions: a new classification of hard tissue lesions of teeth. *Journal of Esthetic Dentistry.* 3 (1): 14–19.

82 Duangthip, D., Man, A., Poon, P.H. et al. (2017). Occlusal stress is involved in the formation of non-carious cervical lesions. A systematic review of abfraction. *American Journal of Dentistry.* 30 (4): 212–220.

83 Poiate, I.A., Vasconcellos, A.B., Poiate Junior, E., and Dias, K.R. (2009). Stress distribution in the cervical region of an upper central incisor in a 3D finite element model. *Brazilian Oral Research.* 23 (2): 161–168.

84 Ahmed, N., Megalan, P., Suryavanshi, S. et al. (2019). Effect of bracket slot and archwire dimension on posterior tooth movement in sliding mechanics: a three-dimensional finite element analysis. *Cureus.* 11 (9): e5756.

85 Gurbanov, V., Bas, B., and Oz, A. (2020). Evaluation of stresses on temporomandibular joint in the use of class II and III orthodontic elastics: a three-dimensional finite element study. *Journal of Oral and Maxillofacial Surgery: Official Journal of the American Association of Oral and Maxillofacial Surgeons.* 78 (5): 705–716.

86 Lei, T., Xie, L., Tu, W. et al. (2012). Development of a finite element model for blast injuries to the pig mandible and a preliminary biomechanical analysis. *The Journal of Trauma and Acute Care Surgery.* 73 (4): 902–907.

87 Soares, P.V., Santos-Filho, P.C., Martins, L.R., and Soares, C.J. (2008). Influence of restorative technique on the biomechanical behavior of endodontically treated maxillary premolars. Part I: fracture resistance and fracture mode. *The Journal of Prosthetic Sentistry.* 99 (1): 30–37.

88 Soares, P.V., Santos-Filho, P.C.F., Queiroz, E.C. et al. (2008). Fracture resistance and stress distribution in endodontically treated maxillary premolars restored with composite resin. *Journal of Prosthodontics: Official Journal of the American College of Prosthodontists.* 17 (2): 114–119.

89 Zelic, K., Milovanovic, P., Rakocevic, Z. et al. (2014). Nano-structural and compositional basis of devitalized tooth fragility. *Dental Materials: Official Publication of the Academy of Dental Materials.* 30 (5): 476–486.

90 Shen, L., Huang, H., Yu, J. et al. (2009). Effects of periodontal bone loss on the natural frequency of the human canine: a three-dimensional finite element analysis. *Journal of Dental Sciences* 4: 81–86.

91 Magne, P. (2010). Virtual prototyping of adhesively restored, endodontically treated molars. *The Journal of Prosthetic Dentistry.* 103 (6): 343–351.

92 Filipovic, N. (2008). *PAK Software for Static and Dynamics Finite Element Analysis.* Kragujevac, Serbia: University of Kragujevac.

93 Kojic, M., Filipovic, N., Stojanovic, B., and Kojic, N. (2008). *Computer Modeling in Bioengineering - Theoretical Background, Examples and Software,* 1e. London: Wiley and Sons.

94 Chuang, S.F., Chang, C.H., and Chen, T.Y. (2011). Contraction behaviors of dental composite restorations--finite element investigation with DIC validation. *Journal of the Mechanical Behavior of Biomedical Materials.* 4 (8): 2138–2149.

95 Sano, H., Ciucchi, B., Matthews, W.G., and Pashley, D.H. (1994). Tensile properties of mineralized and demineralized human and bovine dentin. *Journal of Dental Research.* 73 (6): 1205–1211.

96 Nalla, R.K., Kinney, J.H., Marshall, S.J., and Ritchie, R.O. (2004). On the in vitro fatigue behavior of human dentin: effect of mean stress. *Journal of Dental Research.* 83 (3): 211–215.

97 Craig, R.G. and Peyton, F.A. (1958). Elastic and mechanical properties of human dentin. *Journal of Dental Research.* 37 (4): 710–718.

98 Carvalho, R.M., Santiago, S.L., Fernandes, C.A. et al. (2000). Effects of prism orientation on tensile strength of enamel. *The Journal of Adhesive Dentistry.* 2 (4): 251–257.

99 Haines, D.J. (1968). Physical properties of human tooth enamel and enamel sheath material under load. *Journal of Biomechanics.* 1 (2): 117–125.

100 Chuang, S.F., Chang, C.H., Yaman, P., and Chang, L.T. (2006). Influence of enamel wetness on resin composite restorations using various dentine bonding agents: part I-effects on marginal quality and enamel microcrack formation. *Journal of Dentistry.* 34 (5): 343–351.

101 Ichim, I., Kieser, J.A., and Swain, M.V. (2007). Functional significance of strain distribution in the human mandible under masticatory load: numerical predictions. *Archives of Oral Biology.* 52 (5): 465–473.

102 Kruzic, J.J. and Ritchie, R.O. (2008). Fatigue of mineralized tissues: cortical bone and dentin. *Journal of the Mechanical Behavior of Biomedical Materials.* 1 (1): 3–17.

103 Arola, D., Bajaj, D., Ivancik, J. et al. (2010). Fatigue of biomaterials: hard tissues. *International Journal of Fatigue.* 32 (9): 1400–1412.

104 Bajaj, D., Sundaram, N., Nazari, A., and Arola, D. (2006). Age, dehydration and fatigue crack growth in dentin. *Biomaterials.* 27 (11): 2507–2517.

105 Kwon, Y.H., Jeon, G.H., Jang, C.M. et al. (2006). Evaluation of polymerization of light-curing hybrid composite resins. *Journal of Biomedical Materials Research Part B, Applied Biomaterials.* 76 (1): 106–113.

106 Nalla, R.K., Imbeni, V., Kinney, J.H. et al. (2003). In vitro fatigue behavior of human dentin with implications for life prediction. *Journal of Biomedical Materials Research Part A.* 66 (1): 10–20.

107 Ritchie, R.O. (1999). Mechanisms of fatigue-crack propagation in ductile and brittle solids. *International Journal of Fracture* 100: 55–83.

108 Giannini, M., Soares, C.J., and de Carvalho, R.M. (2004). Ultimate tensile strength of tooth structures. *Dental Materials: Official Publication of the Academy of Dental Materials.* 20 (4): 322–329.

109 van der Bilt, A., Engelen, L., Pereira, L.J. et al. (2006). Oral physiology and mastication. *Physiology & Behavior.* 89 (1): 22–27.

4

Determining Young's Modulus of Elasticity of Cortical Bone from CT Scans

Aleksandra Vulović[1,2] and Nenad D. Filipović[1,2]

[1] *Faculty of Engineering, University of Kragujevac, Kragujevac, Serbia*
[2] *Bioengineering Research and Development Center, BioIRC Kragujevac, Serbia*

4.1   Introduction

Remarkable technological advances achieved in the previous decades have led to significant improvements in all spheres of life. As a result, we have the possibility to analyze the problems that until recently were considered to be unsolvable. Many of these problems are related to the research field of biomedical engineering, where the end goal is to implement engineering methods and skills, in order to analyze and ideally solve the problems that lead to the notable improvement of the life quality. During the technological advances, computers have been significantly improved and can provide us with the ability to analyze various problems, employing a variety of methods, such as the finite element method (FEM).

The FEM belongs to a group of numerical methods that provides us with an ability to analyze and solve complex problems in various fields of engineering. It is based on solving systems of differential equations, which describe complex problems. The main advantage of this method is that it can be used to analyze problems that include complex geometry, such as human bones. Once the geometry of the problem has been developed, the same model can be used for a variety of material properties and the boundary conditions. Because of these characteristics, numerical methods have found great application in the field of

Computational Modeling and Simulation Examples in Bioengineering, First Edition. Edited by Nenad D. Filipovic. © 2022 The Institute of Electrical and Electronics Engineers, Inc. Published 2022 by John Wiley & Sons, Inc.

biomedical engineering, where they have been utilized to analyze numerous problems, such as:

- mechanical behavior of joints [1–6] as well as orthopedic implants [7–12] during routine activities,
- mechanical behavior of blood vessels and stents [13–17] during blood flow through the cardiovascular system as well as for the analysis of plaque progression [18, 19], and
- analysis of respiratory system [20–22] as well as analysis of particles in the lungs and/or numerous types of inhalers [23–25].

The finite element models are becoming widely used in clinical practice. In the interest of following the trend of development and application of the patient-specific therapies/treatments, the number of created personalized finite element models has significantly increased. In order for those models to provide appropriate information for patient- specific therapies/treatment, it is necessary to pay attention to the following elements of the finite element simulation:

- mathematical model that describes the problem that will be analyzed,
- geometry of the problem that is being analyzed,
- material properties that will be assigned to the developed model, and
- boundary conditions (loads, constraints, and contact) that will influence the results of the simulation.

One of the common problems that can arise during the application of the FEM is that the applied material properties are not able to describe the mechanical behavior of the analyzed tissue in the accurate way. This can be especially noticed when analyzing the biomechanics of bone tissues. Bone is considered to be an anisotropic material, whose material properties can significantly vary from person to person. The material properties of bone tissue, available in the literature, are obtained by performing experimental measurements in laboratories. Although the values from the literature will be able to appropriately predict the mechanical behavior of the bone tissue, they will not be able to predict for the person whose model was developed, as the obtained values were measured for someone else. The material properties of a bone tissue depend on a considerable number of parameters (i.e. age, gender, diseases, etc.) so that the most realistic results will be obtained, if the applied material properties belong to the person whose model was created. Considering that experimental measurements are mostly invasive, obtaining patient-specific bone tissue material properties for an alive person from experimental measurement is not an option. One potential solution is based on the analysis of computed tomography images and determining Young's modulus of elasticity from those images.

4.2   Bone Structure

Bone is a dense connecting tissue that builds the human skeleton (also known as skeletal system). All the bones in the human body together with the joints (bones connected to each other by ligaments) make up the human skeleton. It is one of the largest systems in the human body with approximately 15% of the weight. Its primary role is to provide [26]:

- body support and attachment for tendons and muscles needed for body movement and
- protection of internal organs.

Bone consists of three fundamental components – water, organic, and inorganic components. The organic part of a bone consists of collagen and non-collagenous proteins [27]. The inorganic part consists of hydroxyapatite microcrystals. The role of organic components is to provide adequate toughness, while inorganic components affect bone stiffness [28].

The main functions of bones are [27, 29]:

- supporting the body (bones build the skeletal system and keep the organs in place),
- protection of internal organs (i.e. the skull protects the brain; the chest protects the heart and lungs),
- production of blood cells,
- detoxification (bone tissue has the ability to separate heavy metals from the blood),
- maintain pH levels in the blood, and
- storage of minerals (reserves of important minerals for the body, mainly calcium and phosphorus, are created in the bones).

Based on their shapes (Figure 4.1), all bones in the human body can be divided into five groups (OpenStax [30]):

Figure 4.1   Bone shapes.

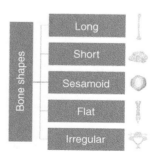

- **long bones** – cylindrical shape; have higher length than width; found in arms, legs, toes, and fingers.
- **short bones** – shaped like a cube, with almost equal thickness, length, and width; found at the root of the feet and hands.
- **flat bones** – thin and often curved bone that often protects internal organs; examples are cranial bones, ribs, and sternum.
- **sesamoid bones** – small and round bone; located in the tendons of the muscles, around the joints of the feet and hands; an example is a patella.
- **irregular bones** – cannot be classified as any other shape; examples are spinal vertebrae and facial bones.

On the microstructural level, bones are organized according to their function. We can distinguish between two bone types – newly formed bone (also known as woven bone) and lamellar bone. While the woven bone is formed rapidly and is very disorganized, the lamellar bone is formed gradually. As a result of slow formation, lamellar bone has properly distributed collagen fibers which cause lamellar bone to be stronger in comparison to the woven bone [27].

On the macroscopic level, bones can be divided into two categories:

- dense cortical bones and
- porous trabecular (spongy or cancellous) bones.

Although most bones contain both types of bone tissue (Figure 4.2), their concentration and distribution primarily depend on the bone function (OpenStax [30]). The majority of the bones in the human body (about 80%) are cortical, while the rest (about 20%) are trabecular bones [31].

Cortical or compact bone is a dense bone tissue, which gives bone strength and can withstand higher compressive forces compared with trabecular bones (OpenStax [30]). The porosity of this type of bone is usually <5%. Cortical bone

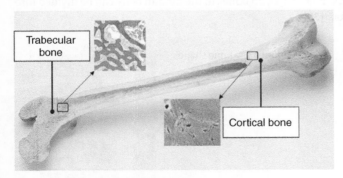

Figure 4.2 Cortical and trabecular.

tissue builds the peripheral part of the flat and short bones, as well as the largest part of the body of long bones. The cortical bone consists of cylindrical structures known as osteons or Haversian systems. Osteon is about 200 μm in diameter and consists of the following elements [31]:

- Haversian canal, which houses blood and lymph vessels, as well as nerve endings;
- Lamellae, which are arranged concentrically around the Haversian canal and are made of a bone matrix (connective tissue made of mineralized substances), and are penetrated by canaliculas;
- Osteoblast cells (bone tissue cells whose primary function is the bone formation), which cover the inner surface of the Haversian canal and synthesize the components of the bone matrix;
- Osteocyte cells (bone cells whose primary function is to maintain the concentration of minerals in the bone matrix), which are circularly distributed between the lamellae; and
- Cement, a layer of matrix different from that in the lamellae in that it does not contain channels.

Trabecular bone is very porous (50–95% porosity), heterogeneous, and anisotropic [32]. It is usually present in cuboid bones, flat bones, spinal vertebrae, and at the ends of long bones. Trabecular bone consists of bone fiber structures known as trabeculae. These structures are established along the stress lines and allow the bone to be more flexible. Due to this structure, trabecular bone is more elastic than cortical bone.

What distinguishes the bone tissue among other types of tissue is the ability to adapt to different loads by growing larger and thicker and to heal microcracks or fractures that can occur [29].

4.3 Young's Modulus of Elasticity of Bone Tissue

4.3.1 Factors Influencing Elasticity Modulus

The mechanical properties of a bone depend on the structure and composition of the bone, which changes continuously. Some factors that affect bone characteristics are [29, 33]:

- Age
- Gender
- Bone type (cortical or trabecular)
- Location in the body

- Porosity
- Mineral content
- Amount of water
- Diseases

The influence of age on the modulus of elasticity of bone is associated with increased bone porosity, mineralization, and accumulation of small bone damage. Based on the results of experimental research, it was concluded that, after the third decade, bone strength under tension and compression is reduced by 2% per decade, while fracture toughness declines by 4% per decade [34]. In men, it is observed that the relation between the modulus of elasticity and age is almost constant until the age of 55, after which (from the age of 55–75) there is a slight increase in the value of modulus. The results indicate that the bone is elastic until the age of 55, after which the elasticity decreases and the bone becomes brittle. A similar conclusion was formulated for women's bones, where bones lose their elasticity after age 55 and become more brittle [35].

Porosity is considered to have the greatest influence on the value changes of elasticity modulus, as porosity leads to lower elasticity modulus. There have been many attempts to propose expressions that will take porosity into consideration [36–40], so far there are no expressions that can accurately calculate elasticity modulus and take into consideration bone anisotropy. Besides porosity, the amount of water in a bone will have a significant effect on the elasticity modulus. Experimental research has shown that elastic modulus of dry bone is higher compared with the wet bone, up to 20% for cortical bone and up to 40% for trabecular bone. It is also demonstrated that wet bones have higher strain to failure and toughness compared with the dry bones [29].

4.3.2 Experimental Calculation of Elasticity Modulus

Young's modulus of elasticity of a bone has been the focus of research for many decades [41–46]. Experimental measurements of elasticity modulus of bone tissues can be performed using several methods. All of these methods can be divided into two groups – noninvasive and invasive methods. Regardless of the method employed, a few simple measurements are performed. These measurements refer to the determination of the amount of water and mineral content in the bone. In addition to these two parameters, the density is also determined. Using known density values for collagen ($1.35 \, g/cm^3$) and hydroxyapatite ($3.15 \, g/cm^3$), the volume fraction of minerals and water can be determined [29].

Bone is an organic material that it is often considered as an engineering material, in order to simplify the analyzed problem. Considering previously mentioned factors that affect elasticity modulus, it is more likely that there will be a variation in values, when measuring bone characteristics, compared with engineering materials. An approach used to calculate Young's modulus of elasticity is to view bone as a composite material, consisting of collagen fibers and an inorganic matrix. This simplified observation of bone allows the Young's modulus of elasticity to be calculated on the basis of the mixture rule (for the loads parallel to the fibers – axial direction) and the inverse mixture rule (for the loads perpendicular to the fibers – transverse direction). Such simplified way of representing bone is not able to calculate Young's modulus of elasticity precisely enough, and as a result, the obtained modulus of elasticity is not optimal for numerical simulations.

Another approach is to use bone and ash density in order to calculate Young's modulus of elasticity of a bone. Authors Carter and Hayes [37] discovered that, the modulus of elasticity and strength of trabecular and cortical bone are related to the cube and square of the apparent density of wet bone, respectively. The first models took into account only the apparent bone density. Further studies have shown that the mechanical properties of bone do not depend solely on the apparent density, but also on the mineral composition [40]. Ash density proved to be the most significant variable:

$$E(MPa) = 10\,500 \bullet \rho_\alpha^{2.57\,\pm\,0.04} \tag{4.1}$$

where ρ_α is ash density.

The previous expression (4.1) can explain over 96% of the statistical variations, in the mechanical behavior of the spine and femurs, for the ash density in range from 0.03 to 1.22 g/cm^3. The authors Keyak, Lee, and Skinner [47] found a relation between Young's modulus of elasticity and ash density for the trabecular bone:

$$E(MPa) = 33\,900 \bullet \rho_\alpha^{2.2}, \rho_\alpha \leq 0.27\,g/cm^3 \tag{4.2}$$

In order to improve the description of the bone mechanical properties, a distinction was made between the volume fraction of bone and the fraction of ash [48]. The apparent bone density (ρ) is expressed as follows:

$$\rho = \frac{BV}{TV}\rho_t = \frac{BV}{TV}(1.41 + 1.29\alpha) \tag{4.3}$$

where

- ρ_t is density of bone tissue that linearly depended on the amount of ash α
- BV is bone volume
- TV is total volume

The previous expressions (4.1)–(4.3) can calculate Young's modulus of elasticity, for the case when bone is considered to be an isotropic material. These expressions do not take into account the influence of structural and microstructural properties on different behaviors in all directions.

Authors Lotz, Gerhart, and Hayes [49] determined Young's modulus of elasticity for the cortical and trabecular femoral bone, in the axial and transverse directions, based on the apparent bone density. The modulus of elasticity of the cortical femur in the axial (4.4) and transverse (4.5) directions can be defined by the following expressions:

$$E(MPa) = 2065 \cdot \rho^{3.09} \tag{4.4}$$

$$E(MPa) = 2314 \cdot \rho^{1.57} \tag{4.5}$$

The modulus of elasticity of the trabecular femoral bone in the axial (4.6) and transverse (4.7) directions can be defined by the following expressions:

$$E(MPa) = 1904 \cdot \rho^{1.64} \tag{4.6}$$

$$E(MPa) = 1157 \cdot \rho^{1.78} \tag{4.7}$$

Table 4.1 shows more examples of empirical expressions for different bone types.

Figure 4.3 shows a comparison between Young's modulus of elasticity obtained using Eqs. (4.12), (4.14), (4.22), and (4.23) from the previous table using apparent density in range from 0 to 1.6 g/cm^3.

Depending on the chosen empirical expression, calculated Young's modulus of elasticity can be significantly different and thus have effect on the obtained results of numerical simulations.

Given the complexity of the bone structure as well as the variety of empirical expression, it is not surprising that reported values of elastic modulus are in a broad range. The values of the elasticity modulus vary from bone to bone. Table 4.2 shows the values of the modulus of elasticity for cortical bones.

The values of the Young's modulus of elasticity shown in the previous two tables show significant differences in the obtained values. The experimentally obtained values of the modulus of elasticity depend on the method of measurement, but also if used sample was wet or dry.

In ideal situation, we would take into consideration that Young's modulus of elasticity has different values in specific bone sections. Including such material properties can significantly improve numerical simulations and lead to obtaining more realistic results. Antic et al. [62] have analyzed mandibular bone considering that both inner and outer sides of the jaw are divided into eight regions. Each region was considered to have orthotropic material properties.

Table 4.1 Overview of empirical expression available in the literature.

Type of bone	Empirical expression	Equation number	References
Vertebra – isotropic modulus of elasticity	E (GPa) $= 2.1\rho - 0.08$	(4.8)	[50]
Proximal femoral bone – isotropic modulus of elasticity	E (GPa) $= 7.541 \frac{BV}{TV} - 0.637$	(4.9)	[51]
Vertebra – isotropic modulus of elasticity	E(GPa) $= 4\,730 \cdot \rho^{1.56}$	(4.10)	[52]
Proximal tibia – isotropic modulus of elasticity	E(GPa) $= 15\,520 \cdot \rho^{1.93}$	(4.11)	[52]
Great trochanter – isotropic modulus of elasticity	$E(GPa) = 15\,010 \cdot \rho^{2.18}$	(4.12)	[52]
Femoral bone neck – isotropic modulus of elasticity	E(GPa) $= 6\,850 \cdot \rho^{1.49}$	(4.13)	[52]
Distal femoral bone – isotropic modulus of elasticity	E(GPa) $= 10.88 \cdot \rho^{1.61}$	(4.14)	[53]
Femoral bone head – isotropic modulus of elasticity	E (GPa) $= 0.573\rho - 0.0094$	(4.15)	[54]
Humerus – orthotropic modulus of elasticity	E_1 (GPa) $= -9.212 + 0.011\rho$ E_2 (GPa) $= -8.540 + 0.011\rho$ E_1 (GPa) $= -6.326 + 0.015\rho$	(4.16)	[55]
Proximal tibia (trabecular bone) – orthotropic modulus of elasticity	E_1(MPa) $= 0.06 \cdot \rho^{1.51}$ E_2(MPa) $= 0.06 \cdot \rho^{1.55}$ E_3(MPa) $= 0.51 \cdot \rho^{1.37}$	(4.17)	[55]
Proximal femoral bone (trabecular bone) – orthotropic modulus of elasticity	E_1(MPa) $= 0.004 \cdot \rho^{2.01}$ E_2(MPa) $= 0.01 \cdot \rho^{1.86}$ E_3(MPa) $= 0.58 \cdot \rho^{1.30}$	(4.18)	[55]
Patella (trabecular bone) – orthotropic modulus of elasticity	E_1(MPa) $= 0.005 \cdot \rho^{1.91}$ E_2(MPa) $= 0.0005 \cdot \rho^{2.21}$ E_3(MPa) $= 0.04 \cdot \rho^{1.68}$	(4.19)	[55]
Proximal humerus (trabecular bone) – orthotropic modulus of elasticity	E_1(MPa) $= 0.06 \cdot \rho^{1.57}$ E_2(MPa) $= 0.07 \cdot \rho^{1.55}$ E_3(MPa) $= 0.32 \cdot \rho^{1.41}$	(4.20)	[55]
Femoral bone – orthotropic modulus of elasticity	E_1 (GPa) $= -6.087 + 0.010\rho$ E_2 (GPa) $= -4.007 + 0.009\rho$ E_1 (GPa) $= -6.142 + 0.014\rho$	(4.21)	[55]
Femoral bone – isotropic modulus of elasticity	E(MPa) $= 14\,664 \cdot \rho_\alpha^{1.49}$	(4.22)	[52]
Femoral bone – isotropic modulus of elasticity	E(GPa) $= 10.5 \cdot \rho_\alpha^{2.29}$	(4.23)	[56]

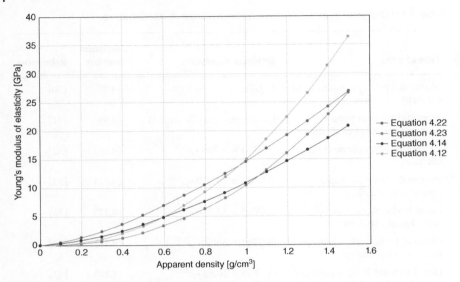

Figure 4.3 Comparison of four empirical expressions.

Table 4.2 Elasticity modulus for cortical bones.

Material property type	Young's modulus of elasticity [GPa]	References
Isotropic (femur)	$E = 16$	[57]
Isotropic (femur)	$E = 14\,371$	[58]
Isotropic (tibia)	$E = 20\,033$	[58]
Isotropic (fibula)	$E = 17\,000$	[59]
Orthotropic (femur)	$E_1 = 12$ $E_2 = 13.4$ $E_3 = 20$	[60]
Orthotropic (femoral bone)	$E_1 = 6.30$ $E_2 = 6.88$ $E_3 = 16$	[57]
Orthotropic (tibia)	$E_1 = 7$ $E_2 = 8.5$ $E_3 = 18.4$	[61]

4.4 Tool for Calculating the Young's Modulus of Elasticity of Cortical Bone from CT Scans

4.4.1 Theoretical Background

Finite element models created the computerized tomography (CT) images that have found great application in the analysis of bone biomechanics. CT is an imaging method, where X-rays and tomography (a method based on a mathematical procedure of image processing) are used. This imaging method involves creating transverse layers of the desired organ, such as bone. Each cross section, known as slice, is of a certain thickness, so the geometry of the organ can be constructed by stacking the obtained sections. When using CT images to develop a finite element model, the imported images are processed automatically or manually, in order to extract the necessary data for creating geometry of the organ. On the imported images, the difference between tissues can be noticed on the basis of colors. The space without tissue is shown in black; less dense tissues, such as muscles, are shown in shades of gray; while solid tissues, such as bone, are shown in shades of white.

In addition to using CT images to create patient-specific finite element models, these images are increasingly being used to obtain information about material properties, namely Young's modulus of elasticity. In order to achieve that, the Hounsfield unit is used.

The Hounsfield unit measures the linear attenuation of tissue within a voxel (volume pixel, whose dimensions are defined based on pixel resolution and slice thickness). Values vary between -1024 and $+3071$, where the value of -1024 corresponds to air, 0 to water, and over $+400$ represents bone. The higher the number, the denser the tissue. The Hounsfield unit is calculated as follows:

$$HU = 1000 \cdot \left(\frac{\mu}{\mu_{H_2O}} - 1 \right) \tag{4.24}$$

where

- μ – coefficient of linear attenuation of tissue in voxel,
- μ_{H_2O} – coefficient of linear attenuation of water.

The calculation of Young's modulus of elasticity is based on the association of the Hounsfield unit with bone density, which has almost linear dependence. The obtained density values are then transformed into the modulus of elasticity, based on the empirical relations of tissue density – modulus of elasticity (expressions presented in Section 3.2). As already shown, the chosen empirical relationship will have great influence on the final results. The reason for this is the different conditions of the experimental measurement of the modulus of elasticity. However, the best confirmation of the calculated modulus of elasticity would be when comparing the calculated value with the measured value in the same place.

In the last couple of years, more and more papers are addressing the determination of material characteristics of bones from images [63–66]. It has been shown that calculating the mean Hounsfield value of an element, based on the average value in the nodes of the element, can lead to modest results when the value of the element is higher or close to the pixel dimension. It is not known whether it is better to adopt the approach, where the mean Hounsfield value is found first, and based on that the Young module, or the approach, where each Hounsfield value of the voxel is first transformed into a voxel of the Young module, and then there is a value for each element. Since the relationship between bone density and Young's modulus is degree, these two approaches will result in different distributions of material characteristics, especially in cortical regions.

Assigning material characteristics to a finite element model, where the value of the characteristics varies from element to element, requires the creation of an additional program, which will be able to connect the position of the element and assign it the average value of the modulus of pixel elasticity. In this case, it is first necessary to generate a model, and then start the calculation of the modulus of elasticity and its assignment [67]. This is still one of the significant problems, especially when analyzing tissues, such as bone, which are anisotropic.

4.4.2 Practical Application

To calculate the material characteristics of a bone, from CT images, the Python programming language, version 3.6, was used. In order for this tool to run, CT images have to be in DICOM format (Digital Imaging and COmmunications in Medicine). Several libraries were necessary to be imported in order to calculate the Young's modulus of elasticity of a bone:

- pydicom – to import DICOM images
- matplotlib – for necessary visualization
- numpy – for necessary mathematical operations
- math – for necessary mathematical operations
- tkinter – for graphical interface

Algorithm that shows steps needed to calculate Young's modulus of elasticity from CT scans is shown in Figure 4.4.

The first thing that the user has to do in order to calculate Young's modulus of elasticity is to import a CT scan that will be used. Open file dialog box opens as soon as the user starts the tool. If the window is closed without choosing a file, text saying that "File not found" will be written on a screen and the user will need to restart the program. If there are some problems with CT scans and tool cannot

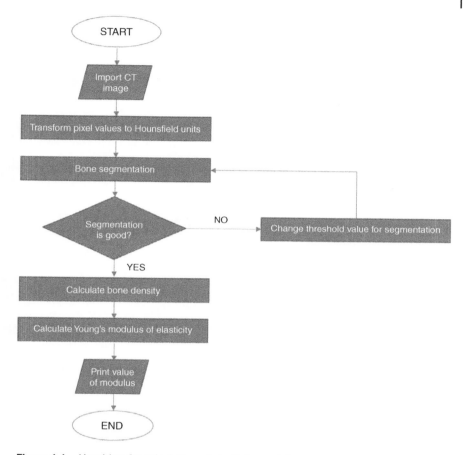

Figure 4.4 Algorithm for calculation of elasticity modulus.

find data it needs from computation of elasticity module, the following message will be written "There is an issue with your CT image. Try another one." Images contain a variety of information (Figure 4.5); however, not all of them are used.

The functions integrated in the tool use the following information:

- Rescale Intercept
- Rescale Slope
- Pixel Data

Without this information, the tool will not be able to work properly, and the user would need to read a different CT scan. After an image has been imported, the

```
(0008, 0008) Image Type                    CS: ['ORIGINAL', 'PRIMARY', 'AXIAL']
(0008, 0012) Instance Creation Date        DA: '20050726'
(0008, 0013) Instance Creation Time        TM: '104015'
(0008, 0016) SOP Class UID                 UI: CT Image Storage
(0008, 0018) SOP Instance UID              UI: 1.2.826.0.1.3680043.2.1125.1.75064541463040.2005072610401589578
(0008, 0020) Study Date                    DA: '20050101'
(0008, 0030) Study Time                    TM: '010100.000000'
(0008, 0050) Accession Number              SH: '1'
(0008, 0060) Modality                      CS: 'CT'
(0008, 0070) Manufacturer                  LO: 'GDCM'
(0008, 0080) Institution Name              LO: 'National Library of Medicine'
(0008, 0081) Institution Address           ST: 'http://www-creatis.insa-lyon.fr/Public/Gdcm'
(0008, 0090) Referring Physician's Name    PN: 'Unknown'
(0008, 1030) Study Description             LO: 'Visible Human Female'
(0008, 103e) Series Description            LO: 'Resampled to 1mm voxels'
(0010, 0010) Patient's Name                PN: 'Eve'
(0010, 0020) Patient ID                    LO: '000-000-001'
(0010, 0030) Patient's Birth Date          DA: '20050101'
(0010, 0032) Patient's Birth Time          TM: '010100.000000'
(0010, 0040) Patient's Sex                 CS: 'F'
(0018, 0050) Slice Thickness               DS: "1.0"
(0018, 5100) Patient Position              CS: 'HFS'
(0020, 000d) Study Instance UID            UI: 1.2.826.0.1.3680043.2.1125.1.75064541463040.2005072610384286421
(0020, 000e) Series Instance UID           UI: 1.2.826.0.1.3680043.2.1125.1.75064541463040.2005072610384286453
(0020, 0010) Study ID                      SH: '1'
(0020, 0011) Series Number                 IS: "1"
(0020, 0013) Instance Number               IS: "505"
(0020, 0032) Image Position (Patient)      DS: [0, 0, 504]
(0020, 0037) Image Orientation (Patient)   DS: [1.000000, 0.000000, 0.000000, 0.000000, 1.000000, 0.000000]
(0020, 0052) Frame of Reference UID         UI: 1.2.826.0.1.3680043.2.1125.1.75064541463040.2005072610384286419
(0020, 1040) Position Reference Indicator  LO: 'SN'
(0028, 0002) Samples per Pixel             US: 1
(0028, 0004) Photometric Interpretation    CS: 'MONOCHROME2'
(0028, 0010) Rows                          US: 512
(0028, 0011) Columns                       US: 512
(0028, 0030) Pixel Spacing                 DS: [1.000000, 1.000000]
(0028, 0100) Bits Allocated                US: 16
(0028, 0101) Bits Stored                   US: 16
(0028, 0102) High Bit                      US: 15
(0028, 0103) Pixel Representation          US: 1
(0028, 1052) Rescale Intercept             DS: "-1024.0"
(0028, 1053) Rescale Slope                 DS: "1.0"
(7fe0, 0000) Group Length                  UL: 524300
(7fe0, 0010) Pixel Data                    OW: Array of 524288 elements
```

Figure 4.5 A list of information contained in a CT scan.

values of each pixel have to be transformed into Hounsfield units. The transformed values are then used to segment a bone and calculate Young's modulus of elasticity. Considering that previous two steps were finished without an issue, the next step includes the segmentation of the bone. The aim of this step is to define boundaries of a bone which will be used to determine the bone density and later the isotropic elasticity modulus of the bone.

The segmentation of the bone depends on the threshold value chosen by the user. Default value is set to 300 and the first segmentation will be performed using that value. After the segmentation, the user will be able to visualize the image of segmented bone in order to determine if whole bone was included. If the automatic segmentation performed well, the user can choose to continue onto the following step. However, if the user is not satisfied with the obtained segmentation, it is possible to perform this step again and to choose new threshold value. The user will be able to change the threshold values for as long as needed. The effect of different threshold values on bone segmentation can be seen in Figure 4.6. The effect of threshold value is demonstrated using femoral bone CT scan, where the body of the femoral bone is shown.

As it can be observed in Figure 4.6, lower threshold values lead to more tissue inclusion, while the higher threshold values reduce the segmented area. This is why it is necessary to carefully choose the threshold as it will have effect on the value of calculated elasticity modulus.

Considering the previous step to be successfully finished, we can start with calculation of bone density. The tool will use Hounsfield unit for each pixel that was part of the segmented bone in order to calculate the bone density as the average Hounsfield unit. This value is necessary for calculation of the Young's modulus of elasticity.

The last step in this process is the calculation of the elasticity modulus. The user will be able to choose between three bone options. The tool can calculate elasticity modulus for the following bones:

- Femur
- Tibia
- Vertebra

The user has to choose one of these three options, and until that is done, the obtained results cannot be shown. After a bone type has been chosen, the user will be able to see values for calculated isotropic Young's modulus of elasticity. The calculation is performed using Eqs. (4.22) for the femur, (4.11) tibia, and (4.10) vertebra.

The calculated value for the chosen CT image will be displayed on the screen (Figure 4.7). Values are given in GPa. Besides the end value, the user will also

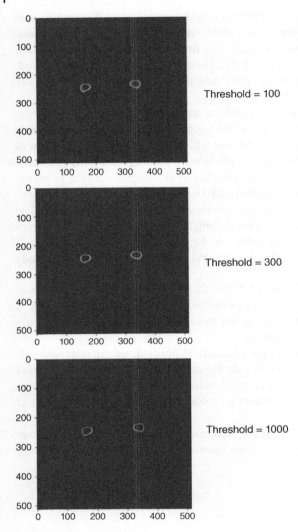

Figure 4.6 Effect of threshold value on the segmentation of the femoral bone body.

Threshold = 100

Threshold = 300

Threshold = 1000

be able to verify the location of the file it was chosen as well as the type of bone used for calculation.

Besides the numerical values, the user will also have a visual representation (Figure 4.8) of the original CT scan (upper half) and the segmented bone used for the calculation of the elasticity modulus (lower half).

The presented tool can be currently used only to calculate Young's modulus of elasticity for isotropic material.

```
Starting program
D:/DICOM_snimci/Kosti/CT_Femur/vhf.608.dcm

Default threshold value is 300. Do you want to change the value? (Y/N)N
Please choose type of bone shown on images (only these are currently available):
Options:
1 .- Femur
2 .- Tibia
3 .- Vertebra
4 .- Exit program

Choose an option: 1
You have chosen Option 1 - Femur

Isotropic material properties:
Young's modulus of elasticity:   15504.142274883412 GPa
```

Figure 4.7   Printing of the calculated values.

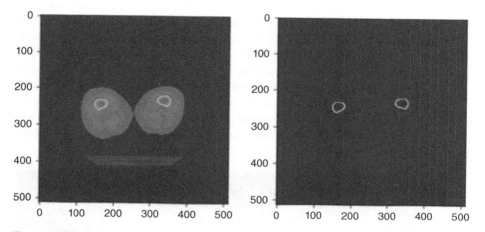

Figure 4.8   Visual representation of the original image and segmented bone used for calculation.

4.5   Numerical Analysis of Femoral Bone Using Calculated Elasticity Modulus

4.5.1   Femoral Bone Model

The aim of the femoral bone geometry reconstruction is to create a realistic anatomical 3D femoral geometry, based on the patient CT images. In order to achieve this, 400 2D images in DICOM format were used. These images belonged to a female patient and were obtained using computed tomography. Characteristics of the used images were:

- Resolution: 512×512 pixels
- Pixel size: 1 mm
- Slice thickness: 1 mm
- The distance between two slices: 1 mm

The first step in the process of femoral bone reconstruction was the segmentation of the femoral bone. This step involved creating a mask, which gathered the information about the contours of the femur, necessary to create the 3D geometry. The contour marking was done in the transverse (horizontal) plane. Figure 4.9 shows one image of the femoral body in the transverse plane, before segmentation.

Femoral segmentation was performed semiautomatically. Two masks were used, so that the segmentation software would automatically mark the contours of the cortical and trabecular bones. Although this significantly reduced the time required to develop the model, it was necessary to manually go through all the images and check that the cortical and trabecular bone were well segmented. An example of a marked cortical part of the femur is presented in Figure 4.10.

After the completion of the segmentation, a calculation of the preliminary 3D model of the femoral bone was performed. The preliminary 3D model was then exported, in order to create a uniform mesh of finite elements.

It is necessary to evaluate the created preliminary model in case there are deficiencies in the surface mech. All defects were removed manually and a uniform

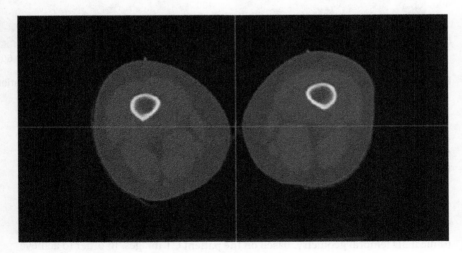

Figure 4.9 Femoral body in the transverse plane before segmentation.

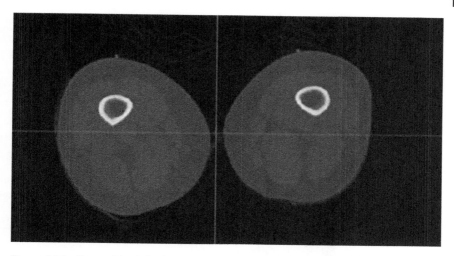

Figure 4.10 Femoral body in the transverse plane after segmentation.

surface mesh was created. The model of the femoral bone, after the refinement of the surface mesh of the cortical and trabecular femoral bone, is presented in Figure 4.11.

The last step in the reconstruction of the femoral bone model is the creating of the volumetric mesh. The total number of nodes and elements for the femoral bone model (including cortical and trabecular part of the femoral bone) was 10 195 and 42 982, respectively.

4.5.2 Material Properties

For the numerical simulation of the mechanical behavior of femoral bone model, isotropic material properties have been used. Both segments of the femoral bone, cortical and trabecular, were considered to be isotropic, linear elastic, and homogeneous. In Table 4.3, Young's elasticity modulus values for the cortical femoral bone are given. Three cases were analyzed; two used calculated elasticity moduli, while one used elasticity modulus adopted from the literature. In all three cases, the Poisson's ratio for the cortical femur was the same – 0.3 [68].

The material properties of the trabecular femoral bone are provided in Table 4.4. The presented properties were the same for all three cases.

4.5.3 Boundary Conditions

The boundary conditions that were used for the numerical simulations were adapted from the literature. For the performed simulations, applied boundary

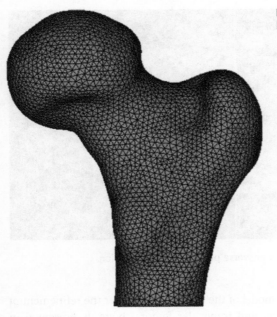

Figure 4.11 Femoral bone model.

Table 4.3 Young's elasticity modulus of the cortical femoral bone.

Case number	Young's modulus of elasticity	References
Case 1	13 079 MPa	Calculated from CT scans
Case 2	12 724 MPa	Calculated from CT scans
Case 3	16 700 MPa	[68]

Table 4.4 Material properties of the trabecular femoral bone.

Young's modulus of elasticity	Poisson's ratio	Reference
600 MPa	0.2	[69]

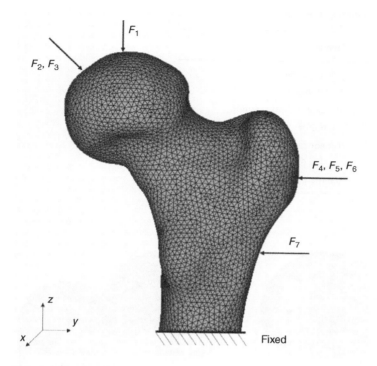

Figure 4.12 Applied boundary conditions for the numerical simulations.

conditions corresponded to the physiological conditions when a person is standing. Figure 4.12 shows the applied constraints and loads for the simulations.

As it can be seen in Figure 4.12, the bottom part of the femoral bone was fixed, while on the rest of the bone model, the forces that represent the effect of weight, contact with the pelvis, and also the impact of muscles were defined. In total, seven forces were applied [70] and their description is given in Table 4.5.

The static finite element analysis of the femoral bone was performed using the software PAK C. This in-house solver has been developed at the Faculty of Engineering, University of Kragujevac, and is used for the numerical simulations of the solid problems such as the analysis of bone biomechanics.

4.5.4 Obtained Results

Figures 4.13–4.18 present the obtained results of numerical simulations (stress and displacement distributions) for three previously defined cases.

Table 4.5 Description of the applied forces.

Force	Name
F_1	Body weight
F_2	Hip contact
F_3	Intersegmental resultant
F_4	Abductor
F_5	Tensor fascia latae, proximal part
F_6	Tensor fascia latae, distal part
F_7	Vastus lateralis

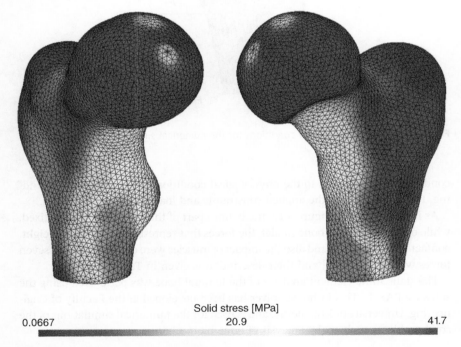

Solid stress [MPa]

0.0667   20.9   41.7

Figure 4.13 Obtained stress distribution – case 1.

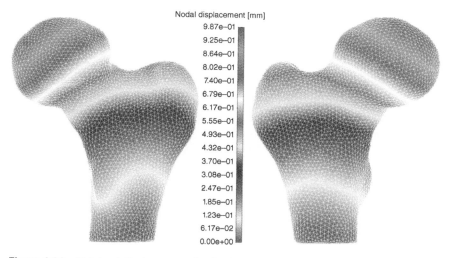

Figure 4.14 Obtained displacement distribution – case 1.

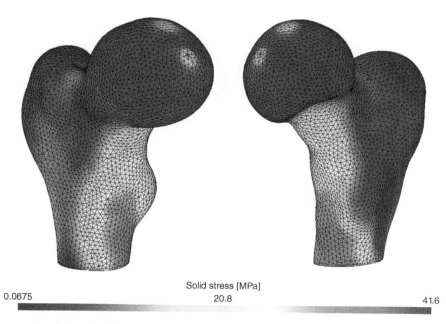

Figure 4.15 Obtained stress distribution – case 2.

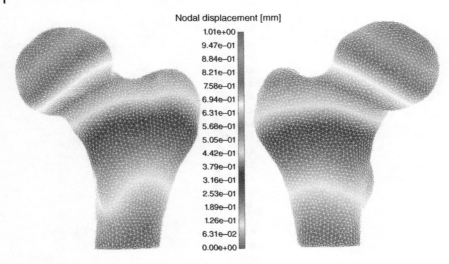

Figure 4.16 Obtained displacement distribution – case 2.

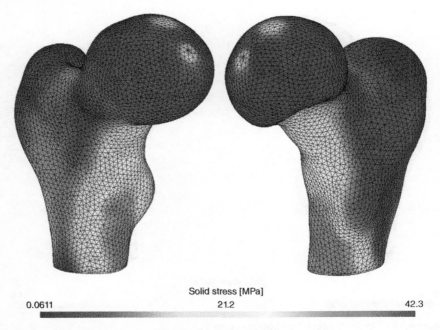

Figure 4.17 Obtained stress distribution – case 3.

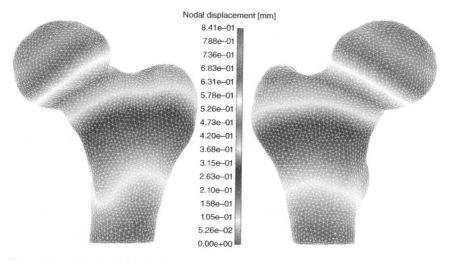

Figure 4.18 Obtained displacement distribution – case 3.

4.5.4.1 Case 1

Figure 4.13 shows the obtained stress distribution for case 1. The obtained stress values for the femoral bone were in the range from 0.0667 to 41.7 MPa.

The maximum stress values can be observed on the body of the femoral bone, as well as on the femoral head where the force F_1 was defined (the effect of body weight). The neck of the femoral bone is a location of high importance due to the fact that the majority of the femoral bone fractures occur on this location. The value of the stress on the neck of the femoral bone did not exceed the value of 26.07 MPa, which was about 15 MPa lower than the maximum calculated stress.

Figure 4.14 shows the displacement distribution for case 1. The calculated femoral bone displacements were in the range from 0 to 0.987 mm.

The maximum displacement values were in the region of the femoral head as it can be seen in Figure 4.14, where the force with the highest values was defined. There was no displacement on the bottom of the model due to defined boundary conditions (the bottom of the femoral body was fixed).

4.5.4.2 Case 2

Figure 4.15 shows the stress distribution for case 2. The calculated stress values were in the range from 0.0675 to 41.6 MPa. As in the previous case (case 1), the maximum stress values were observed at the location where the effect of body weight was defined. Comparison of the stress distribution for cases 1 and 2 showed almost identical distribution. The difference between these two cases was in the range of the calculated stress values. Due to the fact that in this case the applied Young's modulus of elasticity was smaller compared with the Young's modulus value from the first case, the obtained stress values were lower compared with the previously presented result.

Figure 4.16 shows the displacement distribution for case 2. The calculated femoral bone displacements were in the range from 0 to 1.01 mm.

Based on the obtained displacement values, it can be noticed that lower modulus of elasticity led to higher displacement values. As it can be noticed, the displacement distribution presented in Figure 4.16 was identical to the distribution obtained in the previous case.

4.5.4.3 Case 3

The obtained stress distribution for the third case is given in Figure 4.17. It can be noticed that the maximum stress value was higher compared with the previous two cases. All stress values were in the range from 0.0611 to 42.3 MPa.

In the third case, the largest Young's modulus of elasticity was applied (Table 4.3), and by comparison with the previous results, it is possible to observe some differences in the stress distribution. As it can be seen, the location where the maximum stress was calculated has not been changed. However, it can be noticed that the inner body of the femoral bone had larger area with higher stress values compared with the previous two cases. In addition, it can be noticed that the calculated stress values on the head and neck of the femoral bone were higher compared with the previous two cases.

Figure 4.18 shows the displacement distribution for the third case. The calculated displacements of the femoral bone were in range from 0 to 0.841 mm. As in the previous two cases, the displacement distribution was not changed. The only difference between these three cases was regarding the displacement values, which were significantly lower in the third case compared with the previous two.

4.5.4.4 Comparison of the Obtained Results

Table 4.6 shows the comparison of the obtained maximum stress values for all three cases. Row Difference [%] contains information about the difference between

Table 4.6 Comparison of the calculated maximum stress values.

	Case 1	Case 2	Case 3
Maximum calculated stress [MPa]	41.7	41.6	42.3
Difference [%]	−1.42%	−1.65%	—

Table 4.7 Comparision of the calculated maximum displacement values.

	Case 1	Case 2	Case 3
Maximum calculated displacement [mm]	0.987	1.01	0.841
Difference [%]	+17.4 %	+20.1 %	—

the values in the third case (considered to be the absolute value) and the values obtained from the numerical simulations that were using Young's modulus of elasticity calculated from the CT scans.

The results presented in Table 4.6 indicate that the difference between the obtained results for all three cases was not significant. If only stress values are needed, the obtained differences between the results indicated that the calculation of the Young's modulus of elasticity from CT scans can be avoided, and the value of the elasticity modulus from the literature can be adopted. This leads to reduction in time needed for the numerical simulations.

Table 4.7 compares the obtained maximum displacements for all three cases. As in the previous table, the row Difference [%] represents the difference between the values in the third case (considered to be the absolute value) and the values obtained from the simulations that used elasticity modulus calculated utilizing the presented tool.

Unlike the results presented in the previous table (Table 4.6), where the difference between the obtained maximum stress values was quite low, in this case, we have a significantly larger difference between the obtained results (up to 20.1%). Although the obtained stress values indicated that calculation of Young's modulus of elasticity can be omitted for numerical simulation presented in this section, the results for distribution values indicate otherwise.

Figure 4.19 shows the relationship between the applied Young's modulus of elasticity for all three cases and the obtained stress values. The coefficient of determination of linear correlation, based on the obtained values was calculated to be 0.9998. Considering that both cortical and trabecular femoral bone were considered to be linear elastic, the obtained dependence was expected.

Figure 4.20 shows the relationship between the applied Young's modulus of elasticity for all three cases and the obtained stress values.

The coefficient of determination of linear correlation based on the values from numerical simulations and obtained Young's modulus was calculated to be 0.9995. Similar to the previous correlation, the obtained dependence was expected, since the cortical and trabecular femoral bones were defined as a linear elastic material.

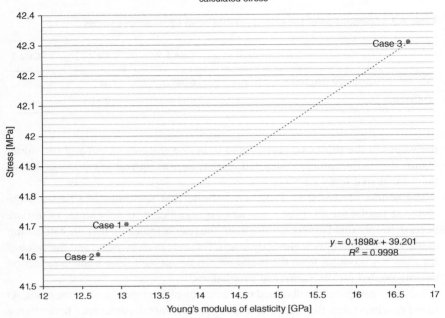

Figure 4.19 Correlation between Young's modulus of elasticity and calculated stress.

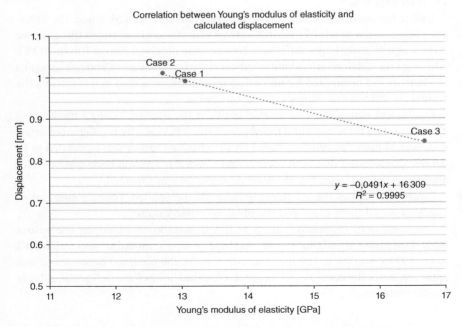

Figure 4.20 Correlation between Young's modulus of elasticity and calculated displacement.

4.6   Conclusion

The technological development achieved in the previous decades has enabled significant progress in the analysis of the human body. The progress is especially noticeable in the research that is focused on the application of the numerical methods as a way to gain novel insights. Nowadays, research in the field of bioengineering is increasingly turning toward the analysis of the patient-specific models, with a goal to provide better patient care and achieve the best possible results during the treatment of patients.

The tool for determining the Young's modulus of elasticity of cortical bone from the patient's CT images is a way to improve the accuracy of numerical simulations. As shown in Section 4.5, the Young's modulus of elasticity can have a considerable impact on the obtained results, and, therefore, can affect the further course of treatment. The importance of determining the realistic Young's modulus of elasticity is easy to notice when analyzing patients with diseases, such as osteoporosis, where bone density decreases over time and thus the value of the modulus of elasticity. This can allow researchers to perform the risk assessments for bone fractures as a consequence of the disease. Moreover, the appropriate value of Young's modulus of elasticity may be significant when analyzing the interaction of orthopedic aids, such as a hip implant, and the corresponding bone.

Based on the presented results for the finite element analysis of femoral bone, it is possible to conclude that, in the case of isotropic, linearly elastic, and homogeneous material, Young's modulus of elasticity has a significantly lower impact on the obtained results, compared with the elasticity modulus taken from the literature. When analyzing the femur, for static load conditions, the value of the elasticity modulus from the literature can be adopted, especially if only the stress distribution is needed as the difference is lower than 5%. In this situation, the use of experimental values significantly reduces the time required to prepare the model for numerical simulations, without significantly affecting the final outcome. It can be assumed that the results would be significantly different if the femoral bone was divided into several units such as the head of the femur, the neck of the femur, and the body of the femur.

The main limitation of the presented tool is that it can currently be employed for the cortical part of the bone. The presented tool will be upgraded in order to provide improved results experience by improving the detection of cortical bone.

Acknowledgements

The research has been carried out with the support of the Ministry of Education, Science and Technological Development, Republic of Serbia, with projects III41007 and OI174028 and funding by contract 451-03-68/2020-14/200107.

References

1 Büchler, P., Ramaniraka, N.A., Rakotomanana, L.R. et al. (2002). A finite element model of the shoulder: application to the comparison of normal and osteoarthritic joints. *Clinical Biomechanics* 17 (9–10): 630–639.

2 Chen, W.M., Park, J., Park, S.B. et al. (2012). Role of gastrocnemius–soleus muscle in forefoot force transmission at heel rise—a 3D finite element analysis. *Journal of Biomechanics* 45 (10): 1783–1789.

3 Cheung, J.T.-M. and Zhang, M. (2005). A 3-dimensional finite element model of the human foot and ankle for insole design. *Archives of Physical Medicine and Rehabilitation* 86 (2): 353–358.

4 Kopperdahl, D.L., Aspelund, T., Hoffmann, P.F. et al. (2014). Assessment of incident spine and hip fractures in women and men using finite element analysis of CT scans. *Journal of Bone and Mineral Research* 29 (3): 570–580.

5 Pena, E., Calvo, B., Martinez, M.A., and Doblare, M. (2006). A three-dimensional finite element analysis of the combined behavior of ligaments and menisci in the healthy human knee joint. *Journal of Biomechanics* 39 (9): 1686–1701.

6 Schmidt, H., Shirazi-Adl, A., Galbusera, F., and Wilke, H.-J. (2010). Response analysis of the lumbar spine during regular daily activities—a finite element analysis. *Journal of Biomechanics* 43 (10): 1849–1856.

7 Bennett, D. and Goswami, T. (2008). Finite element analysis of hip stem designs. *Materials & Design* 29 (1): 45–60.

8 Godest, A.C., Beaugonin, M., Haug, E. et al. (2002). Simulation of a knee joint replacement during a gait cycle using explicit finite element analysis. *Journal of Biomechanics* 35 (2): 267–275.

9 Himmlová, L., Dostálová, T., Kácovský, A., and Konvicková, S. (2004). Influence of implant length and diameter on stress distribution: a finite element analysis. *The Journal of Prosthetic Dentistry* 91 (1): 20–25.

10 Reggiani, B., Leardini, A., Corazza, F., and Taylor, M. (2006). Finite element analysis of a total ankle replacement during the stance phase of gait. *Journal of Biomechanics* 39 (8): 1435–1443.

11 Vulović, A. and Filipović, N. (2019). Computational analysis of hip implant surfaces. *Journal of the Serbian Society for Computational Mechanics* 13 (1): 109–119.

12 Watanabe, Y., Shiba, N., Matsuo, S. et al. (2000). Biomechanical study of the resurfacing hip arthroplasty: finite element analysis of the femoral component. *The Journal of Arthroplasty* 15 (4): 505–511.

13 Auricchio, F., Conti, M., De Beule, M. et al. (2011). Carotid artery stenting simulation: from patient-specific images to finite element analysis. *Medical Engineering & Physics* 33 (3): 281–289.

14 Early, M., Lally, C., Prendergast, P.J., and Kelly, D.J. (2009). Stresses in peripheral arteries following stent placement: a finite element analysis. *Computer Methods in Biomechanics and Biomedical Engineering* 12 (1): 25–33.

15 Gültekin, O., Dal, H., and Holzapfel, G.A. (2016). A phase-field approach to model fracture of arterial walls: theory and finite element analysis. *Computer Methods in Applied Mechanics and Engineering* 312: 542–566.

16 Lally, C., Dolan, F., and Prendergast, P.J. (2005). Cardiovascular stent design and vessel stresses: a finite element analysis. *Journal of Biomechanics* 38 (8): 1574–1581.

17 Migliavacca, F., Petrini, L., Colombo, M. et al. (2002). Mechanical behavior of coronary stents investigated through the finite element method. *Journal of Biomechanics* 35 (6): 803–811.

18 Filipovic, N., Rosic, M., Tanaskovic, I. et al. (2011). ARTreat project: three-dimensional numerical simulation of plaque formation and development in the arteries. *IEEE Transactions on Information Technology in Biomedicine* 16 (2): 272–278.

19 Sadat, U., Li, Z.-Y., Young, V.E. et al. (2010). Finite element analysis of vulnerable atherosclerotic plaques: a comparison of mechanical stresses within carotid plaques of acute and recently symptomatic patients with carotid artery disease. *Journal of Neurology, Neurosurgery & Psychiatry* 81 (3): 286–289.

20 Bahmanzadeh, H., Abouali, O., Faramarzi, M., and Ahmadi, G. (2015). Numerical simulation of airflow and micro-particle deposition in human nasal airway pre-and post-virtual sphenoidotomy surgery. *Computers in Biology and Medicine* 61: 8–18.

21 Koullapis, P., Kassinos, S.C., Muela, J. et al. (2018). Regional aerosol deposition in the human airways: the SimInhale benchmark case and a critical assessment of in silico methods. *European Journal of Pharmaceutical Sciences* 113: 77–94.

22 Wang, Y. and Elghobashi, S. (2014). On locating the obstruction in the upper airway via numerical simulation. *Respiratory Physiology & Neurobiology* 193: 1–10.

23 Kopsch, T., Murnane, D., and Symons, D. (2016). Optimizing the entrainment geometry of a dry powder inhaler: methodology and preliminary results. *Pharmaceutical Research* 33 (11): 2668–2679.

24 Tong, Z.B., Zheng, B., Yang, R.Y. et al. (2013). CFD-DEM investigation of the dispersion mechanisms in commercial dry powder inhalers. *Powder Technology* 240: 19–24.

25 Vulović, A., ŠuŠterŠič, T., Cvijić, S. et al. (2018). Coupled in silico platform: computational fluid dynamics (CFD) and physiologically-based pharmacokinetic (PBPK) modelling. *European Journal of Pharmaceutical Sciences* 113: 171–184.

26 Su, N., Yang, J., Xie, Y. et al. (2019). Bone function, dysfunction and its role in diseases including critical illness. *International Journal of Biological Sciences* 15 (4): 776.

27 Burr, D.B. and Akkus, O. (2014). Bone morphology and organization. In: *Basic and Applied Bone Biology* (eds. D.B. Burr and M.R. Allen), 3–25. Academic Press.

28 Sun, X., Lijia, L., Guo, Y. et al. (2018). Influences of organic component on mechanical property of cortical bone with different water content by nanoindentation. *AIP Advances* 8: 035003.

29 Novitskaya, E., Chen, P.Y., Hamed, E. et al. (2011). Recent advances on the measurement and calculation of the elastic moduli of cortical and trabecular bone: a review. *Theoretical and Applied Mechanics* 38 (3): 209–297.

30 OpenStax College (2013). *Anatomy and Physiology*. OpenStax College.

31 Clarke, B. (2008). Normal bone anatomy and physiology. *Clinical Journal of the American Society of Nephrology* 3: S131–S139.

32 Oftadeh, R., Perez-Viloria, M., Villa-Camacho, J.C. et al. (2015). Biomechanics and mechanobiology of trabecular bone: a review. *Journal of Biomechanical Engineering* 137 (1): 0108021–01080215.

33 Chittibabu, V., Santa Rao, K., and Govinda Rao, P. (2016). Factors affecting the mechanical properties of compact bone and miniature specimen test techniques: a review. *Advances in Science and Technology Research Journal* 10 (32): 169–183.

34 Morgan, E.F., Unnikrisnan, G.U., and Hussein, A.I. (2018). Bone mechanical properties in healthy and diseased states. *Annual Review of Biomedical Engineering* 20: 119–143.

35 Havaldar, R., Pilli, S.C., and Putti, B. (2014). Insights into the effects of tensile and compressive loadings on human femur bone. *Advanced Biomedical Research* 3: 101.

36 Bonfield, W. and Clark, E.A. (1973). Elastic deformation of compact bone. *Journal of Materials Science* 8 (11): 1590–1594.

37 Carter, D.R. and Hayes, W.C. (1977). The compressive behavior of bone as a two-phase porous structure. *The Journal of Bone and Joint Surgery* 59A: 954–962.

38 Dong, X.N. and Guo, X.E. (2004). The dependence of transversely isotropic elasticity of human femoral cortical bone on porosity. *Journal of Biomechanics* 37 (8): 1281–1287.

39 Gibson, L.J. (1985). The mechanical behaviour of cancellous bone. *Journal of Biomechanics* 18 (5): 317–328.

40 Schaffler, M.B. and Burr, D.B. (1988). Stiffness of compact bone: effects of porosity and density. *Journal of Biomechanics* 21: 13–16.

41 Cuppone, M., Seedhom, B.B., Berry, E., and Ostell, A.E. (2004). The longitudinal Young's modulus of cortical bone in the midshaft of human femur and its correlation with CT scanning data. *Calcified Tissue International* 74 (3): 302–309.

42 Currey, J.D. (1988). The effect of porosity and mineral content on the Young's modulus of elasticity of compact bone. *Journal of Biomechanics* 21 (2): 131–139.

43 Katz, J.L. (1980). Anisotropy of Young's modulus of bone. *Nature* 283 (5742): 106–107.

44 Keller, T.S., Mao, Z., and Spengler, D.M. (1990). Young's modulus, bending strength, and tissue physical properties of human compact bone. *Journal of Orthopaedic Research* 8 (4): 592–603.

45 Rho, J.Y., Ashman, R.B., and Turner, C.H. (1993). Young's modulus of trabecular and cortical bone material: ultrasonic and microtensile measurements. *Journal of Biomechanics* 26 (2): 111–119.

46 Rho, J.Y., Tsui, T.Y., and Pharr, G.M. (1997). Elastic properties of human cortical and trabecular lamellar bone measured by nanoindentation. *Biomaterials* 18 (20): 1325–1330.

47 Keyak, J.H., Lee, I., and Skinner, H.B. (1994). Correlations between orthogonal mechanical properties and density of trabecular bone: use of different densitometric measures. *Journal of Biomechanical Materials and Research* 28: 1329–1236.

48 Hernandez, C.J., Beaupre, G.S., Keller, T.S., and Carter, D.R. (2001). The influence of bone volume fraction and ash fraction on bone strength and modulus. *Bone* 29 (1): 74–78.

49 Lotz, J.C., Gerhart, T.N., and Hayes, W.C. (1991). Mechanical properties of metaphyseal bone in the proximal femur. *Journal of Biomechanics* 24: 317–329.

50 Kopperdahl, D.L. and Keaveny, T.M. (1998). Yield strain behavior of trabecular bone. *Journal of Biomechanics* 31 (7): 601–608.

51 Ciarelli, T., Fyhrie, D., Schaffler, M., and Goldstein, S. (2000). Variations in three-dimensional cancellous bone architecture of the proximal femur in female hip fractures and in controls. *Journal of Bone and Mineral Research* 15 (1): 32–40.

52 Morgan, E.F., Bayraktar, H.H., and Keaveny, T.M. (2003). Trabecular bone modulus–density relationships depend on anatomic site. *Journal of Biomechanics* 36 (5): 897–904.

53 Kaneko, T.S., Bell, J.S., Pejcic, M.R. et al. (2004). Mechanical properties, density and quantitative ct scan data of trabecular bone with and without metastases. *Journal of Biomechanics* 37 (4): 523–530.

54 Li, B. and Aspden, R.M. (1997). Composition and mechanical properties of cancellous bone from the femoral head of patients with osteoporosis or osteoarthritis. *Journal of Bone and Mineral Research* 12 (4): 641–651.

55 Rho, J.Y., Hobatho, M.C., and Ashman, R.B. (1995). Relations of mechanical properties to density and CT numbers in human bone. *Medical Engineering & Physics* 17 (5): 347–355.

56 Keller, T.S. (1994). Predicting the compressive mechanical behavior of bone. *Journal of Biomechanics* 27 (9): 1159–1168.

57 Krone, R., & Schuster, P. (2006). An investigation on the importance of material anisotropy in finite-element modeling of the human femur. SAE Technical Paper Series, No. 2006-01-0064.

58 Soni, A., Chawla, A., & Mukherjee, S. (2007). Effect of muscle contraction on knee loading for a standing pedestrian in lateral impacts. *Proceedings of 20th ESV Conference*, Lyon, France (18–21 June 2007). Paper Number 07-0458.

59 Tseng, J.-G., Huang, B.W., Liang, S.-H. et al. (2014). Normal mode analysis of a human fibula. *Life Science Journal* 11 (12): 711–718.

60 Prochor, P. (2017). Finite element analysis of stresses generated in cortical bone during implantation of a novel Limb cortical bone during implantation of a novel Limb. *Biocybernetics and Biomedical Engineering* 37 (2): 255–262.

61 Taheri, E., Sepehri, B., and Ganji, R. (2012). Mechanical validation of perfect tibia 3D model using computed tomography scan. *Engineering* 4 (12): 877–880.

62 Antic, S., Vukicevic, A., Milasinovic, M. et al. (2015). Impact of the lower third molar presence and position on the fragility of mandibular angle and condyle: a three-dimensional finite element study. *Journal of Cranio-Maxillofacial Surgery* 43 (6): 870–878.

63 Duchemin, L., Bousson, V., Raossanaly, C. et al. (2008). Prediction of mechanical properties of cortical bone by quantitative computed tomography. *Medical Engineering & Physics* 30 (3): 321–328.

64 Helgason, B., Taddei, F., Pálsson, H. et al. (2008). A modified method for assigning material properties to FE models of bones. *Medical Engineering & Physics* 30 (4): 444–453.

65 Inacio, J.V., Malige, A., Schroeder, J.T. et al. (2019). Mechanical characterization of bone quality in distal femur fractures using pre-operative computed tomography scans. *Clinical Biomechanics* 67: 20–26.

66 Taddei, F., Schileo, E., Helgason, B. et al. (2007). The material mapping strategy influences the accuracy of CT-based finite element models of bones: an evaluation against experimental measurements. *Medical Engineering & Physics* 29 (9): 973–979.

67 Chen, G., Schmutz, B., Epari, D. et al. (2010). A new approach for assigning bone material properties from CT images into finite element models. *Journal of Biomechanics* 43 (5): 1011–1015.

68 Dhanopia, A. and Bhargava, M. (2017). Finite element analysis of human fractured femur bone implantation with PMMA thermoplastic prosthetic plate. *Procedia Engineering* 173: 1658–1665.

69 Perez, A., Mahar, A., Negus, C. et al. (2008). A computational evaluation of the effect of intramedullary nail material properties on the stabilization of simulated femoral shaft fractures. *Medical Engineering & Physics* 30 (6): 755–757.

70 Chalernpon, K., Aroonjarattham, P., and Aroonjarattham, K. (2005). Static and dynamic load on hip contact of hip prosthesis and Thai femoral bones. *International Journal of Mechanical and Mechatronics Engineering* 9 (3): 251–255.

5

Parametric Modeling of Blood Flow and Wall Interaction in Aortic Dissection

Igor B. Saveljic[1,2] and Nenad D. Filipovic[2,3]

[1] *Institute for Information Technologies University of Kragujevac, Kragujevac, Serbia*
[2] *Department of Computer Fluid Dynamics, Bioengineering Research and Development Center, BIOIRC Kragujevac, Serbia*
[3] *Faculty of Engineering, University of Kragujevac Kragujevac, Serbia*

5.1   Introduction

The term *cardiovascular disease* refers to a group of diseases that affect the heart or blood vessels. All cardiovascular diseases occur as a consequence of narrowed or blocked blood vessels. These blood vessel obstructions can cause chest pain, heart attack, or stroke.

Atherosclerosis is a pathological process in which lipids, parts of dead cells, and calcium accumulate (deposit) in the artery wall. Deposited lipids in the wall of a blood vessel come from blood lipoproteins, primarily cholesterol. Atherosclerosis develops gradually, usually affecting large and medium-sized arteries. Mild forms of atherosclerosis usually do not give any symptoms, while over time a blood clot (thrombus) can form at the sites of thickening of the arteries. Coronary atherosclerosis is the most common cause of ischemic heart disease and acute coronary syndromes (unstable angina pectoris, myocardial infarction, and sudden cardiac death). In cerebrovascular diseases, the situation is similar; the process of atherosclerosis develops in the blood vessels that supply the brain with blood, resulting in narrowing or complete obstruction of the blood vessels.

Aortic dissection is a serious medical condition in which the inner layer of the aortic wall ruptures. Then, the blood rushes through the cleft, leading to stratification (dissection) of the inner and middle layers of the aortic wall. If the blood penetrates the outer layer of the wall, aortic dissection is often fatal. This condition is relatively rare. It most often occurs in men in the seventh and eighth decade of life. Symptoms of aortic dissection can mimic the symptoms of other diseases,

Computational Modeling and Simulation Examples in Bioengineering, First Edition. Edited by Nenad D. Filipovic. © 2022 The Institute of Electrical and Electronics Engineers, Inc. Published 2022 by John Wiley & Sons, Inc.

which often leads to delays in making the right diagnosis. However, when detected and treated early, the chances of survival increase significantly.

Aortic dissection is characterized by stratification of the aortic wall, which begins with a small crack in the intimate layer and results in filling of the media space with blood, thus creating another lumen – false lumen [1, 2]. The development and progression of aortic dissection is associated with several of the most important risk factors, such as hypertension [1, 3], genetic factors [4], gender, and age. This condition of the aorta can lead to several serious complications, such as aortic rupture, cardiac tamponade, hypotension, and organ ischemia, very often with a fatal outcome [2].

Numerous methods are used to diagnose aortic dissection, two of which are the most common. The first method, transesophageal echocardiography (TEE), with a percentage of application of 33%, is used when the existence of aortic dissection is initially suspected, as well as in the case of defining the mechanisms of aortic regurgitation [5, 6]. The second method, computed tomography (CT), with an application rate of 61%, is used to obtain information on the degree of aortic involvement [6, 7].

The numerical method, i.e. the finite element method, is used to analyze the flow of blood and other hemodynamic variables of the aorta and its branches. Numerous studies have examined changes in healthy aortas in this way [8, 9], in dissected aortas [10–12], and in aortas with aneurysms [13–15].

Studies [16, 17] assume that the flow in the aorta is laminar, since it is a large blood vessel where the mean flow is small, so the Reynolds number is relatively small [4, 18]. Regarding blood characteristics, most of the numerical studies assume it Newtonian blood [11, 16, 18], based on neglecting the effect of particles in large arteries and that the shear rate is high in large arteries.

A significant number of numerical simulations have been used to predict the progression of aortic dissection [4, 19–21], thrombus formation [18, 22], various treatment strategies, including fenestration and stent grafts [5, 18, 23]. Great attention has been paid to the entrance slits, location, size, and influence on the further propagation of the false lumen [10, 24, 25]. In addition, the number of recurrent gaps and their impact on the fate of the false lumen have been studied [26]. Research has led to the fact that even the shape of the inlet gap can have an impact on the way the false lumen propagates [18, 27].

In this chapter, continuum mechanics is applied for the analysis of aortic dissection and appropriate models of fluid flow in a healthy aorta and through aortic dissection. The theoretical basis of the work is the basic fluid mechanics and deformable body mechanics. The basic method used in the paper is the finite element method. The finite element method is one of the most common numerical methods used in engineering practice. This method is based on the physical discretization of the continuum.

5.2 Medical Background

5.2.1 Circulatory System

The human circulatory system is closed and consists of blood, blood vessels, and heart; lymph, lymph vessels, and lymph glands (Figure 5.1). This system enables the organism to exchange oxygen and nutrients, and the products of metabolism from cells and tissues are secreted into it. The circulatory system consists of two subsystems: the pulmonary (small bloodstream) and the systemic (large bloodstream) circulatory circuit. The pulmonary cycle includes the blood vessels in the lungs and the blood vessels that connect the lungs and the heart. The systemic circuit includes all other blood vessels and organs in the body. The main difference between these circulatory subsystems is in the composition of the blood that passes through them.

The right side of the heart supplies blood to the pulmonary circulatory circuit, while the left side of the heart supplies blood to the systemic circulatory circuit. The blood of these two subsystems never mixes (except in the case of a heart septal defect). Both subsystems consist of a large number of capillaries that form a network of blood vessels in which the exchange of molecules takes place. The oxygenated blood leaves the pulmonary capillaries and is further transported to the organs through the vascular system. There is an exchange of matter in the organs, i.e. oxygen from "fresh" blood is exchanged for carbon dioxide, which is released from the cells. Blood that carries carbon dioxide is transported through the systemic circulatory system [28].

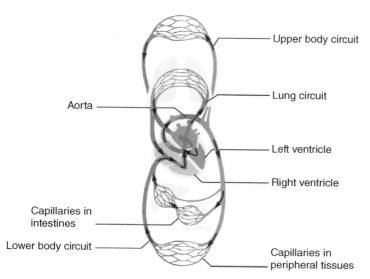

Figure 5.1 Schematic representation of the cardiovascular system [28].

The shape and structure of the walls of blood vessels vary depending on the pressure and the nature of the pulse wave. For this reason, the walls of the arteries are considerably thicker. The walls of the arteries and veins below the heart are generally thicker than those above, due to the greater hydrostatic pressure in the lower regions when the body is in an upright position. The systemic arteries have thicker walls than the pulmonary ones, reflecting a significant difference in pressure. Arteries in the skull and vertebral canals ("closed box" system) have very thin walls in relation to their diameter; the thickness of the outer and middle tunic is reduced, which is a situation that is also found in the canals of long bones. In places where they are subjected to significant pressures or other mechanical forces, such as coronary and carotid arteries, the vessels often have thick, muscular walls, because of a thicker middle tunic. Elastic arteries usually have walls one-tenth the thickness of their caliber; this ratio in muscular arteries is about a quarter, in arterioles it is about a half, and in veins it is one-tenth or less.

5.2.2 Aorta

The aorta is the main and largest blood vessel in the body. It starts from the base of the left heart chamber and is a "channel" that delivers oxygen-rich blood to all tissues of the human body. Immediately after exiting the heart, the ascending aorta begins and forms an arch from which three main branches first emerge: the brachiocephalic trunk, the left common carotid artery, and left subclavian artery (Figure 5.2). These three branches play a role in supplying blood to the upper extremities and the brain. After the arch, the aorta descends into the torso and is called the descending aorta. Two parts can be distinguished in this segment: the thoracic aorta, located above the diaphragm; and the abdominal aorta, located below the diaphragm.

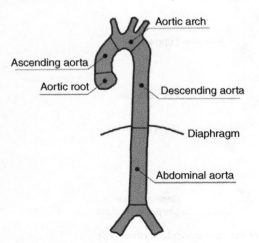

Figure 5.2 Anatomy of the aorta [21].

Aortic arch

Ascending aorta

Aortic root

Descending aorta

Diaphragm

Abdominal aorta

The thoracic aorta represents the upper part of the descending part of the aorta. It begins at the height of the fourth thoracic vertebra and descends downward, forward, and slightly to the right, so that in its course, it gradually approaches the middle plane of the vertebrae. The end is located at the height of the twelfth thoracic vertebra, where it passes through the aortic opening of the diaphragm and passes into the abdominal aorta. From the thoracic aorta, smaller and numerous lateral branches stand out, which are mostly intended for the nutrition of the thoracic organs.

The abdominal aorta is the largest artery in the abdominal cavity that continues to the thoracic aorta, and begins at the point where the descending part of the thoracic aorta passes through an opening in the diaphragm at the level of the lower edge of the body of the twelfth thoracic vertebra. It then descends through the retroperitoneal space in the midline in front of the vertebral body all the way to the level of the body of the fourth lumbar vertebra. At that level, it divides into two common iliac arteries. During its journey through the abdomen, it gives its visceral and parietal branches. Visceral branches serve to nourish the organs of the abdominal cavity.

5.2.3 Structure and Function of the Arterial Wall

The structure of the arterial wall varies along the vascular tree, as well as across age, sex, and disease. However, arteries can be categorized according to two main types:

- elastic arteries, such as the aorta, common carotid artery, common iliac artery, and main pulmonary artery, and
- muscular arteries, such as coronary artery, femoral artery, renal artery, and cerebral artery.

At the microscopic level, the arterial wall consists of a layered structure composed of three concentric zones (intimate tunic, media tunic, and adventitial tunic), separated by elastic membranes and containing first elastin, collagen fibers, fibroblast, endothelial and smooth muscle cells housed in amorphous, gel-like environment consisting mainly of water. It is important to note that the histology of each arterial segment changes with location according to what physiological function it has in the circulatory system. Of course, there are no sudden variations between large and small arteries.

The composite structure of the arterial wall (Figure 5.3), as already mentioned, consists of three clearly separated base layers:

- intima (lat. Tunica intima),
- media (lat. Tunica media), and
- adventitia (lat. Tunica externa).

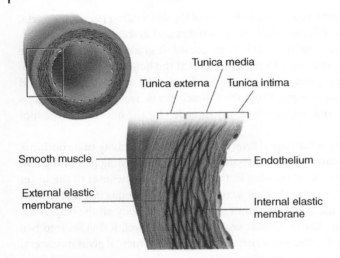

Figure 5.3 The structure of the wall of the main components of healthy elastic arteries. *Source:* Henry Gray/John William Parker 1918/Public Domain.

The mechanical properties of all three layers are directly conditioned by the concentration and structural arrangement of constituents, primarily collagen fibers, which in biological tissues have a similar role as steel reinforcements in reinforced concrete structures and are, therefore, responsible for the anisotropic properties of blood vessels [29].

The intima is the inner layer of the blood vessel, which practically forms a single layer of endothelial cells that serve as a barrier that prevents the formation of blood clots. Young and healthy arteries have a very thin layer of intimacy, while with aging and development of the disease its thickness increases significantly and can amount to up to 40% of the total thickness of the blood vessel. The families of collagen fibers within this layer have practically no dominant direction of supply – they are scattered over the volume of the tissue. It is known that pathological changes of the intima are associated with atherosclerosis, one of the most common diseases of blood vessels manifested by accumulation of cells, fat, calcium, fibrin, and tissue debris, which ultimately leads to the formation of fibrous plaque of complex geometry and biochemical structure. In the advanced stages of the disease, the central layer of the arterial wall – the medium – can be affected. Therefore, the mechanical characteristics of healthy and diseased arteries differ significantly.

The medium is the central layer of the artery. It consists of a complex three-dimensional network of densely packed bundles of collagen fibers, elastin and smooth muscle cells, spirally arranged in relation to the axis of the blood vessel. It is separated from the intima by a thin elastic membrane. The media plays a crucial role in the contraction of the arterial wall and in adjusting the volume of the blood vessel to the variable values of blood pressure. It is the bearer of biomechanical properties of blood vessels in the area of lower and medium value blood pressure, primarily in the circular direction. Numerous analyses of the media microstructure in healthy and atherosclerotic blood vessels show that there are no significant changes in the arrangement of collagen fiber bundles.

Adventitia is the outer layer of the arterial wall. It consists of cells that produce collagen and elastin and of spirally twisted thick bundles of collagen fibers arranged in two dominant directions, whereby the density of these collagen fiber bundles is significantly higher on the outside of the artery. Adventitia is inhomogeneous in the radial direction of the blood vessel. As for the thickness of the adventitia, it depends on the type of artery and the physiological function of the blood vessel. In the area of small and medium loads, it is significantly more elastic than the medium, in which it has a small influence on the overall mechanical characteristics of the blood vessel. Regarding the area of high pressures, due to large deformations of the blood vessel in the radial direction, i.e. increase in the diameter, there is a tightening of spirally twisted collagen fibers, and thus a very large increase in adventitious stiffness, which protects the artery from overload and possible rupture [30].

The structure of the medium ensures great strength, durability, and the ability of the wall to withstand loads both along the longitudinal and in the circumferential directions. All mechanisms that weaken the aortic wall, such as atherosclerosis, traumatic or hereditary disease, cause increased wall stress that can cause aortic dilatation, creating an aneurysm that can be complicated by aortic dissection or rupture. From a mechanical perspective, the media is the most important layer in a healthy artery. Dissections usually occur in the media or between the media and advent layers.

5.2.4 Aortic Dissection

Aortic dissection is a disease characterized by the penetration of blood between the layers of the aortic wall and the formation of a false lumen, and aortic rupture often results in fatal bleeding. Aortic dissection most often occurs as a consequence of long-term arterial hypertension and atherosclerosis. It also occurs as part of hereditary connective tissue disorders that lead to aortic wall weakening. The hemodynamic properties of the cardiovascular system expose the aorta to significant traumatic forces. Systolic aortic pressure depends on vascular resistance, stroke volume, and aortic flexibility. During systole, blood enters the aorta and the pressure in it rises. The aorta expands under the influence of increased pressure due to the elastic properties of the aortic media. In diastole, the pressure inside the aorta decreases with anterograde blood flow. As the aorta ages, fragmentation of elastin occurs and the mass ratio of collagen to elastin increases, increasing wall stiffness. Furthermore, atherosclerosis causes atrophy of the media, or death of muscle cells that also contribute to the elasticity of the aorta. This progressive loss of elasticity results in increased stress within the aortic wall, and thus an increased possibility of dissection.

5.2.5 History of Aortic Dissection

In 1728, the English physician Frank Nicholls (1699–1778) noticed that aneurysms could rupture or only rupture in the inner layers, allowing the outer lumen to expand. This was the first official case of aortic dissection. The French physician and inventor of the stethoscope, Rene-Theophile-Hyacinthe Laënnec (1791–1826), was the first to introduce the term *aortic dissection* with the aim of describing the rupture of the intima distal to the aortic valve. Thomas Bevill Peacock (1812–1882), a famous English cardiologist, published his research on 80 patients in which he described three stages of this disease:

- splitting of the intima layer,
- spreading of the dissected hematoma with possible rupture, and
- lumen recanalization.

Aortic dissection was exclusively diagnosed posthumously, until angiography appeared in 1939. The famous American surgeon Michael Ellis DeBakey (1908–2008) performed the first successful dissection operation on the descending part of the aorta in 1955. In 1996, the International Registry of Acute Aortic Dissection was established with the aim of collecting data on etiological factors, clinical characteristics, treatment, and hospital outcomes of patients with aortic dissection worldwide.

Diseases of the aortic wall contribute to a high degree of cardiovascular mortality. By applying new diagnostic methods, such as magnetic resonance imaging (MRI) and CT, much earlier and better diagnosis of aortic diseases is possible. Therefore, there has been a change in the approach to the treatment of patients because diagnoses and decisions on further treatment are made faster. A number of factors can alter the structure of the aortic wall, making the aorta more susceptible to dissection. All the mechanisms that weaken the aortic wall, especially the media layer, are the cause of increased wall stress that can cause aortic dilatation and aneurysm formation. This can be additionally complicated by aortic dissection or rupture.

5.2.6 Classification of Aortic Dissection

Dissection usually begins with a rupture (called entry tear) of the intima, allowing blood to enter the tear that continues to spread at the border of the outer and middle third of the media. In this way, a new lumen is created within the media,

Figure 5.4 CT scan shows the false and true lumen. *Source:* [21] Igor B. Saveljic 2020, p.137–178/with permission of Elsevier.

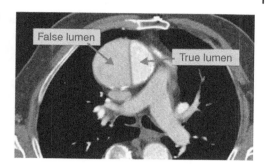

the so-called false lumen (Figure 5.4). Usually, the tear occurs transversely and does not cover the entire circumference of the aorta [31]. The false lumen represents the blood-filled space between the separated layers of the aorta. The rupture can spread in proximity to the heart, sometimes even to the coronary arteries and rupture into the pericardial cavity. Distal dilation can lead to the iliac arteries. Moreover, in the distal part of the dissected aorta, a new rupture called the re-entry tear may occur, which returns the blood to the lumen.

Aortic dissection is classified according to anatomical localization and duration. Depending on the duration of symptoms, aortic dissection is classified as acute and chronic. Acute dissection is a dissection that is diagnosed two weeks after the onset of symptoms, and when a complicated, life-threatening situation occurs. Chronic dissection is a dissection that lasts longer than this period. Although such a division appears to be arbitrary, it is based on research that has shown that 74% of deaths due to aortic dissection occur within the first 14 days [32].

There are three classification schemes for aortic dissection based on the site of entry tear as well as proximal or distal involvement of the aorta. According to research [33], 65% of aortic ruptures occur in the ascending part of the aorta, 20% in the descending part, 10% in the aortic arch, and 5% in the abdominal aorta. American surgeon DeBakey proposed three classification schemes (Figure 5.5).

The Stanford classification divides aortic dissection into two types: type A, which affects only the ascending part of the aorta, and type B, which affects only the descending part of the aorta. Acute aortic dissection type A is fatal, with a mortality rate 1–2% per hour after the first appearance of symptoms. The characteristic symptoms of this type of dissection are a sudden strong, stabbing pain in the front of the chest (in about 85% of cases), or behind the chest (46%). In addition, there is also abdominal pain (22%), sudden loss of consciousness (13%), weakened pulse (20%), and not infrequently stroke (6%) [33, 35]. Acute dissection type A always requires surgical treatment. Treatment using only drugs is associated with a mortality rate of almost 24% to 24 hours, 30% to 48 hours, 40% to 7 days, and 50% to 1 month. Even after surgery, the mortality rate is 10% to 24 hours, 13% to 7 days, and almost 20% to 30 days [36]. Current ascending aortic surgical techniques aim

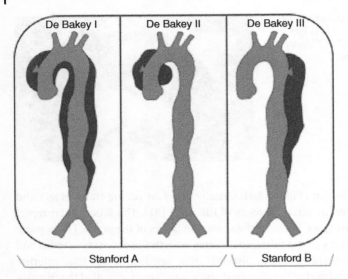

Figure 5.5 Classification of aortic dissection. *Source:* Adopted from [34].

to replace or repair the aortic root and aortic valve. Elimination of the remaining false lumens and repair of aortic dissection on the descending aorta are not in the foreground. Replacement or repair of the ascending aorta do not completely eliminate the flow and pressure in the distal part of the false lumen.

Acute aortic dissection type B is less dangerous than type A, but there are not many differences between them in the clinical picture. The most common symptoms of acute aortic dissection type B are back pain (64%) and chest pain (63%), as well as sudden abdominal pain (43%) and hypertension [36]. The mortality rate of patients with minor complications in type B dissection is 10% in 30 days. In case of renal insufficiency or visceral ischemia, as well as leg ischemia, urgent surgical repair of the aorta is required. The mortality rate in these cases is 20% per day and 25% for one month. The two most common sites of entry tear are the aortic root, for type I and II dissections, and a few inches below the left subclavian artery, for the third type of DeBakey dissection.

Population studies estimate that the annual incidence of acute aortic dissection ranges between 2.9 and 3.5 per 100 000 [37]. Risk factors for this pathological condition include age, hypertension, and structural abnormalities of the aortic wall. According to a study by the International Registry of Aortic Dissection, men are more often affected than women with a 4 : 1 ratio. In 60% of the cases, it is Stanford type A dissection. The incidence of this type is highest in the population between 50 and 60 years of age, while type B mainly affects the slightly older population with a peak incidence between 60 and 70 years of age. Hypertension occurs in 70% of cases, according to the IRAD database [38].

5.2.7 Diagnostic Techniques

Acute aortic dissection is generally suspected already on the basis of anamnesis and clinical examination. According to the analysis of 250 patients with acute chest or back pain (128 of whom had dissection), it was found that 96% of acute dissections can be identified based on a combination of the following three signs: sudden chest or abdominal pain, aortic enlargement on X-rays, peripheral pulsation and/or blood pressure disorders (difference between left and right arm pressure greater than 20 mmHg). The main objectives of diagnostic methods are rapid confirmation of the presence (or absence) of acute aortic dissection, classification of the type of dissection, identification and location of the intimate tear, and confirmation of the presence of true/false lumen and thrombus.

5.2.7.1 Aortography

Retrograde aortography was the first accurate diagnostic method for assessing patients with suspected dissection aorta. Diagnosis of aortic dissection based on aortography was first established and published by Robb and Steinberg in 1939 and became a routine method in the 1960s. When it comes to the diagnosis of aortic dissection in the thorax, the sensitivity of the aortography ranges between 86 and 88% and the specificity between 75 and 94%, but a false-negative finding may be present in case a false lumen thrombosis has occurred [39, 40, 41]. Aortography is less commonly used today in the process of diagnosing dissection because of its invasive nature and risk for developing contrast nephropathy. Moreover, the preparation of the aortography teams and the necessary equipment can last for too long, which certainly affects the increase in mortality [42].

5.2.7.2 Computed Tomography

CT is a radiological method of recording, which, in addition to X-rays, uses tomography, a method based on the mathematical procedure of recording images using modern computers. The first-generation devices gave a rough picture and the length of the examination lasted about 30 minutes. Spinal CT, a third-generation apparatus, has the ability to simultaneously move the table on which the patient lies and be exposed to X-rays while giving accurate images and shorter imaging times. Many factors affect the recording results. The biggest role in that is played by the collimation of X-rays, the characteristics of the table, as well as the choice of the reconstruction interval. The basic technical characteristics of CT devices are 3 mm collimation and table motion 6 mm/s [43, 44]. According to IRAD data, CT is used in the diagnosis of approximately 75% of patients. CT is also available as a noninvasive diagnostic test, with a sensitivity of 83–95% and a specificity of 87–100% when diagnosing acute aortic dissection [45]. According to the protocol, the entire aorta is imaged and, in most cases, this is enough to make a diagnosis and show the true and false lumen and the place of rupture of the intima and the entry of blood into the blood vessel wall. Such an accurate presentation is of great help for further intervention planning. A thrombus can also be seen on CT scans, if it has occurred. It mainly refers to the false lumen because it is located in it, but in patients who have an aneurysmal enlargement, the thrombus may also be located in the right lumen. In dissection of the aorta descendens, the false lumen is in more than 90% of cases larger in diameter than the true lumen, which is also important for their differentiation. Three-dimensional CT may be helpful in planning therapy and interventions in such patients, but axial imaging still provides the best insight into the topographic relationships of the right and false lumens and the possible compromise of aortic branches. Compared with other diagnostic tests, CT provides a very good insight into anatomical relationships, which is extremely important for further surgical or endovascular procedures and is the most reliable for further follow-up of patients [46].

5.2.7.3 Echocardiography

Echocardiography is a noninvasive diagnostic method that uses ultrasonic waves to show the heart and blood vessels. The sensitivity of transthoracic echocardiography (TTE) ranges between 35 and 80%, and the specificity between 40 and 95%. Although it is a noninvasive and accessible diagnostic test that does not radiate, it also has its drawbacks. Some of them are narrow intercostal spaces that prevent good visualization, lower efficiency in obese people, and people with emphysema of the lungs. Compared with this method, TEE is preferred, also due to the close anatomical relationships of the esophagus and aorta. The sensitivity of TTE echocardiography is from 77 to 80%, while the specificity ranges from 93 to 96% for the ascending aorta [47]. Like all other ultrasound examinations, it is not harmful to health because it does not radiate. Moreover, this diagnostic test can localize the site of intima rupture; detect a false lumen and blood flow in it, i.e. a thrombus; detect involvement of coronary arteries or aortic branches and the degree of regurgitation and pleural effusion. If the color Doppler technique is additionally used, it significantly reduces the percentage of false-positive results and increases the accuracy of the diagnostic test. The main limitation in TEE is the inability to visualize the distal aorta and aortic arch well due to air in the trachea and left bronchus [48].

5.2.7.4 Magnetic Resonance

The MRI is a noninvasive and precise diagnostic method that gives a picture of the health of certain organs. It is painless and harmless to the patient, so the examination can be repeated several times. Among all the abovementioned diagnostic tests, MRI has the highest sensitivity and specificity in the diagnosis of aortic dissection, even between 95 and 100%. This examination can detect the site of intima rupture, the spread of dissection within the aortic wall, and involvement of the branches. It is also possible to distinguish the true from the false lumen. MR can show involvement of aortic branches with sensitivity and specificity from 90 to 96% [49]. Much of the literature [49, 50] shows that this method can be successfully used in the diagnosis of aortic dissection. MRI is contraindicated in patients who have pacemakers or other metal implants [48].

5.2.7.5 Intravascular Ultrasound

Intravascular ultrasound examination is performed by introduction of catheter into the aorta via conventional conductor wires, with catheter-guide or without catheter-guide. This procedure can be done in less than 10 minutes. The use of intravascular ultrasound is very useful in completing the information obtained by angiography in patients with aortic dissection. Given that performed during cardiac catheterization, this technique can overcome most of the potential disadvantages of conventional angiography. Intravascular ultrasound allows direct representation of the structure of the wall of blood vessels from aortic lumen. In the literature, data on sensitivity and accuracy of almost 100% can be found. This technique reveals thrombosis in a false lumen with much greater precision and sensitivity than TEE [51, 52]. Moreover, intravascular ultrasound can show a change in the aortic wall due to bleeding in the media layer due to the increase in wall thickness [53].

5.2.8 Treatment of Acute Aortic Dissection

Treatment of aortic dissection aims to prevent death and ischemic damage to the abdominal organs. Acute Stanford type A dissection is considered an urgent surgical condition. On the other hand, when it comes to acute Stanford type B dissection, conservative therapy is preferred, except in case of complications, when surgical or endovascular intervention is approached [54, 55]. Based on numerous researches [2, 33, 55, 56], medication can play an important role in treating aortic dissection. Almost two-thirds of approximately 85% of patients who were discharged from the hospital felt well without complications after aggressive antihypertensive drug therapy [57], thus avoiding the risk of emergency surgery.

5.2.8.1 Drug Therapy

The goal of drug treatment of acute aortic dissection is to lower blood pressure and reduce the rate of contractions, i.e. the frequency of the left ventricle, because it has been observed that this reduces the traumatic effect on the aorta and the spread of dissection along the wall. A patient with aortic dissection should be cared for as soon as possible in an intensive care unit and morphine should be administered for analgesia and blood pressure reduced to

100–120 mmHg [7]. Initial therapies for blood pressure control are intravenous beta-blockers. They also lower the heart rate below 60 beats per minute. An infusion of propranolol or labetalol may be used. In comparison, esmolol has several advantages in acute therapy due to the shorter half-life and faster action. It is also preferred in patients who do not tolerate beta-blockers, for example due to asthma. Calcium channel blockers, verapamil, and diltiazem are used in therapy if the patient has a contraindication to beta-blockers [1]. The transition from intravenous to oral therapy is possible only when blood pressure control is achieved.

Acute aortic dissection of the ascending aorta can be fatal, where the mortality rate ranges from 1 to 2% per hour after the onset of the first symptoms [33]. Acute type A dissection is a surgical emergency. Drug-only dissection treatment is associated with a mortality rate of approximately 20% within 24 hours of the first symptoms, 30% within 48 hours, 40% within 7 days, and 50 within 1 month. Even after surgery, the mortality rate is 10% in 24 hours, 13% in 7 days, and almost 20% in a month [35, 36].

5.2.8.2 Surgical Treatment

Surgical or endovascular intervention is used in case of complications of Stanford type B aortic dissection. Indications for such an approach are persistent hypertension or pain that does not respond to drug therapy, spreading of the dissection along the wall, occlusion of one of the main aortic branches resulting in organ ischemia, enlargement dissecting aortic diameter, and rupture [58]. Since such invasive therapeutic methods solve only complicated cases, it is not surprising that mortality in this group of patients is higher compared with those treated with drugs. According to IRAD data, in-hospital mortality for operated patients is 32%, while mortality in the conservatively treated group is 10%. However, these are data only on the short-term outcome. When it comes to the long-term outcome, mortality is similar in both cases [59].

The goal of surgical treatment of aortic dissection type A (type I, II) is prevention of rupture or development of a pericardial effusion that can lead to cardiac tamponade. It is also necessary to eliminate aortic regurgitation and prevent myocardial ischemia. In type B dissection (type III), the main goal is to prevent aortic rupture. In patients with dissection aortic type I, it is necessary to perform a resection of the part intimacy affected by cleavage. If dissection is limited to the ascending aorta with or without aortic arch involvement, resection of the entire cleft of the intima is possible. Complete recovery is rarely achieved in dissections of type A and B (types I–III). There are various approaches to the treatment of acute care aortic dissection, which provide a wide range of methodological possibilities.

5.3 Theoretical Background

This part of the chapter describes the basic principle of fluid and solid modeling. An overview of the equations that enable solving the problem of modeling blood as a fluid and a deformable solid – the aortic wall – is given.

5.3.1 Continuum Mechanics

Continuum mechanics deals with the study of the motion of solid, liquid, and gaseous bodies at the macroscopic level, defining the term *continuous medium*, which assumes that matter completely, fills the space occupied by the body [60]. The term *continuum* implies that physical quantities are defined in a geometric point without volume, using limits, and thus in the mechanics of the continuum, the application of the infinitesimal calculus is enabled. The basic physical quantities defined by continuum mechanics are displacement, velocity, acceleration, deformation, force, mass, density, and stress. The laws on which the mechanics of the continuum is based are the law of conservation of mass, the law of momentum, and the first and second laws of thermodynamics.

5.3.1.1 Lagrange and Euler's Formulation of the Material Derivative

In continuum mechanics, the quantities observed are a function of space and time. There are two choices for observing variable-sized positions. One is Lagrange's description in which size is observed together with the motion of a material particle, i.e. some variable f is a function of material coordinates

$$
\begin{aligned}
f &= f(a_1, a_2, a_3, t) \\
x_i &= a_i + u_i(a_1, a_2, a_3, t)
\end{aligned}
\tag{5.1}
$$

where a_i are the initial coordinates of the material point at time $t = 0$, x_i are the coordinates at time t, and u_i are the components of the displacement of the material point, as shown in Figure 5.6. This way of describing the motion of particles is applicable to the motion of solids because the relative displacements between material points are generally small.

Another way of describing the motion of particles is Euler's, which is applied to fluids, in which deformations during fluid motion are significant. Thus, in Euler's description, a point in space and a quantity related to a material point that is at that point in space at a given moment are practically observed.

$$
\begin{aligned}
a_i &= x_i - u_i(x_1, x_2, x_3, t) \\
f &= f(x_1, x_2, x_3, t)
\end{aligned}
\tag{5.2}
$$

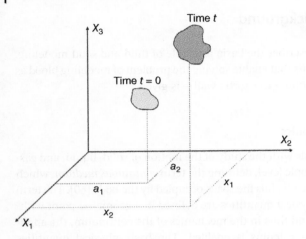

Figure 5.6 Coordinate definitions.

Since the material particle moves in space and time, it is necessary to determine the rate of change of the physical quantity at the material point. In the case of Lagrange's description, the rate of change of the physical quantity f is simply the derivative over time, $\dot{f}$ at a_i = const. This is Lagrange's formulation of the material derivative. In Euler's description, the procedure for calculating the excerpt is as follows. During the interval Δt, the material point crosses the path from position x_i to $x_i + \Delta x_i$ and the quantity f we observe changes by Δf.

If, assuming that f is a continuous function, Δf develops into a Taylor series, we obtain

$$\Delta f = \delta f + \frac{1}{2}\delta^2 f + \dots \tag{5.3}$$

where δf is calculated as

$$\delta f = \frac{\partial f}{\partial x_i}\Delta x_i + \frac{\partial f}{\partial t}\Delta t \tag{5.4}$$

The rates of change of magnitude f at a material point are defined as

$$\frac{\mathrm{D}f}{\mathrm{D}t} = \lim_{\Delta t \to 0} \frac{\Delta f}{\Delta t} \tag{5.5}$$

whereas, since it is a matter of observing a material particle, the derivative $\frac{\mathrm{D}}{\mathrm{D}t}$ is called a *material derivative*. The velocity of the material particle itself is defined as

$$\mathbf{v} = \lim_{\Delta t \to 0} \frac{\Delta \mathbf{r}}{\Delta t} \equiv \frac{\mathrm{D}\mathbf{r}}{\mathrm{D}t}, \quad v_j = \lim_{\Delta t \to 0} \frac{\Delta x_j}{\Delta t} = \frac{\mathrm{D}x_j}{\mathrm{D}t} \tag{5.6}$$

If Eq. (5.3) is replaced by (5.5), based on (5.4) and (5.6) and ignoring the members of the second order in the Taylor series, we obtain

$$\frac{Df}{Dt} = \frac{\partial f}{\partial t} + v_i \frac{\partial f}{\partial x_i} \tag{5.7}$$

Thus, a material derivative of a size consists of a local derivative of a size with a fixed point in space, and a convective part that occurs due to the movement of the particle. If the integral of a scalar function f per unit mass is defined as

$$\int_{mass} f \, dm = \int_{volume} f\rho dV \tag{5.8}$$

then the material derivative of the total integral has the form:

$$\frac{D}{Dt} \int_V (f\rho)dV = \int_V \frac{\partial(f\rho)}{\partial t} dV + \int_S (f\rho)v_n dS \tag{5.9}$$

where the first term on the right side of Eq. (5.9) represents the total increase in the quantity f within the control volume V, and the second term is the external flux of transport of the physical quantity f through the boundary in the direction of normal through the control surface S. Since $(f\rho)v_n dS = (f\rho)v_i n_i dS$ by using Gauss's theorem [61] to convert the surface into a volume integral, we finally obtain

$$\frac{D}{Dt} \int_V (f\rho)dV = \int_V \left\{ \frac{\partial}{\partial t}(f\rho) + \frac{\partial}{\partial x_i}(\rho f v_i) \right\} dV \tag{5.10}$$

The previous equation represents Reynolds' transport theorem, and denotes the material derivative of the volume integral for the case when there is a motion of mass through volume V.

5.3.1.2 Law of Conservation of Mass
By applying Reynolds's theorem to the law on the maintenance of mass, it is obtained [28]

$$\frac{Dm}{Dt} = \frac{D}{Dt} \int_V \rho dV = \int_V \left(\frac{D\rho}{Dt} + \rho \frac{\partial v_i}{\partial x_i} \right) dV = 0 \tag{5.11}$$

For an arbitrary control volume V, the equation of continuity in the point is obtained

$$\frac{D\rho}{Dt} + \rho \frac{\partial v_i}{\partial x_i} = 0 \tag{5.12}$$

After this, Reynolds's theorem is reduced to the following form:

$$\frac{\mathrm{D}}{\mathrm{D}t}\int_V (f\rho)\mathrm{d}V = \int_V \left(\frac{\partial f}{\partial t} + v_i \frac{\partial f}{\partial x_i}\right)\rho \mathrm{d}V = \int_V \rho \frac{\mathrm{D}f}{\mathrm{D}t}\mathrm{d}V \tag{5.13}$$

5.3.1.3 Navier–Stokes Equations

The Navier–Stokes equations describe the motion of viscous incompressible fluids. They represent a system of partial differential equations derived from the second Newtonian law. If we imagine the volume V of a fluid at time t, the external forces are represented as the surface forces $\mathbf{f}^S$ per unit area, and the volume $\mathbf{f}^B$ per mass unit. Based on the equation of the moment of change in the amount of motion, it is obtained

$$\frac{\mathrm{D}}{\mathrm{D}t}\int_V \rho v \mathrm{d}V = \int_V \mathbf{b}\mathrm{d}V + \int_S \mathbf{p}\mathrm{d}S \tag{5.14}$$

By applying Eq. (5.13), which represents the equation of maintaining the mass, Eq. (5.14) is reduced to the following form:

$$\int_V \rho \frac{\mathrm{D}\mathbf{v}}{\mathrm{D}t}\mathrm{d}V = \int_V \mathbf{f}^B \mathrm{d}V + \int_S \mathbf{f}^S \mathrm{d}S \tag{5.15}$$

Using the Gaussian and Koshi theorem [62] on the transformation of a surface into a volume integral, we obtain

$$\int_V \rho \frac{\mathrm{D}v_i}{\mathrm{D}t}\mathrm{d}V = \int_V f_i^B \mathrm{d}V + \int_V \frac{\partial \sigma_{ij}}{\partial x_j}\mathrm{d}V \tag{5.16}$$

Subsequently, constituent relations for Newton's fluid are introduced

$$\sigma_{ij} = -p\delta_{ij} + 2\mu \dot{e}_{ij} \tag{5.17}$$

where p represents the fluid pressure, μ the dynamical viscosity, and $\dot{e}$ the deformation rate tensor. The deformation rate tensor is determined as

$$\dot{e}_{ij} = \frac{1}{2}\left(\frac{\partial v_i}{\partial x_j} + \frac{\partial v_j}{\partial x_i}\right) \tag{5.18}$$

When Eqs. (5.17) and (5.18) are replaced by Eq. (5.16), it is obtained

$$\int_V \rho \frac{\mathrm{D}v_i}{\mathrm{D}t}\mathrm{d}V = \int_V f_i^B \mathrm{d}V + \int_V \left(-\frac{\partial p}{\partial x_i} + \mu \left(\frac{\partial^2 v_i}{\partial x_j \partial x_j} + \frac{\partial^2 v_j}{\partial x_j \partial x_i}\right)\right)\mathrm{d}V \tag{5.19}$$

where the addition is performed according to the repeated index ($j = 1, 2, 3$).

Since the control volume V is arbitrary, a differential form of Eq. (5.19) can be written, so that in the end, the standard form of the Navier–Stokes equations for incompressible viscous flow is obtained

$$\rho\left(\frac{\partial v_i}{\partial t} + v_j\frac{\partial v_i}{\partial x_j}\right) = -\frac{\partial p}{\partial x_i} + \mu\left(\frac{\partial^2 v_i}{\partial x_j\partial x_j} + \frac{\partial^2 v_j}{\partial x_j\partial x_i}\right) + f_i^B \tag{5.20}$$

5.3.1.4 Equations of Solid Motion

One of the basic principles in continuum mechanics is the principle of virtual work. This principle is applied in many numerical programs as a basis for deriving the necessary relations. We observe the body in equilibrium with the given boundary conditions (Figure 5.7) that are affected by the given external loads. Virtual displacements and virtual deformations at a material point are $\delta\mathbf{u}$ and $\delta\mathbf{e}$. Suppose that the field of virtual displacements is given and that the external loads are immutable. Virtual displacements at points where external forces act are $\delta\mathbf{u}^{(C)}$ and $\delta\mathbf{u}^{(D)}$, and virtual displacements at support points are limited. Virtual movements correspond to the equilibrium state of the body under given loads. Stress and displacements are given on the surfaces S_σ and S_u, respectively. The assumption is that the virtual displacements are infinitesimal and satisfy the given boundary conditions.

If it starts from the equilibrium, equations are derived from [63]

$$\frac{\partial\sigma_{ij}}{\partial x_i} + F_j^V = 0 \tag{5.21}$$

with the application of border conditions

$$\sigma_{ij}n_i - F_j^S = 0 \tag{5.22}$$

Figure 5.7 Schematic representation of a deformable body under the action of external forces.

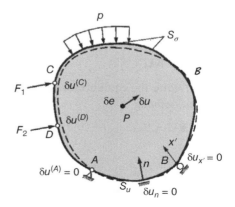

the equality of virtual works of internal and external forces arises

$$\delta W_{\text{int}} = \delta W_{\text{ext}} \tag{5.23}$$

where

$$\delta W_{\text{int}} = \int_V \sigma_{ij} \delta e_{ij} \mathrm{d}V \tag{5.24}$$

virtual work of internal forces on virtual deformations, and

$$\delta W_{\text{ext}} = \int_V F_k \delta u_k \mathrm{d}V + \int_{S^\sigma} F_k^S \delta u_k^S \mathrm{d}V + \sum_i F_k^{(i)} \delta u_k^{(i)}, \quad k = 1, 3 \tag{5.25}$$

virtual operation of external forces on virtual displacements.

Equations (5.24) and (5.25), written in matrix form, are

$$\delta W_{\text{int}} = \int_V \delta \mathbf{e}^{\mathrm{T}} \boldsymbol{\sigma} \mathrm{d}V \tag{5.26}$$

$$\delta W_{\text{ext}} = \int_V \delta \mathbf{u}^{\mathrm{T}} \mathbf{F}^V \mathrm{d}V + \int_{S^\sigma} \delta \mathbf{u}^{\mathrm{T}} \mathbf{F}^S \mathrm{d}V + \sum_i \delta \mathbf{u}^{\mathrm{T}} \mathbf{F}^{(i)} \tag{5.27}$$

The virtual deformations corresponding to the virtual displacements are

$$\delta e_{kj} = \frac{1}{2} \left(\delta \left(\frac{\partial u_k}{\partial x_j} \right) + \delta \left(\frac{\partial u_j}{\partial x_k} \right) \right) \tag{5.28}$$

Applying the principles of virtual work and constituent relations for linear elastic material in matrix form:

$$\boldsymbol{\sigma} = \mathbf{C}\mathbf{e} \tag{5.29}$$

using the isoparametric concept of interpolation [64] within the final elements, on the basis of which the coordinates and displacements at any point within the element

$$\mathbf{x} = \mathbf{N}\mathbf{X} \tag{5.30}$$

$$\mathbf{u} = \mathbf{N}\mathbf{U} \tag{5.31}$$

we can write the equilibrium equation of a finite element, which we give here without deriving

$$\mathbf{K}\mathbf{U} = \mathbf{F}_{\text{ext}} \tag{5.32}$$

where

K – element stiffness matrix,
B – moving in nodes, containing statements of interpolation functions,

C – elastic constituent matrix,
e = **BU** – matrix deformation,
U – movements in nodes,
X – coordinates of nodes,
N – matrix of interpolation functions,
F$_{\text{ext}}$ – external forces in the element nodes.

Equation (5.32) is obtained from the principle of virtual work and represents the equality of internal and external forces. In a general case, the internal forces are nonlinear displacement functions. By linearization, Eq. (5.32) can be written in the form [65, 66]:

$$^{t+\Delta t}\mathbf{K}^{(i-1)}\Delta\mathbf{U}^{(i)} = {}^{t+\Delta t}\mathbf{F}_{\text{ext}} - {}^{t+\Delta t}\mathbf{F}_{\text{int}}^{(i-1)} \tag{5.33}$$

where

$\Delta\mathbf{U}^{(i)}$ – vector of incremental displacement vector,
$^{t+\Delta t}\mathbf{F}_{\text{int}}^{(i-1)} = \int_V {}^{t+\Delta t}\mathbf{B}^{\text{T}(i-1)t+\Delta t}\boldsymbol{\sigma}^{(i-1)}dV$ – the internal force vector at the end of the load step.

The equations of motion of a material system can be written using the principles of virtual work, taking into account the effects of inertial forces. The elementary volume inertial force is

$$d\mathbf{F}^{\text{in}} = -\ddot{\mathbf{u}}dm = -\ddot{\mathbf{u}}\rho dV \tag{5.34}$$

When taking into account the influence of inertial forces, the virtual work of external forces is

$$\delta W_{\text{ext}} = \int_V \delta\mathbf{u}^{\text{T}}(\mathbf{F}^V - \rho\ddot{\mathbf{u}})dV + \int_{S^\sigma} \delta\mathbf{u}^{\text{T}}\mathbf{F}^S dV + \sum_i \delta\mathbf{u}^{\text{T}}\mathbf{F}^{(i)} \tag{5.35}$$

Time differentiation Eq. (5.31) gives us interpolations for velocity and acceleration of points

$$\dot{\mathbf{u}} = \mathbf{N}\dot{\mathbf{U}} \tag{5.36}$$
$$\ddot{\mathbf{u}} = \mathbf{N}\ddot{\mathbf{U}} \tag{5.37}$$

where

$\dot{\mathbf{u}}$ – the velocity of the material point in the element,
$\ddot{\mathbf{u}}$ – the acceleration of material point in the element,
$\dot{\mathbf{U}}$ – the velocity in nodes,
$\ddot{\mathbf{U}}$ – the acceleration in nodes.

In matrix form, Eq. (5.36) can be represented in the form:

$$\mathbf{M\ddot{U}} + \mathbf{KU} = \mathbf{F} \tag{5.38}$$

Equation (5.38) can be written in the following form:

$$\mathbf{M\ddot{U}} + {}^{t+\Delta t}\mathbf{K}^{(i-1)}\Delta\mathbf{U}^{(i)} = {}^{t+\Delta t}\mathbf{F}_{\text{ext}} - {}^{t+\Delta t}\mathbf{F}_{\text{int}} \tag{5.39}$$

In linear solid analysis, the basic assumption is that solid displacements are infinitesimally small and that the material is linearly elastic. In addition, the assumption is that the nature of boundary conditions remains unchanged under the influence of external loads. Under these assumptions, the finite element equilibrium equation is derived for static analysis (5.32). Equation (5.32) refers to the linear analysis of solids because displacement $\mathbf{U}$ is a linear function of external forces $\mathbf{F}_{\text{ext}}$. In the case when the displacements are not linearly dependent on the load, nonlinear analysis is applied.

5.3.2 Solid–Fluid Interaction

The blood flow through blood vessels represents fluid flow through a tube/solid of deformable walls. Modeling such a physical process requires the determination of the interaction of fluids and solids. During the last 30 years, a lot of effort has been invested in the development of computational mechanics and the field of analysis of solids and fluids. The development of computer technology expands the application of the method of solving the problem of fluids and solids in many engineering but also in other multidisciplinary areas. Special numerical disciplines for solid problems (Computational Structure Dynamics [CSD]) and fluid problems (Computational Fluid Dynamics [CFD]) have been developed. The solid–fluid interaction requires simultaneous numerical solutions of both the fluid domain and the domain of solids.

5.4 Blood Flow in the Arteries

The aorta and arteries have low resistance to blood flow compared with arterioles and capillaries. When a ventricular contraction occurs, the volume of blood is abruptly expelled into the arterial blood vessels. Because the outflow into the arterioles is relatively slow due to their high resistance to flow, the arteries become inflated to receive additional blood volume. During diastole, elastic twitching of the arteries pushes blood into the arterioles. In this way, the elastic properties of the arteries help to convert the pulsating blood flow from the heart into a more continuous flow through the rest of the circulation. *Hemodynamics* is a term used to describe the mechanisms that affect the dynamics of blood circulation. In order to make an accurate model of blood flow in the arteries, it is necessary to include realistic characteristics such as:

- the flow is pulsating,
- arteries are elastic and narrow,
- the geometry of the artery is complex and includes narrowing, rounding, as well as branching, and
- in small arteries, the viscosity depends on the diameter of the blood vessel and the shear rate.

The general time-dependent fluid equations in a right cylindrical tube are represented by the continuity equation and the Navier–Stokes equations in cylindrical coordinates:

$$\frac{\partial v}{\partial r} + \frac{v}{r} + \frac{\partial u}{\partial z} = 0 \tag{5.40}$$

$$\frac{\partial u}{\partial t} + v\frac{\partial u}{\partial r} + u\frac{\partial u}{\partial z} = F_z - \frac{1}{\rho}\frac{\partial P}{\partial z} + \frac{\mu}{\rho}\left(\frac{\partial^2 u}{\partial r^2} + \frac{1}{r}\frac{\partial u}{\partial r} + \frac{\partial^2 u}{\partial z^2}\right) \tag{5.41}$$

$$\frac{\partial v}{\partial t} + v\frac{\partial v}{\partial r} + u\frac{\partial v}{\partial z} = F_t - \frac{1}{\rho}\frac{\partial P}{\partial r} + \frac{\mu}{\rho}\left(\frac{\partial^2 v}{\partial r^2} + \frac{1}{r}\frac{\partial v}{\partial r} - \frac{v}{r^2} + \frac{\partial^2 v}{\partial z^2}\right) \tag{5.42}$$

where u is the axial component of the fluid velocity, v the radial component of the fluid velocity, r the radial coordinate, z the axial component, ρ the density of the fluid, and μ the viscosity.

5.4.1 Stationary Flow

The simplest model of stationary laminar flow in a circular cylinder is known as *Hagen–Poiseuille flow* [67]. For the axisymmetric flow in a circular tube of inner radius R_0 and length l, the boundary conditions are as follows:

$$u(r = R_0) = 0 \quad \text{and} \quad \frac{\partial u}{\partial r}(r = 0) = 0 \tag{5.43}$$

For a uniform pressure gradient (ΔP) along the pipe, a parabolic Poiseuille solution is obtained

$$u(r) = -\frac{\Delta P}{4\mu l}\left(R_0^2 - r^2\right) \tag{5.44}$$

The previous equation indicates that the pressure gradient ΔP requires the creation of a volumetric flow $Q = uA$ with a proportional increase in Q. Therefore, the vascular resistance R counts as

$$R = \frac{\Delta P}{Q} \tag{5.45}$$

Arteries are adaptable blood vessels, and their wall stiffness increases with deformation. Because of their ability to dilate, blood vessels can function as a storehouse of blood volume under pressure. In that sense, they function as capacitive elements. The linear relationship between the volume V and pressure defines the capacity of the stored element, or vascular capacity

$$C = \frac{dV}{dP} \tag{5.46}$$

It should be noted that the capacity decreases with increasing pressure and with the age of blood vessels. Another useful expression of arterial capacity per unit length is C_u. Knowing the wall thickness h, modulus of elasticity E, and inner radius R_0, the mentioned expression is calculated as

$$C_u \equiv \frac{dC}{dz} \approx \frac{2\pi R_0{}^3}{hE} \tag{5.47}$$

5.4.2 Oscillatory (Pulsating) Flow

The flow of blood in the large arteries takes place by the work of the heart, that is, its contractions, because of which this flow is called oscillatory or pulsating. The simplest model of pulsating flow was developed by the British mathematician John Ronald Womersley [68] for a fully developed oscillatory flow of an incompressible fluid in a rigid right cylinder. The problem is defined for a sinusoidal pressure gradient composed of sine and cosine

$$\frac{\Delta P}{l} = K e^{i\omega t} \tag{5.48}$$

where the oscillation frequency is $\omega/2\pi$. When we insert the previous equation in (5.41), we obtain

$$\frac{\partial^2 u}{\partial r^2} + \frac{1}{r}\frac{\partial u}{\partial r} - \frac{1}{v}\frac{\partial u}{\partial t} = -\frac{K}{\mu}e^{i\omega t} \tag{5.49}$$

By separating the variables, the following solution is obtained:

$$u(r,t) = W(r) \bullet e^{i\omega t} \tag{5.50}$$

By inserting the previous equation in (5.49), the Bessel equation is obtained

$$\frac{d^2 W}{dr^2} + \frac{1}{r}\frac{dW}{dr} + \frac{i^3 \omega \rho}{\mu}W = -\frac{A}{\mu} \tag{5.51}$$

where $W(r)$ counts as

$$W(r) = \frac{K}{\rho}\frac{1}{i\omega}\left\{1 - \frac{J_0\left(\frac{r}{\sqrt{\frac{\omega}{v}}} \cdot i^{3/2}\right)}{J_0\left(\frac{R}{\sqrt{\frac{\omega}{v}}} \cdot i^{3/2}\right)}\right\} \tag{5.52}$$

where J_0 is a Bessel function of zero order of the first kind, $v = \mu/\rho$ is the kinematic viscosity, while the α dimensional parameter is known as the Womersley number, and is represented as

$$\alpha = {}^{R_0}\sqrt{\frac{w}{v}} \tag{5.53}$$

where R_0 is the radius of the tube, ω is the angular frequency, and v is the kinematic viscosity. The Womersley number represents the ratio of inertial (nonstationary) forces and viscous forces. When the mentioned number has a small value, viscous forces are dominant and velocity profiles are parabolic. If α has a large value, the inertial forces take over the dominance and the velocity profile becomes flatter, that is, it has a blunt shape. At higher frequencies, the amplitudes of motion decrease, which leads to a phase difference of 90° between the pressure gradient and the flow.

5.4.3 Flow in Curved Pipes

Arteries and veins are not generally straight tubes, but some have curved structures, especially the aorta, which has a complex three-dimensional curved geometry. To understand the effect of bent blood flow, we will observe a simple example of a stationary laminar flow in a bent tube (Figure 5.8).

Figure 5.8 Schematic representation of the distorted axial velocity profile and secondary motion developed in laminar flow in a curved tube.

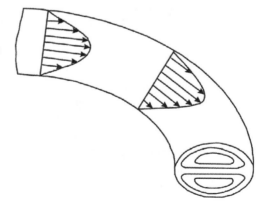

When a stationary fluid flow enters a bent tube in a horizontal plane, all its elements are exposed to centripetal acceleration normal to their original directions. This force is obtained from the pressure gradient in the bending plane, which is more or less uniform in the cross section. Therefore, all fluid elements are exposed to approximately the same lateral acceleration, and elements that move faster with higher inertia will change their directions less than elements that move less. The end result is that faster-moving elements, which initially occupy the fluid core near the center, penetrate faster to the outside of the curve along the diametrical plane and are replaced by internal circular motions of slower parts of the fluid located near the wall. Thus, the total current field is composed of a single passive distorted axial component over which the secondary flow of two counter-rotating vortices is superior. An analytical solution of a fully developed, stationary vortex flow in curved tubes of circular cross section was developed by the British mathematician Dean in 1927, who expressed the ratio of centrifugal inertial forces and viscous forces over Dean's number

$$\text{De} = \text{Re}\sqrt{\frac{r}{R_{\text{curve}}}} \tag{5.54}$$

where Re is the Reynolds number, r is the radius of the pipe, and R_{curve} radius is the curvature of the pipe. As the number De increases, the maximum axial velocity is more inclined toward the outer wall. Blood flow in the aorta is complex and topics such as inlet flow from the aortic valves, pulsating flow, and their influence on shear wall tension have been the subject of numerous experimental and numerical studies [69, 70].

5.4.4 Blood Flow in Bifurcations

The arterial system is a complex asymmetric branching system that delivers blood to all organs and tissues. Simplified arterial bifurcation can be represented by two curved tubes arising from a straight tube. The pattern of blood flow after the fork can be shown as in Figure 5.9.

The boundary layer is generated on the inner side of the wall downstream of the flow division at maximum speed near the boundary layer. As in the case of flow in curved pipes, the maximum speed is inclined toward the outer curve, which is the inner wall of the fork. A large number of experiments have been conducted to obtain a flow profile in branching branches, then energy loss as well as shear stress levels in these branching zones [70–72]. Of great interest are carotid bifurcation and bifurcation of the lower extremities because their blockage can lead to stroke, i.e. the inability to walk.

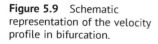
Figure 5.9 Schematic representation of the velocity profile in bifurcation.

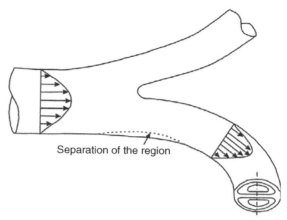

Separation of the region

5.5 Numerical Simulations

Numerical simulations, conducted in this chapter, aim to describe the behavior of blood within the aortic wall, both in healthy and dissected aorta. A parametric model of a healthy aorta was developed, and then three typical types of aortic dissection were made by computer manipulation. Figure 5.10 shows a parametric model of a healthy aorta and the dimensions used in the modeling.

In order to create a 3D mesh of finite elements, gmsh was used, after which tetrahedral elements were converted into brick elements, using our developed software. For example, the mesh of healthy aorta consists of 120 308 elements for lumen/blood domain, and 92 900 for solid/wall domain. Figure 5.11 shows geometries of parametric model for four examples of the modeled aorta.

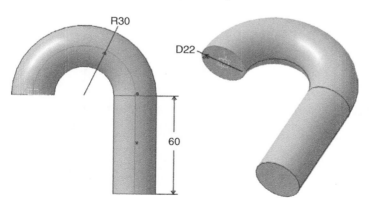

Figure 5.10 Geometric dimensions of the aorta.

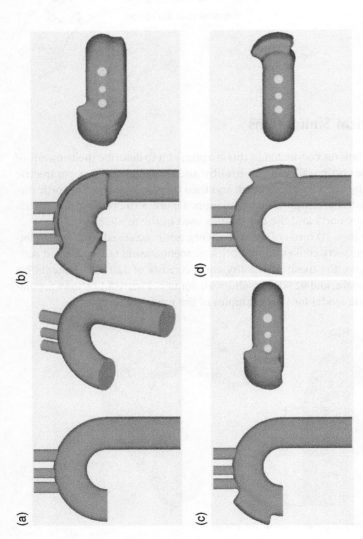

Figure 5.11 Geometries of parametric models: (a) healthy aorta; (b) dissected aorta type I; (c) dissected aorta type II; (d) dissected aorta type III according to DeBakey.

All numerical simulations were performed using the PAKSF solver [63]. A parabolic flow profile is assigned at the entrance of the ascending part of the aorta (Figure 5.12), while the constant value of the zero pressure is given at the outputs [73].

The blood is modeled as a Newtonian fluid, suitable for larger arteries. The blood density was $\rho = 1.05$ g/cm^3, dynamic viscosity $\mu = 0.0037$ Pa s. It is assumed that blood flow is laminar in large blood vessels, since it is assumed that the low mean velocity results in low Reynolds number values [27, 74]. A one-layered wall model was used in the present study, based on research by a group of authors [75]. They used noninvasive ultrasonic methods to determine the Young's modulus of 3321 human arteries and showed the value to be 983–701 kPa. In the present study, the mean Young's modulus of aorta is assumed to be 850 kPa with a Poisson coefficient of 0.45. The calculation time step was set to 0.01. All the results presented here are for the maximum systole. Size of entry tear was 5 mm. Figure 5.13 shows numerical results conducted on the healthy aorta. It can be seen that the maximum developed velocity, shown in longitudinal cross section, through the aorta is 2.03 m/s, while the pressure is 1.75e04 Pa.

Figure 5.14 shows distribution of wall shear stress and von Mises stress of the healthy aorta. The maximum value of the wall shear stress of the fluid domain is 8.8 Pa, while the maximum stress at the wall domain is 3.5e05 Pa, which are considered as normal aortic values.

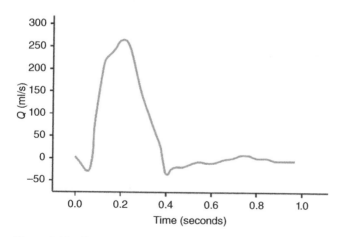

Figure 5.12 The time-dependent pulsatile waveform at the ascending aorta.

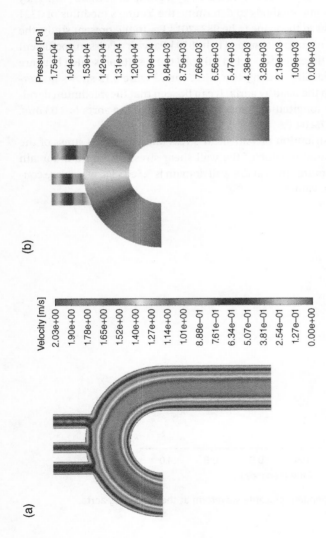

Velocity [m/s]

2.03e+00
1.90e+00
1.78e+00
1.65e+00
1.52e+00
1.40e+00
1.27e+00
1.14e+00
1.01e+00
8.88e−01
7.61e−01
6.34e−01
5.07e−01
3.81e−01
2.54e−01
1.27e−01
0.00e+00

(a)

Pressure [Pa]

1.75e+04
1.64e+04
1.53e+04
1.42e+04
1.31e+04
1.20e+04
1.09e+04
9.84e+03
8.75e+03
7.666+03
6.56e+03
5.47e+03
4.38e+03
3.28e+03
2.19e+03
1.09e+03
0.00e+03

(b)

Figure 5.13 Example 1 – distribution of velocity (a) and pressure (b) through the healthy aorta.

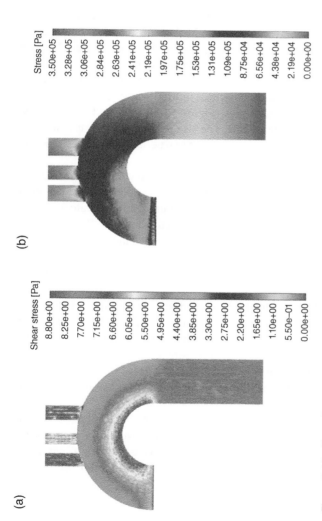

Figure 5.14 Example 1 – distribution of shear stress on fluid domain (a) and von Mises stress (b) on solid domain of the healthy aorta.

In the case when the aorta is dissected and there are thickenings on certain parts, the simulation results are different. Figure 5.15 shows distribution of velocity and pressure through the dissected aorta in case of type II according to DeBakey.

From Figure 5.15a, it can be seen that the fluid penetrates through the entry tear and creates a flow in that part of the dissection. Figure 5.15b shows pressure distribution and maximal value of 1.61e04 Pa in the zone of proximal part of false lumen. This value of pressure is worrying because the size of the entry tear of 5 mm soon leads to further stratification and increase of flow in the false lumen.

Figure 5.16a shows wall shear stress of 2.17 Pa in the entry tear zone, while Figure 5.16b shows the maximal values of von Mises stress around the dissected parts, which indicates a further direction of expansion of the false lumen. The size of the entry tear is 5 mm and there is no return circulation.

Figure 5.17a shows that a significant amount of fluid passes through the entry tear in the first layer of the aortic wall. This leads to stratification of the aortic layer with increasing pressure in the false lumen (Figure 5.17b), 1.37e04 Pa in the proximal part and 8.95e03 Pa in the distal part. The size of the entry tear is 5 mm and the size of the distal return tear is 7 mm. This type of true/false lumen relationship is called communicating circulation.

Figure 5.18 shows the distribution of wall shear stress and von Mises stress of the dissected aorta, type I (DeBakey). The ascending and descending parts of the aorta have a similar distribution as in the healthy aorta, but significant values of wall shear stress of 2.67 Pa in the proximal part of the dissection and 1.84 Pa in the distal part were recorded (Figure 5.18a). When the von Mises stress in the solid domain is observed (Figure 5.18b), a zone of higher values is seen, and a peak of the highest stress value of 3.5e05 Pa. This stress value represents the potential location of the aortic rupture.

Figure 5.19 shows the distribution of velocity and pressure through the dissected aorta in the case of type III, according to DeBakey. The size of the entry tear is 5 mm and there is no return circulation. Through the dissected part of the aorta, the highest velocity of 0.32 m/s was recorded (Figure 5.19a). The maximum value of the pressure in the false lumen is 8.04 Pa (Figure 5.19b).

Figure 5.20 shows distribution of wall shear stress and von Mises stress of the dissected aorta, type I.

From Figure 5.20a, it can be seen that a wall shear stress value of 1.25 Pa was recorded in the entry tear zone. Regarding the result of the stress on the wall of the false lumen (Figure 5.20b), elevated values of the range from 1.79e05 to 3.18e05 Pa are observed, which indicate the direction of further stratification.

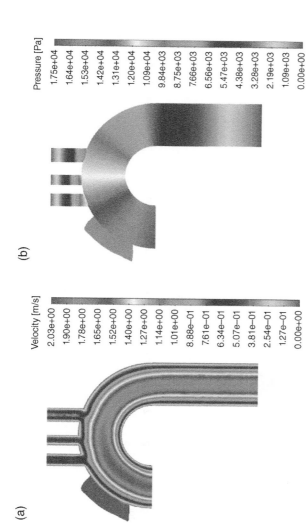

Velocity [m/s]

2.03e+00
1.90e+00
1.78e+00
1.65e+00
1.52e+00
1.40e+00
1.27e+00
1.14e+00
1.01e+00
8.88e-01
7.61e-01
6.34e-01
5.07e-01
3.81e-01
2.54e-01
1.27e-01
0.00e+00

Pressure [Pa]

1.75e+04
1.64e+04
1.53e+04
1.42e+04
1.31e+04
1.20e+04
1.09e+04
9.84e+03
8.75e+03
7.66e+03
6.56e+03
5.47e+03
4.38e+03
3.28e+03
2.19e+03
1.09e+03
0.00e+00

(a)

(b)

Figure 5.15 Example 2 – distribution of velocity (a) and pressure (b) through the dissected aorta, type II.

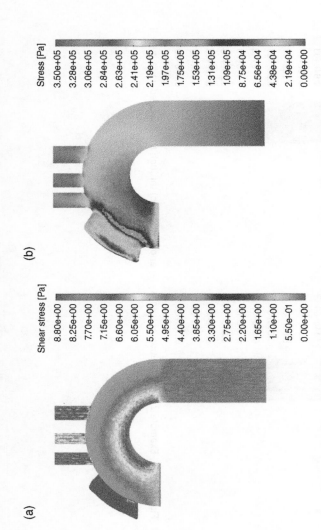

Figure 5.16 Example 2 – distribution of wall shear stress on fluid domain (a) and von Mises stress (b) on solid domain through the dissected aorta, type II.

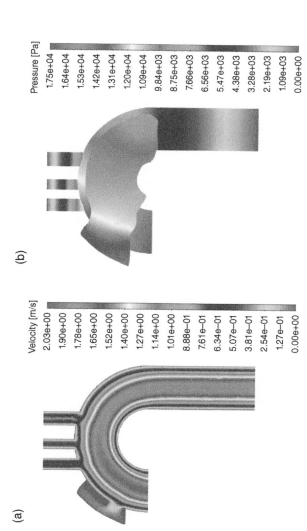

Figure 5.17 Example 3 – distribution of velocity (a) and pressure (b) through the dissected aorta, type I.

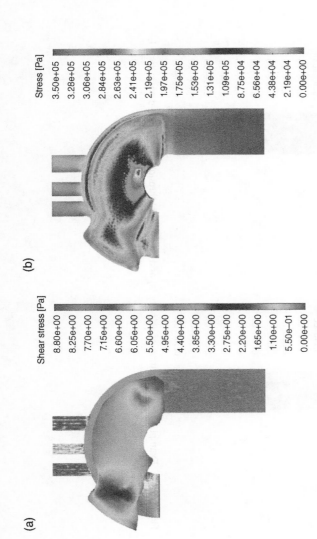

Figure 5.18 Example 3 – distribution of wall shear stress on fluid domain (a) and von Mises stress (b) on solid domain through the dissected aorta, type I.

Shear stress [Pa]

8.80e+00
8.25e+00
7.70e+00
7.15e+00
6.60e+00
6.05e+00
5.50e+00
4.95e+00
4.40e+00
3.85e+00
3.30e+00
2.75e+00
2.20e+00
1.65e+00
1.10e+00
5.50e-01
0.00e+00

(a)

Stress [Pa]

3.50e+05
3.28e+05
3.06e+05
2.84e+05
2.63e+05
2.41e+05
2.19e+05
1.97e+05
1.75e+05
1.53e+05
1.31e+05
1.09e+05
8.75e+04
6.56e+04
4.38e+04
2.19e+04
0.00e+00

(b)

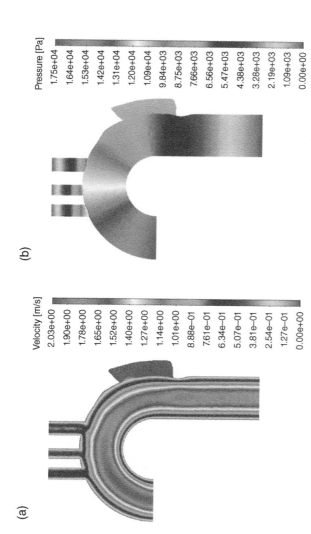

Figure 5.19 Example 4 – distribution of velocity (a) and pressure (b) through the dissected aorta, type III.

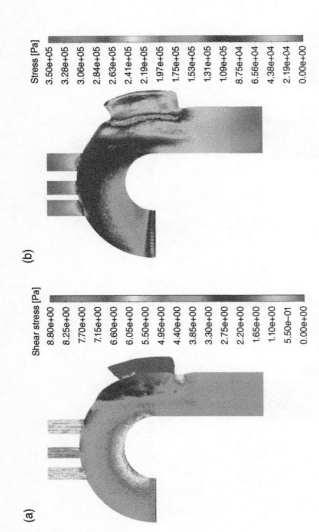

Figure 5.20 Example 4 – distribution of wall shear stress on fluid domain (a) and von Mises stress (b) on solid domain through the dissected aorta, type III.

5.6 Conclusions

Heart diseases are the leading cause of death in both developed and developing countries. Practically, they are so widespread that it is increasingly difficult to find the part of the planet where the population is protected from these diseases. What is most worrying is that the number of patients is constantly growing and this trend continues. Modern lifestyle, daily stress exposure, reduced physical activity, irregular nutrition, and obesity are some of the causes of hypertension, which is one of the most significant risk factors for the development of cardiovascular diseases.

Aortic dissection or dissecting aortic aneurysm is a disorder of the aorta that is characterized by a longitudinal rupture of its wall, whereby blood enters the fissure and, consequently, a second, false lumen of the aorta is formed. Some classify aortic dissection as aortic aneurysm, while others single it out as a special disease. An aneurysm is an enlargement of blood vessels (aorta) due to congenital or acquired weakness of their wall. Aortic dissections make up about 15–20% of all aneurysms.

In this chapter, three-dimensional simulations were presented in order to determine the hemodynamic behavior of blood in the healthy aorta, and in the aorta with the presence of acute dissection. The aim of the simulated exercises was to determine the numerical relationship between the true and false lumens of acute aortic dissection for three types of this disease. It can be concluded that numerical simulations can help doctors understand the whole process and determine what factors can contribute to reducing the mortality rate. By using computer manipulation and the dicom images obtained from a doctor, it is possible to perform an accurate numerical simulation of a patient's condition, and to conduct a virtual operation, and then run the simulation again to predict the course of the disease before the actual operation. In this way, the potential positions of the aortic rupture can be determined. It is of great importance to detect the sites with high stress values because they represent the points with the highest potential for developing a re-entry or aortic rupture, which can lead to a fatal outcome. The advantages of this approach are numerous and, taking into consideration the fact that it saves a lot of time, the postoperative outcomes are improved. These four presented examples can also be used for educational purposes. By changing the input function, blood density, and modulus of elasticity of the aortic wall, changes in fluid velocity, fluid pressure, its impact on the aortic wall via shear stress and stress in the dissected aortic wall can be monitored.

References

1 Tsai, T.T., Nienaber, C.A., and Eagle, K.A. (2005). Acute aortic syndromes. *Circulation* **112**: 3802.
2 Golledge, J. and Eagle, K.A. (2008). Acute aortic dissection. *The Lancet* **372** (9632): 55–66.

3 Thrumurthy, S.G., Karthikesalingam, A., Patterson, B.O. et al. (2012). The diagnosis and management of aortic dissection. *British Medical Journal* **344**: d8290. https://doi.org/10.1136/bmj.d8290.

4 Tse, K.M., Chiu, P., Lee, H.P., and Ho, P. (2011). Investigation of hemodynamics in the development of dissecting aneurysm within patient-specific dissecting aneurismal aortas using computational fluid dynamics (CFD) simulations. *Journal of Biomechanics* **44** (5): 827–836.

5 Ince, H. and Nienaber, C.A. (2007). Diagnosis and management of patients with aortic dissection. *Heart* **93** (2): 266–270.

6 Braverman, A.C. (2011). Aortic dissection: prompt diagnosis and emergency treatment are critical. *Cleveland Clinic Journal of Medicine* **78** (10): 685–696.

7 Erbel, R., Alfonso, F., Boileau, C. et al. (2001). Diagnosis and management of aortic dissection. *European Heart Journal* **22**: 1642.

8 Shahcheranhi, N., Dwyer, H.A., Cheer, A.Y. et al. (2002). Unsteady and three-dimensional simulation of blood flow in the human aortic arch. *Journal of Biomechanical Engineering* **124** (4): 378–387.

9 Wen, C.Y., Yang, A.S., Tseng, L.Y., and Chai, J.W. (2010). Investigation of pulsatile flowfield in healthy thoracic aorta models. *Annals of Biomedical Engineering* **38** (2): 391–402.

10 Karmonik, C., Bismuth, J., Davies, M.G. et al. (2011). A computational fluid dynamics study pre-and post-stent graft placement in an acute type B aortic dissection. *Vascular and Endovascular Surgery* **45** (2): 157–164.

11 Tang, A.Y.S., Fan, Y., Cheng, S.W.K., and Chow, K.W. (2012). Biomechanical factors influencing type B thoracic aortic dissection: computational fluid dynamics study. *Engineering Applications of Computational Fluid Mechanics* **6**: 622–632.

12 Karmonik, C., Partovi, S., and Muller-Eschner, M. (2012). Longitudinal computational fluid dynamics study of aneurysmal dilatation in a chronic DeBakey type III aortic dissection. *Journal of Vascular Surgery* **56** (1): 260.e1–263.e1.

13 Borghi, A., Wood, N.B., Mohiaddin, R.H., and Xu, X.Y. (2008). Fluid–solid interaction simulation of flow and stress pattern in thoracoabdominal aneurysms: a patient-specific study. *Journal of Fluids and Structures* **24** (2): 270–280.

14 Tan, F.P.P., Borghi, A., Mohiaddin, R.H. et al. (2009). Analysis of flow patterns in a patient-specific thoracic aortic aneurysm model. *Computers and Structures* **87** (11–12): 680–690.

15 Filipovic, N., Nikolic, D., Saveljic, I. et al. (2013). Computer simulation of thromboexclusion of the complete aorta in the treatment of chronic type B aneurysm. *Computer Aided Surgery* **18** (1–2): 1–9.

16 Cheng, S.W., Lam, E.S., Fung, G.S. et al. (2008). A computational fluid dynamic study of stent graft remodeling after endovascular repair of thoracic aortic dissections. *Journal of Vascular Surgery* **48** (2): 303–309, discusion 309–310.

17 Lam, S.K., Fung, G.S.K., Cheng, S.W.K., and Chow, K.W. (2008). A computational study on the biomechanical factors related to stent-graft models in the thoracic aorta. *Medical and Biological Engineering and Computing* **46** (11): 1129–1138.

18 Fan, Y., Cheng, S.W.K., Qing, K.X., and Chow, K.W. (2010). Endovascular repair of type B aortic dissection: a study by computational fluid dynamics. *Journal of Biomedical Science and Engineering* **3**: 900–907.

19 Jung, J., Hassanein, A., and Lyczkowski, R.W. (2006). Hemodynamic computation using multiphase flow dynamics in a right coronary artery. *Annals of Biomedical Engineering* **34** (3): 393–407.

20 Cheng, Z., Tan, F.P.P., and Riga, C.V. (2010). Analysis of flow patterns in a patient-specific aortic dissection model. *Journal of Biomechanical Engineering* **132** (5): 051007. https://doi.org/10.1115/1.4000964.

21 Saveljic, I. (2016). Numerical solving of relationship between true and false lumen in acute aortic dissection. Ph.D. Thesis. Kragujevac: Faculty of Engineering, University of Kragujevac, Serbia.

22 Hou, G., Tsagakis, K., and Wendt, D. (2010). Three-phase numerical simulation of blood flow in the ascending aorta with dissection. *Proceedings of the 5th European Conference on Computational Fluid Dynamics (ECCOMAS CFD '10)*, Lisbon, Portugal (14–17 June 2010).

23 Filipovic, N., Saveljic, I., Nikolic, D. et al. (2015). Numerical simulation of blood flow and plaque progression in carotid–carotid bypass patient specific case. *Computer Aided Surgery* **20** (1): 1–6.

24 Karmonik, C., Bismuth, J., and Redel, T. (2010). Impact of tear location on hemodynamics in a type B aortic dissection investigated with computational fluid dynamics. *Proceedings of the 32nd Annual International Conference of the IEEE Engineering in Medicine and Biology Society (EMBS '10)*, Buenos Aires, Argentina (August 31–September 4, 2010).

25 Weiss, G., Wolner, I., Folkmann, S. et al. (2012). The location of the primary entry tear in acute type B aortic dissection affects early outcome. *European Journal of Cardiothoracic Surgery* **42** (3): 571–576.

26 Wan Ab Naim, W.N., Ganesan, P.B., and Sun, Z. (2014). The impact of the number of tears in patient-specific Stanford type B aortic dissecting aneurysm: CFD simulation. *Journal of Mechanics in Medicine and Biology* **14** (2): 1450017.(20 p.). doi: 10.1142/S0219519414500171.

27 Chen, D., Müller-Eschner, M., and Tengg-Kobligk, H.V. (2012). A patient-specific study of type-B aortic dissection: evaluation of truefalse lumen blood exchange. *Biomedical Engineering Online* **12**: 1–16.

28 Filipovic, N., 1999. Numerical analysis of coupled problems: deformable body and fluid flow. Ph.D. Thesis. Faculty of Mech. Engrg., University of Kragujevac, Serbia.

29 https://en.wikipedia.org/wiki/Tunica_media.

30 Rutherford, R.B. (2000). *Vascular Surgery, Section 4, Chapter 18*, 313–329. Philadelphia: Saunders.

31 Khan, I.A. and Nair, C.K. (2002). Clinical, diagnostic, and management perspectives of aortic dissection. *Chest* **122**: 311–328.

32 Crawford, E.S. (1990). The diagnosis and management of aortic dissection. *JAMA* **264**: 2537–2541.

33 Hagan, P.G., Nienaber, C.A., Isselbacher, E.M. et al. (2000). The International Registry of Acute Aortic Dissection (IRAD): new insights into an old disease. *JAMA* **283**: 897–903.

34 Reul, G.J., Cooley, D.A., Hallman, G.L. et al. (1975). Dissecting aneurysm of the descending aorta. *The Archives of Surgery* **110**: 632–640.

35 Mehta, R.H., O'Gara, P.T., Bossone, E., and Nienaber, C.A. (2002). Acute type A aortic dissection in the elderly: clinical characteristics, management, and outcomes in the current era. *Journal of the American College of Cardiology* **40**: 685–692.

36 Suzuki, T., Mehta, R.H., Ince, H. et al. (2003). Clinical profiles and outcomes of acute type B aortic dissection in the current era: lessons learned from the International Registry of Aortic Dissection (IRAD). *Circulation* **108**: 312–317.

37 Meszaros, I., Mórocz, J., Szlávi, J. et al. (2000). Epidemiology and clinicopathology of aortic dissection. *Chest* **117**: 1271–1278.

38 Januzzi, J.L. (2004). Characterizing the young patient with aortic dissection: results from the International Registry of Aortic Dissection (IRAD). *Journal of the American College of Cardiology* **43**: 665–669.

39 Erbel, R., Bednarczyk, I., Pop, T. et al. (1990). Detection of dissection of the aortic intima and media after angioplasty of coarctation of the aorta. An angiographic, computer tomographic, and echocardiographic comparative study. *Circulation* **81**: 805–814. https://doi.org/10.1161/01.cir.81.3.805.

40 Erbel, R., Engberding, R., Daniel, W. et al. (1989). Echocardiography in diagnosis of aortic dissection. *Lancet* **1**: 457–461.

41 Erbel, R., Oelert, H., and Meyer, J. (1993). Influence of medical and surgical therapy on aortic dissection evaluated by transesophageal echocardiography. *Circulation* **87**: 1604–1615.

42 Deeb, G.M., Williams, D.M., and Bolling, S.F. (1997). Surgical delay for acute type A dissection with malperfusion. *The Annals of Thoracic Surgery* **64**: 1669–1675.

43 Sommer, T., Fehske, W., and Holzknecht, N. (1996). Aortic dissection: a comparative study of diagnosis with spiral CT, multiplanar transesophageal echocardiography, and MR imaging. *Radiology* **199**: 347–352.

44 Nienaber, C.A. and von Kodolitsch, Y. (1997). Diagnostic imaging of aortic diseases. *Radiologe* **37**: 402–409.

45 Hartnell, G. and Costello, P. (1993). The diagnosis of thoracic aortic dissection by noninvasive imaging procedures. *The New England Journal of Medicine* **328**: 1637; author reply 1638.

46 LePage, M.A. et al. (2001). Aortic dissection: CT features that distinguish true lumen from false lumen. *AJR American Journal of Roentgenology* **177**: 207–211.

47 Mintz, G.S., Kotler, M.N., Segal, B.L., and Parry, W.R. (1979). Two dimensional echocardiographic recognition of the descending thoracic aorta. *The American Journal of Cardiology* **44**: 232–238.

48 Erbel, R., Mohr-Kahaly, S., and Oelert, H. (1990). Diagnostic strategies in suspected aortic dissection: comparison of computed tomography, aortography and transesophageal echocardiography. *American Journal of Cardiac Imaging* **4**: 157–172.

49 Prince, M.R., Narasimham, D.L., Jacoby, W.T. et al. (1996). Three-dimensional gadolinium-enhanced MR angiography of the thoracic aorta. *AJR American Journal of Roentgenology* **166**: 1387–1397.

50 Krinsky, G.A., Rofsky, N.M., and DeCorato, D.R. (1997). Thoracic aorta: comparison of gadolinium-enhanced three dimensional MR angiography with conventional MR imaging. *Radiology* **202**: 183–193.

51 Yamada, E., Matsumura, M., Kyo, S., and Omoto, R. (1995). Usefulness of a prototype intravascular ultrasound imaging in evaluation of aortic dissection and comparison with angiographic study, transesophageal echocardiography, computed tomography, and magnetic resonance imaging. *The American Journal of Cardiology* **75**: 161–165.

52 Wolff, K.A., Herold, C.J., Tempany, C.M. et al. (1991). Aortic dissection: atypical patterns seen at MR imaging. *Radiology* **181**: 489–495.

53 Alfonso, F., Goicolea, J., Aragoncillo, P. et al. (1995). Diagnosis of aortic intramural hematoma by intravascular ultrasound imaging. *The American Journal of Cardiology* **76**: 735–738.

54 Elefteriades, J.A., Lovoulos, C.J., Coady, M.A. et al. (1999). Management of descending aortic dissection. *The Annals of Thoracic Surgery* **67**: 2002–2005.

55 Augoustides, J.G.T. and Andritsos, M. (2010). Innovations in aortic disease: the ascending aorta and aortic arch. *Journal o Cardiothoracic and Vascular Anesthesis* **24**: 198–207.

56 Trimarchi, S., Nienaber, C.A., and Rampoldi, V. (2006). Role and results of surgery in acute type B aortic dissection: insights from the International Registry of Acute Aortic Dissection (IRAD). *Circulation* **114**: I-357–I-364.

57 Svensson, L.G., Kouchoukos, N.T., and Miller, D.C. (2008). Expert consensus document on the treatment of descending thoracic aortic disease using endovascular stent-grafts. *The Annals of Thoracic Surgery* **85**: S-1–S-41.

58 DeSanctis, R.W., Doroghazi, R.M., Austen, W.G., and Buckley, M.J. (1987). Aortic dissection. *The New England Journal of Medicine* **317**: 1060.

59 Doroghazi, R.M., Slater, E.E., DeSanctis, R.W. et al. (1984). Long-term survival of patients with treated aortic dissection. *Journal of the American College of Cardiology* **3**: 1026.

60 Kojić, M., Slavković, R., Živković, M. et al. (1995). Analiza tačnosti rešenja strujanja viskoznog nestišljivog fluida sa prenosom toplote metodom konačnih elemenata, 21. *Jugoslovenski kongres teorijske i primenjene mehanike, str. 83-88, Niš.*

61 Connor, J.J. and Brebbia, C.A. (1976). *Finite Element Techniques for Fluid Flow.* Bristol: Newnes-Butterworths.

62 Kojić, M., Slavković, R., Živković, M., and i Grujović, N. (1998). *Metod Konačnih Elemenata I.* Mašinski fakultet u Kragujevcu.

63 Kojic, M., Filipovic, N., Stojanovic, B., and Kojic, N. (2008). *Computer Modeling in Bioengineering – Theoretical Background, Examples and Software.* Chichester, England: Wiley.

64 Bathe, K.J. (1996). *Finite Element Procedures.* Englewood Cliffs, N.J.: Prentice-Hall, Inc.

65 Kojic, M. and Bathe, K.J. (2005). *Inelastic Analysis of Solids and Structures.* Heildelberg, Berlin: Springer-Verlag.

66 Zivkovic, M. (2006). *Nelinearna analiza konstrukcija.* Srbija: Mašinski fakultet u Kragujevcu.

67 Kutz, M. (2003). *Standard Handbook of Biomedical Engineering & Design.* McGRAW-HILL https://www.accessengineeringlibrary.com/content/book/9780071356374.

68 Womersley, J.R. (1955). Method for the calculation of velocity, rate of flow and viscous drag in arteries when the pressure gradient isknown. *Journal of Physiology* **127**: 553–563.

69 Pedley, T.J. (1980). *The Fluid Mechanics of Large Blood Vessels.* Cambridge, United Kingdom: Cambridge University Press.

70 Liepsch, D. (1993). Fundamental flow studies in models of human arteries. *Frontiers of Medical and Biological Engineering* **5**: 51–55.

71 Pedley, T.J. (1995). High Reynolds number flow in tubes of complex geometry with application to wall shear stress in arteries. *Symposia of the Society for Experimental Biology* **49**: 219–241.

72 Perktold, K., Hofer, M., Rappitsch, G. et al. (1998). Validated computation of physiologic flow in realistic coronary artery branch. *Journal of Biomechanics* **31**: 217–228.

73 Shi, Y., Zhu, M., Yu, C. et al. (2016). The risk of stanford type—a aortic dissection with different tear size and location: a numerical study. *BioMedical Engineering OnLine* **15** (Suppl 2): 128.

74 Morris, L., Delassus, P., and Walsh, M. (2004). A mathematical model to predict the in vivo pulsatile drag forces acting on bifurcated stent grafts used in endovascular treatment of abdominal aortic aneurysms (AAA). *Journal of Biomechanics* **37**: 1087–1095.

75 Riley, W.A., Barnes, R.W., Evans, G.W., and Burke, G.L. (1992). Ultrasonic measurements of the elastic modulus of the common carotid artery: the Atherosclerosis Risk in Communities (ARIC) Study. *Stroke* **23**: 952–956.

6

Application of AR Technology in Bioengineering

Dalibor D. Nikolic and Nenad D. Filipovic

Department of Technical-Technological Sciences, Institute of Information Technologies, University of Kragujevac, Kragujevac, Serbia

6.1 Introduction

The breakthrough of technological progress in certain areas often has direct and lasting effects on our daily lives, and this is especially pronounced in medical practice. A good example of these achievements is virtual reality (VR), which is already widely used in areas such as games, entertainment, design, simulation training, and other similar fields. Augmented reality (AR), which emerged as a product of VR, is also evolving rapidly. Unlike VR, which takes place within a completely virtual world, AR technology brings components, i.e. virtual objects and connects them to the real environment. This will make AR users feel an authentic new environment. With the advancement of computer technology and smartphone technology, the application of AR technology is also expanding. The application of AR technologies can be seen in video games, industrial design, entertainment, education, and many other aspects. Engineers often develop software tools that have direct application in clinical practice [e.g. software for calculating virtual fractional flow reserve (FFR) [1, 2], arterial plaque progression and processing [3], myocardial bridge [4], etc.]. However, because of difficult interpretation in the clinical setting, these tools are not used enough. AR technologies have the potential to improve these tools and facilitate their application in the clinic. Biomedical engineering also takes advantage of the opportunities provided by AR and introduces this technology to biomedical and clinical practice.

Computational Modeling and Simulation Examples in Bioengineering, First Edition. Edited by Nenad D. Filipovic. © 2022 The Institute of Electrical and Electronics Engineers, Inc. Published 2022 by John Wiley & Sons, Inc.

6.2   Review of AR Technology

By definition, AR is an interactive experience of a real-world environment where the objects that reside in the real world are enhanced by computer-generated virtual objects. The first definition of AR was proposed by the researcher Azuma in his paper [5]. In his article entitled "Augmented Reality Research," AR is treated as a system with the following characteristics:

1) Combines real and virtual;
2) Interactive in real time;
3) Registered in 3-D.

From this definition, it can be concluded that AR does not intend to create a completely virtual world, but to include some virtual content within the real world. Figure 6.1 shows an example of AR application and creation of an interactive book, in which a virtual object (heart) appears on an actual object (book).

6.2.1   Augmented Reality Devices

AR technology has evolved over the years and has gone through several stages of development. With the development and improvement of hardware, three types of AR hardware systems can be found on the market at the moment: monitor-based AR, mobile-based display, and head-mounted display (HMD).

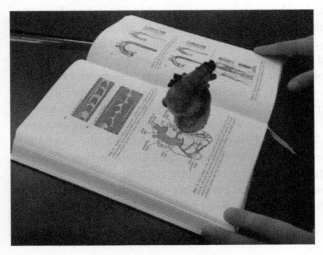

Figure 6.1   A simple example of augmented reality. *Source:* [6] Kojic et al., (2008)/with permission of John Wiley and Sons.

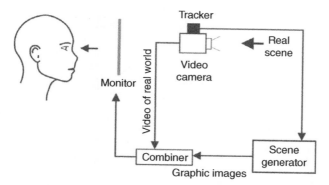

Figure 6.2 Schematic representation of AR monitor system.

6.2.2 AR Screen Based on the Monitor

This is one of the oldest forms of displaying AR-virtual objects, generated by a computer, superimposed on the real environment that the camera records in real time, and then displays on the user's computer monitor. Figure 6.2 shows a schematic representation of this system.

6.2.3 AR Screen Based on Mobile Devices

With the development of smartphones and the improvement of hardware, the integration of AR software on these devices has occurred naturally, so that today the appropriate AR applications can be easily applied on a smartphone. At the moment, applications marked with QR markers are most often used on mobile phones. With the growth of the Android market, Google's development team has developed an ARcore platform that has enabled the development and application of marker-less AR applications. Some of mobile phone AR-based simulations are developed for the real machine controls [7].

Figure 6.3 presents the AR effect shown on a smartphone. When the user points the camera of the mobile phone in the background, he/she can see the real object and the virtual one at the same time.

6.2.4 Head Mounting Screen

A HMD is a type of display device worn by the user, also known as AR display glasses. First, helmet display (HMD) was created in 1968 [8]. The optical perspective helmet display was developed by professor Ivan Sutherland. This device makes it possible to superimpose simple graphics constructed by computers in real time on real scenes. Modern HMD devices can be divided into two categories shown in

Figure 6.3 Augmented reality application on mobile phone. *Source*: UNIBOA/Unsplash.

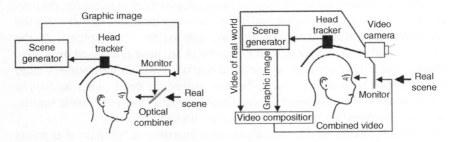

Figure 6.4 Schematic diagrams: SHDM device (left); VHDM device (right).

Figure 6.4: one is a cheaper, optically transparent HMD device so-called optical see-through HMD (SHMD) and the other is a video HMD device (VHMD). The principles of operation of these two types of devices are shown in Figure 6.4 (left and right).

As we mentioned, AR represents the overlapping of virtual content into real space or landscape. In order to position these virtual objects in the desired place, it is necessary to define their location in some way (known as "registration"), which means that the AR system must follow the relevant targets (markers) in the input video and set up virtual objects to the appropriate position. The system monitors the position of these target objects on the video and, if they move, the virtual objects will also change their position, i.e. virtual object will follow the target locations. In this way, the impression is gained that the virtual object is part of the real environment and this effect is called a "combination of virtuality and reality."

The main supporting technologies of the AR system contain algorithms that include tracking and positioning, interaction, and fusing (merging). These three technologies are a basic condition for a useful and functional AR system.

Registration (marker detection) is the first step of the AR system – displaying virtual objects in the exact location in the real world. This process requires spatial data (location) obtained from the current scene from the video file. The location of the virtual object, as well as the location of the observer, is relative, which means that the location system in the AR system must detect the position of the observer, i.e. the point of view of the observer. Some systems even detect the direction of movement of the observer's (e.g. head); the AR system can decide how to reconstruct the coordinate system in accordance with the viewer's field of view and display a virtual object. This process is called tracking. In the AR system, tracking is the basis of registration, which includes the detection of observers, as well as the acquisition of spatial location data from stage markers, additional interactive tools, or other equipment. A simple schematic representation of video detection (the most common method of tracking and registration) is shown in Figure 6.5.

Interaction within AR is one of the primary goals of AR technology. Its role is to achieve a more natural interaction between users and virtual objects in real-life scenarios. There are several types of AR application interaction techniques and some of them are Spatial Point, Command, and some of the special tools developed for AR.

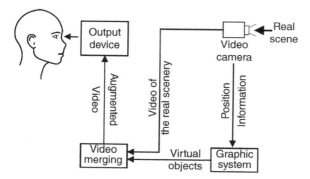

Figure 6.5 Schematic representation – the fundamental principle of AR system.

In order to get a more realistic look of the AR scene, it is necessary for the system to detect not only accurate registration based on the tracking and positioning process, but also the effects of shadows and light and occlusion between virtual objects and real objects. The occlusion between virtual objects and real objects is very difficult to detect and developers pay a lot of attention to this problem. The problem of occlusion refers to the fact that, in reality, it happens that a real object on the scene partially or completely covers a virtual object.

Currently, in most augmented reality systems, only virtual objects overlap with real objects, so the real scene has always been clogged with virtual objects. It is difficult to correctly estimate the position relationship between real and virtual objects. This complex issue in the AR system has not yet been fully resolved. There are some relevant algorithms proposed by individual research institutions. Huazhong University of Science and Technology proposed a series of algorithms [9] based on the latest advances in digital image processing, such as pattern recognition, computer vision (CV), and nonlinear optimization theory. All of those proposed algorithms have realized semi-automatic real-time occlusion of non-rigid objects for the AR system.

6.2.5 AR in Biomedical Engineering

Biomedical engineering is an interdisciplinary science and a combination of basic science, medicine, and engineering. In essence, biomedical engineering deals with solving problems related to prevention and treatment of diseases using engineering theory and methods. This discipline is closely related to modern information technologies, so the advancement of computer technology has a huge impact on the development of biomedical engineering and, thus, on clinical practice.

AR technology, as a specific branch of VR technology, brings many new features. Although AR technology is constantly developing, its significant influence on the biomedical field is already noticed.

Currently, the application of AR technology is in a development phase, but there are a lot of research and experimental applications of AR technology in biomedical engineering, on different types of medical simulations, anatomy of the human body and surgical education, etc.

Some of the applications of AR technology are directly related to clinical practice. For example, using the AR technology, engineers can connect a three-dimensional virtual model of an organ reconstructed by CT or MRI and fuse them with the appropriate real organ of the patient during surgery. Moreover, in this way, AR provides the surgeon with new information about the internal anatomy of the organ, e.g. about the position of blood vessels that cannot be seen with the naked eye.

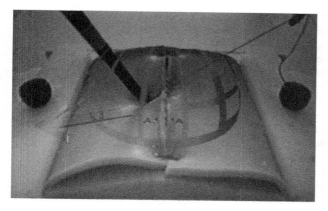

Figure 6.6 Dome and arrow during the suturing task to guide the trainee in the proper direction when pulling the knot tight. *Source*: [11] Botden et al., 2009, p.2131–2137/with permission of Springer Nature.

The Biomedical Engineering College of Capital Medical University [10] has developed a medical augmented reality (MAR) system based on a real model of a female phantom doll training that is connected to a set of virtual data obtained from real patients using MRI technology. This AR-based system is used for training surgical simulation of brain cysts, cyst tissues, probes, skulls, and brain lobes, in which MRI medical images are interactively displayed on a real model (phantom). Botden from the Catharina Hospital of the Netherlands [11] in his paper proposed a safe way to train surgeons for laparoscopic surgery called ProMIS. In this paper, AR is used as a tool to train suturing skills in laparoscopic surgery (Figure 6.6). In his study, Oostema [12] tried to prove that AR-based surgery training had advantages such as better performance that can be assessed using unique computer indicators, which instructors can analyze after the intervention. This study attempted to determine whether the results obtained with the AR hybrid simulator correlated with laparoscopic surgical skill.

The basic structure of an AR-based biomedical engineering system consists of three steps: the first step is to create a virtual model of a tissue or an organ, then the second step is to initialize the camera system and real objects, and the third step is to combine a real environment with virtual objects.

In the first step, a digital medical model of one of the patient's organs should be created. This is usually achieved by segmenting multi-slice DICOM images obtained from MRI or CT machines. This step can be called the preparation of virtual AR system objects. Then, in the second step, the camera calibration task will be performed and all the markers will be initialized before user interaction. This setting information will be stored as a matrix in a specific file. At the beginning of the

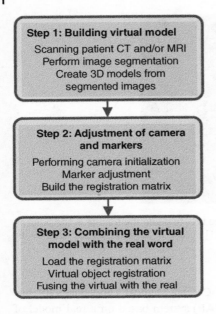

Step 1: Building virtual model

Scanning patient CT and/or MRI
Perform image segmentation
Create 3D models from
segmented images

**Step 2: Adjustment of camera
and markers**

Performing camera initialization
Marker adjustment
Build the registration matrix

**Step 3: Combining the virtual
model with the real word**

Load the registration matrix
Virtual object registration
Fusing the virtual with the real

Figure 6.7 Schematic representation of the flow of an AR-based biomedical engineering system.

process, the system will load the camera matrix setup data from the file and set the camera parameters, and then activates the initialized process to load all the markers. In this way, the camera starts working on capturing real objects and scenes.

In the third step, the virtual models (created in the first step) will be combined with the actual objects on the computer screen or HMD. The people who use this AR system will get the impression that they can see through the human body, so the preparation for surgery will be more intuitive and accurate.

A schematic representation of the flow of an AR-based biomedical engineering system is shown in Figure 6.7.

Over time, a number of tools for the development AR were developed. In this chapter, we will mention only some of them.

ARKit is an AR development platform launched by Apple in 2017. Developers can use this suite of tools to develop AR applications for the iPhone and iPad. The advantage of ARKit is that it allows developers to develop AR applications that can support up to two devices that share the same virtual objects. This makes the AR experience more interesting.

ARCore is a software platform developed by Google's AR application building team, which is very similar to Apple's ARKit. The ARcore can take advantage of cloud-based software and hardware to bring digital objects to the real world. Its main functions are motion capture, perception of the environment as well as the perception of light sources.

Vuforia is currently the most popular SDK. The main recognition function supports Android, iOS, and UVP. Also, different SDKs are available depending on different platforms. The developers can choose any environment such as Android Studio, Xcode, Visual Studio, and Unity and use the Vuforia development tool according to their needs.

The Wikitude SDK reconstructs its propositions using a development framework for image recognition and tracking and geolocation technology, including image recognition and tracking, 3D model display, video overlay, and location-based AR. In 2017, Wikitude introduced Simultaneous Localization and Mapping – SLAM technology, which enables recognition and tracking of objects, as well as marker-less real-time tracking.

6.3 Marker-based AR Simple Application, Based on the OpenCV Framework

In this part of the chapter, we will present the basics of marker-based AR applications based on the OpenCV framework. It will be presented how to create a standard real-time project using OpenCV (for desktop) that uses the ArUco marker AR method. For understanding this part, a minimum of basic knowledge of C++ is required. Also, for source code testing, a software environment Visual Studio (2019) or later version and OpenCV 4 library with integrated modules OpenGL and ArUco are required.

Advantages of marker-based AR systems are:

- Cheap (fast) detection algorithm.
- Very robust against lighting changes.

However, the main weaknesses are:

- Do not work if the marker is partially covered.
- The marker must generally be in black and white.
- In order to be easily detected, the markers must be square-shaped.
- Do not have a beautiful aesthetic appearance.
- Do not have anything in common with objects from the real world.

ArUco [13] markers are most common marker types which have been used for a while in AR, camera calibration, and camera pose estimation. Those markers were originally developed by Garrido-Jurado et al. [13], and published in their paper "Automatic generation and detection of highly reliable fiducial markers under occlusion." The name ArUco stands for the Augmented Reality University of

Figure 6.8 Some examples of ArUco markers.

Cordoba in Spain, the place where they were developed. The example of ArUco markers is presented in Figure 6.8.

Pose estimation is very important in many CV applications. This is a process that requires finding correspondences between points in the real environment and their 2D image projection. Usually, this is a difficult step, and thus it is common to use synthetic or fiducial markers to make it easier. An ArUco marker is a square fiducial marker. Using this marker is very simple, just print them and place them on the object or scene being imaged. ArUco is a binary square marker with a black background and white boundaries and a white generated pattern inside of his that uniquely identifies it. The black boundary helps to make their detection easier. They can be easily generated in a variety of sizes. Depending on object size and scene, these markers can be smaller or larger for successful detection. In some cases, very small markers are not being detected, but simply increasing their size will facilitate detection.

The general idea is that we can print these markers and put them in some places in the real world. Then, we can use application to photograph or record the real world and detect these markers uniquely. These markers are widely applied. In a robotics application, we can put these markers along the path of the warehouse or on storage boxes, so that a robot equipped with a camera can trace the route and find something. The robot with a camera mounted when detecting one of these markers can know the exact location in the warehouse because each marker has a unique ID. In the example **ArUcoExample01** whose source code we shared with this chapter, we place the printed markers at the corners of a real picture in the right word. Then, when we uniquely identify the markers and find the position of picture corners, we are able to replace the real picture with an AR object like an arbitrary video or image. This new picture (AR object) will have the correct perspective distortion if we move the camera.

6.3.1 Generating ArUco Markers in OpenCV

ArUco markers can be very easily generated by using OpenCV. The ArUco module included in OpenCV frameworks has a total of 25 versions of predefined dictionaries of markers. All the markers in a dictionary have the same number of bits or blocks (4×4, 5×5, 6×6, or 7×7), and each dictionary contains an exact number of markers (50, 100, 250, or 1000). Function for generating ArUco markers is:

```
//Create Aruco markers
void createArucoMarkers()
{
    Mat outputMarker;
    Ptr<aruco::Dictionary> myMarkerDictionary = aruco::
    getPredefinedDictionary
                    (aruco::PREDEFINED_DICTIONARY_NAME::DICT_4X4_50);
    for (int i = 0; i < 50; i++)
    {
        aruco::drawMarker(myMarkerDictionary, i, 500, outputMarker, 1);
        ostringstream convert;
        string imgName = "4x4ArucoMarker_";
        convert << imgName << i << ".jpg";
        imwrite(convert.str(), outputMarker);
    }
}
```

The `drawMarker` function allows us to print a marker of choosing. We use for loop to print all markers from the dictionary (in our case 50). The first parameter represents the dictionary, the second parameter is marker id (in this case i) from the collection of 50 markers with IDs from 0 to 49. The third parameter in `drawMarker` function defines the size of the generated marker (500 × 500 pixels). The next parameter represents the object that would store the generated marker. And the last parameter represents the thickness parameter – how many blocks should be added as a boundary around the generated binary pattern. In our example, a boundary of one bit would be added one block boundary around the 4 × 4 generated pattern and will produce an image with 5 × 5 bits in a 500 × 500 pixel image. An example of the generated marker is presented in Figure 6.9. Usually, in most applications, we need to use multiple markers, print them, and put them on the scene.

Figure 6.9 Example of ArUco marker.

The first step in the development stage of marker-base AR is to detect the markers.

```
// Load the dictionary that was used to generate the markers.
Ptr<Dictionary> dictionary = getPredefinedDictionary(DICT_4X4_50);
//Ptr<Dictionary> dictionary = getPredefinedDictionary(DICT_6X6_250);

/* Declare the vectors that would contain the detected marker corners and
the rejected marker candidates*/
vector<vector<Point2f>> marker_Corners, rejected_Candidates;

// Initialize the detector parameters - default values
Ptr<DetectorParameters> parameters = DetectorParameters::create();

// Detect the markers
cv::aruco::detectMarkers(frame, dictionary, marker_Corners, marker_IDs,
parameters, rejected_Candidates);
```

To detect marker's function for detection, we must know which marker dictionary we are using. Therefore, in the beginning, we load the same dictionary to our detection function that we used to generate the markers. An initial set of parameters will be detected using `DetectorParameters::create()`. Detection parameters can be adjusted including the adaptive threshold value. However, in most cases, the default parameter will work well.

The function `detectMarkers` is used to detect and locate all corners of the markers. In the case of successful marker detection, all the four corner points of the marker are detected. These four detected corner points are stored as a vector of points, in order from top left, top right, bottom right, and bottom left. Also, multiple markers in the image are stored in a `vector` of vector of points. The function `detectMarkers` is used to detect and locate all corners of the markers.

The first input parameter is the image of our scene. The second parameter is the dictionary of markers (used to generate the markers). The third parameter is an object `marker_Corners` where successfully detected markers will be stored, and `marker_IDs` will store markers ids. Next, the input is `DetectorParameters` object, which is initialized earlier and is passed as a parameter. The last input parameter `rejected_Candidates` will store the rejected candidates. It is very important while cutting and placing the markers in a scene, that we retain some white border around the marker's black boundary. This will make detection much easier. The ArUco markers are mainly developed to solve the problem of camera pose estimation. They can be used for various applications including AR.

In this chapter, we provide **ArUcoExample01** C++ project, which will be used for developing an AR application. This application lets us overlay one specific part of an existing image or video with any new image. For testing purposes, we pick a scene at home with a large picture on the wall. The idea is that AR object (new image) replaces the existing picture in the frame to see how they look up on the wall. The next step is to try to insert an image in the frame of the video file. For this purpose, we prepare (print, and cut) four large ArUco markers. Then, we put them on the wall (simulate corners of the picture) (Figure 6.10). We take the picture of the wall and record a small video. In the case of the video file, we process each frame of the video individually (like an image) in order.

Detected markers represent four corresponding set of points in the input image and corner points of the AR object (new image) are used to compute a

Figure 6.10 ArUco markers on the wall – preparation AR scene.

homography. Basically, homography is a transformation (3 × 3 matrix) that maps the points from one image to the corresponding points in the other image. Homography 3 × 3 matrix can be written as:

$$H = \begin{bmatrix} h_{00} & h_{01} & h_{02} \\ h_{10} & h_{11} & h_{12} \\ h_{20} & h_{21} & h_{22} \end{bmatrix}$$

If we write the first set of corresponding points as a (x_1, y_1) in the first image, and set of corresponding points (x_2, y_2) in the second image, then, the Homography H will map them in the following way:

$$\begin{bmatrix} x_1 \\ y_1 \\ 1 \end{bmatrix} = H \begin{bmatrix} x_2 \\ y_2 \\ 1 \end{bmatrix} = \begin{bmatrix} h_{00} & h_{01} & h_{02} \\ h_{10} & h_{11} & h_{12} \\ h_{20} & h_{21} & h_{22} \end{bmatrix} \begin{bmatrix} x_2 \\ y_2 \\ 1 \end{bmatrix}$$

Simply, homography allows us to warp one image onto the other. In **ArUcoExample01** application, the homography matrix is used to warp the new AR object (new image) into the quadrilateral defined by points obtained from markers' positions in our captured image.

```
// Compute homography
Mat h = cv::findHomography(pts_src, pts_dst);

// Warped imageMat warpedImage;
// Warp source image to destination
cv::warpPerspective(im_src, warpedImage, h, frame.size(), INTER_CUBIC);

Mat mask = Mat::zeros(frame.rows, frame.cols, CV_8UC1);
cv::fillConvexPoly(mask, pts_dst, Scalar(255, 255, 255), LINE_AA);

// Erode the mask
Mat element = getStructuringElement( MORPH_RECT, Size(5,5));
cv::erode(mask, mask, element);
// Copy the warped image into the wall
Mat imOut = frame.clone();

warpedImage.copyTo(imOut, mask);
```

The new image (AR object) corner points were used as the source points (pts_src), and corner points from the corresponding picture on the wall were obtained from marker positions and used as the destination points(pts_dst). The OpenCV function `findHomography` was used to compute the homography function H between the source and destination points. In case when video file was

Figure 6.11 An example of using ArUco marker – AR images on a real wall.

used, this process will be repeated on each frame. In Figure 6.11, Homography results are presented.

Example application **ArUcoExample01** can be run in the next ways:

- Run on a single image:

```
ArUcoExample01.exe --image=test.jpg
```

- Run on a recorded video file:

```
ArUcoExample01.exe --video=test.mp4
```

ArUcoExample02 application is created to familiarize the user with the process of generating the ArUco marker explained.

Example application **ArUcoExample02** can be executed by the next commands:

- Generate 50 ArUco markers 4x4:

```
ArUcoExample02.exe --marker
```

- Run live web camera calibration:

```
ArUcoExample02.exe --calibration
```

- Run web camera and detect the printed 4x4 ArUco marker:

```
ArUcoExample02.exe -- webcam
```

The `ArUcoExample02.exe -marker` command will call `createAruco-Markers()` function which creates ArUco markers in the form of a JPG file. With the current settings, the function will create 50 4 × 4 markers. This can be changed by changing the dictionary `DICT_4X4_50` with some of the predefined markers dictionaries in the ArUco Library. **Note**: if different types of markers are generated, e.g. 6 × 6 or 8 × 8, the dictionary must be changed in all detection functions.

Also, within this example, functions have been created that allow the webcam to be calibrated in real time, by executing command `ArUcoExample02.exe --calibration`.

Starting of the calibration launches the webcam screen and pressing the SPACE button takes a photo of the chessboard pattern. The pattern of a chessboard in pdf format size A4 that the user needs to perform calibration is located within the source code. For successful calibration, more than 15 images of the chessboard pattern from different angles and distances are necessary. After the images are taken, the user starts the camera calibration process by pressing the enter key, and the calibration parameters will be saved within the file.

On the end, `ArUcoExample02.exe --webcam` command first call function `loadCameraCalibrationFile` to load camera calibration parameters from `CameraCalibrationFile` which is made by the previous calibration process, then call `WebCamMonitoringArUco` function which launches the webcam and detects markers in its field of vision. The axes of the local coordinate system are plotted on each detected marker. If the calibration is poorly performed, this axis will not correctly mark the position of the marker. An example of the use of the ArUco marker AR in biomedical engineering is shown in the Figure 6.12.

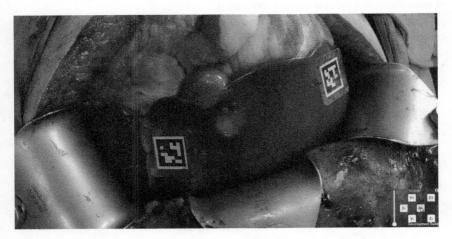

Figure 6.12 ArUco markers – biomedical application.

6.4 Marker-less AR Simple Application, Based on the OpenCV Framework

In this part of the chapter, we will present the basics of marker-less AR applications **ArExample03** based on the OpenCV framework. Through source code of this application, it will be presented how to create a standard real-time project using OpenCV (for desktop) that uses the marker-free AR method, and instead of using printed square markers, the real environment will be used as input. For understanding of this part, a minimum of basic knowledge of C++ is required. Also, for testing example, a software environment Visual Studio (2019) or later version, and OpenCV 4 library with integrated OpenGL are required.

Advantages of marker-based AR systems are:

- Cheap (fast) detection algorithm.
- Very robust against lighting changes.

However, the main weaknesses are:

- Do not work if the marker is partially covered.
- The marker must generally be in black and white.
- In order to be easily detected, the markers must be square (shape).
- Do not have a beautiful aesthetic appearance.
- Do not have anything in common with things from the real world.

So, markers are a very good thing to start working with AR, but for a more beautiful and natural work, an AR system without markers is necessary. This (No Marker) AR technique is based on recognizing objects that exist in the real world and treating them as a marker.

Excellent examples of targets for marker-less AR are company logos, toys, album thumbnails, and so on. Essentially, any object that has sufficient descriptive and discriminative information in relation to the rest of the subject on the scene can be a target for markers-less AR.

The main advantages of marker-less AR applications are:

- They can easily be used to detect objects from the real environment.
- They will work even if part of the target object is covered.
- The target object can be of any shape and texture.

Marker-less AR systems can use real images as well as real objects to position the 3D space camera and present some of eye-catching effects on top of the actual image. The basis without marker-less AR system cannot work are algorithms for object detection and image recognition. Unlike markers, whose shapes are fixed and recognizable, real objects cannot be defined in this way. It is very common for objects to have a complex shape and modified pose assessment algorithms are required to find their correct 3D transformations. Marker-less AR system requires a lot of processing power for very complex calculations, so it is very demanding to use them on a mobile device, often is unable to ensure smooth operation without "glitching." In this chapter, an example will be presented of how to use marker-less AR on a desktop computer. We used Microsoft's development environment, but the code can also be used to develop multiform applications. For this purpose, it is necessary to adapt the project **ArExample03** for CMake build system or something similar.

6.4.1 Use Feature Descriptors to Find the Target Image in a Video

Image recognition is a CV technique that searches an input image in order to find a specific bitmap pattern. A good image recognition algorithm should be able to detect a pattern even if it is rotated or scaled, or has a different brightness than the target image. So, the key question is how does CV compare the target image with other images?

Since the pattern can be affected by perspective transformation, it is not possible to directly compare the pixels of the target image and the test image. Therefore, in this case, the Feature Points and the Feature descriptions are usable. This feature is usually defined as the "interesting" part of the image. These features are used as a starting point for a numerous of CV algorithms. Feature point is a part of the image defined by a center point, radius, and orientation.

The method of finding areas of interest in the input image is called a feature detection. Numerous algorithms have been developed to detect characteristics based on methods that search for blobs, edges, or corners. This chapter will show the method for detecting corners. For corners detecting, it is necessary to detect the edges in the image. The "Edge detection algorithm" analyzing image gradient and searches for his rapid changes. This is done by finding extremums of the first derivative of the image gradients in the X and Y directions. Feature-point orientation is computed as a direction of dominant image gradient in a particular area of interest. In the case, when the image is scaled or rotated, the feature-detection algorithm is used to recalculate orientation of dominant gradient. Therefore, regardless of image rotation, the orientation of feature points is not changed and those features are called rotation invariant. The size of feature point is helpful to overcome the scaling problem. Depending on the approach, feature-detection algorithms use fixed-size features, or calculate the optimal size for each key point separately. Knowing of this characteristic allows the algorithm to find the same feature points on scaled images. In this way is features scale invariant. OpenCV library has integrated several algorithms for feature-detection and they are derived from the base class cv::FeatureDetector. Any of those of the feature-detection algorithms can be created on two ways:

Through the explicit call of the wished feature detector class constructor:

```
cv::Ptr<cv::FeatureDetector> detector = cv::Ptr<cv::FeatureDetector>
                        (new cv::SurfFeatureDetector());
```

Or by creating a feature detector by algorithm name:

```
cv::Ptr<cv::FeatureDetector> detector = cv::FeatureDetector::create
("SURF");
```

For detection of feature points, simply call the detect method:

```
std::vector<cv::KeyPoint> keypoints;
detector->detect(image, keypoints);
```

All detected feature points will be placed in the keypoints container, also each key point contains its center, radius, angle, and score, and some correlation factor with the "strength" or "quality" of the feature point. Depending on used feature detection algorithm, this score will be calculated in a different way. So, it is only valid to compare scores of the key points detected by the same detection algorithm.

For achieving the best results in pattern detection, the detector should calculate key point orientation and size, in this way key points become invariant to rotation and scale. In the OpenCV library, the most famous and robust keypoint detection algorithms are used in SIFT and SURF feature detection. They are patented, and cannot be used free for commercial use. But ORB or FREAK algorithm is very good and free replacement. ORB detection is an updated FAST feature detector. The FAST detector is very good and fast but does not calculate the orientation or the size of the key point. Its successor ORB algorithm estimates key point orientation, but not and the feature size. This deficiency could be overcome by using some cheap programming tricks.

Why does feature point matters so much in image recognition?

Let us say that we work with images, with a minimum resolution of 640 × 480 pixels and a color depth of 24 bits per pixel, in that case we have 912KB of data for each image. Searching and comparing pixels with pixels takes too long, and it is necessary to solve their position, i.e. rotation and scaling. This is not easy to do on the strongest computer. Therefore, the using of feature points is necessary; a similar technique is used in biometric fingerprinting. By discovering a sufficient number of key points and comparing them, we can be sure that algorithms find a good image. Moreover, by using angle-based detectors, we will have a large number of key points because they return edges, corners, and other sharp figures. Therefore, to find correspondences between two images (target and working image), we only need to match the key points. From the area defined by the key point, a vector called descriptor will be extracted. These forms represent the feature points. Many methods of extraction of the descriptor from the feature point can be used, and each of them have own strengths and weaknesses. SIFT and SURF descriptor-extraction algorithms use a lot of CPU power but provide robust descriptors and good distinctiveness. In this chapter, we freely use the ORB descriptor-extraction algorithm because we are also using the same type of algorithm for a feature detector. It is a practice for the developer to use feature detector and descriptor extractor from the same algorithm. In this way, we can be sure that they fit each other perfectly. Feature descriptor is represented in a form of vector with fixed size (e.g. 16 or more elements). Therefore, image with a resolution of 640×480 pixels will have around 1500 feature points. Then, it will require $1500 * 16 * $ size of (float) = 96 KB (for SURF algorithm). This is 10 times smaller than the original image data and it is much easier for us to operate with descriptors rather than bitmaps. For two feature descriptors, we will introduce a similarity score—a factor which defines the level of similarity between two vectors. Usually, its L2 norm or hamming distance (depend on feature descriptor used).

OpenCV library have integrated a several feature descriptor-extraction algorithms and they are derived from the base class `cv::DescriptorExtractor`. In the similar way like feature-detection algorithms, they can be created by explicit constructor calls or by either specifying their name.

To define a pattern of some object, we will use a class called Pattern, which holds a trained image, list of all features and extracted descriptors, as well as 2D and 3D correspondences for initial pattern position:

```
/*
    Store the target image data and computed descriptors of target pattern
*/
struct ARPattern
{
    cv::Size                    size;
    cv::Mat                     frame;
    cv::Mat                     grayImg;
    std::vector<cv::KeyPoint>   keypoints;
    cv::Mat                     descriptors;

    std::vector<cv::Point2f>    points2d;
    std::vector<cv::Point3f>    points3d;
};
```

The finding of frame-to-frame correspondences can be explained as the search of the nearest neighbor from one set of descriptors for every element of another set.

This is called the "matching" procedure. OpenCV have two main algorithms for descriptor matching:

```
cv::BFMatcher // Brute-force matcher
cv::FlannBasedMatcher // Flann-based matcher
```

The brute-force matcher (as his name suggests) looks for each descriptor in the first set and the closest descriptor in the second set by trying each one (slow search).

`cv::FlannBasedMatcher` uses the fast search algorithm based on approximation of nearest neighbor and finding correspondences.

The result of descriptor matching is a list of correspondences between two sets of descriptors. The first set of descriptors is called the train set because it corresponds to our target pattern image, and the second set, the query set, belongs to the image where we search for the pattern. The more patterns to image correspondences exist, the greater the chances are that the target pattern is present on the image. In order to increase the matching speed, it is necessary to train the matching earlier by calling the match function. This training phase is used to optimize the performance of cv::FlannBasedMatcher. For these needs, train classes will make index trees for input train descriptors. This will increase the matching speed for large data sets (e.g. finding matches on hundreds of images). In the case of cv:: BFMatcher, this train class will do nothing

```
void ARPatternDetector::train(const ARPattern& pattern)
{
    // Store the pattern object
    m_pattern = pattern;
    // Clear old train data:
    m_matcher->clear();
    std::vector<cv::Mat> descriptors(1);
    descriptors[0] = pattern.descriptors.clone();
    m_matcher->add(descriptors);
    // After adding train data perform actual train:
    m_matcher->train();
}
```

To match query descriptors, in OpenCV, `cv::DescriptorMatcher:` can use one of the methods:

```
void match(const Mat& queryDescriptors,
    vector<DMatch>& matches, const vector<Mat>& masks = vector<Mat>());
//this method will finds a simple list of the best matches

void knnMatch(const Mat& queryDescriptors,
    vector<vector<DMatch> >& matches, int k,
```

```
    const vector<Mat>& masks = vector<Mat>(), bool compactResult = false);
//this method will find K nearest matches for each descriptor

void radiusMatch(const Mat& queryDescriptors,
    vector<vector<DMatch> >& matches, maxDistance,
    const vector<Mat>& masks = vector<Mat>(), bool compactResult = false);
/*this method will find correspondences whose distances are not larger
than the defined distance*/
```

Mismatch during the matching phase is normal. Two types of matching errors can be identified:

- False-positive matches – correspondence of characteristic points is wrong.
- False-negative matches – no point matches even though characteristic points are observed in both images.

We cannot deal with false-negative matches because the matching algorithm did not recognize them for some reason. The problem is false-positive matches and the main goal is to minimize their number. To reject those wrong correspondences, a cross-match technique will be used. The main idea is that target train descriptors match a set of queries and opposite that set queries to match with target train, only the common matches for these two types of matches will be used. This technique usually gives the best results with a minimum number of deviations if there are enough matches. Cross-match filter is available in the cv::BFMatcher class. To enable them simply when creating cv::BFMatcher set second argument to true.

```
//Example
  cv::Ptr<cv::DescriptorMatcher> matcher(new cv::BFMatcher(cv::
NORM_HAMMING, true));
```

In Figure 6.13 is presented the result of matching by using cross-checks filter.

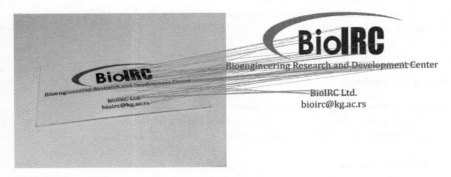

Figure 6.13 Matching results by using cross-checks filter.

Also, a very well-known technique for eliminating deviations is the so-called ratio test. First, we use KNN-matching with $K = 2$. In this way, the two closest descriptors are returned for each match. However, only if the distance ratio between the first and second match is large enough, the match will return (usually the ratio threshold is near two).

```
void ARPatternDetector::getMatches(const cv::Mat& queryDescriptors, std::
vector<cv::DMatch>& matches)
{
    matches.clear();

    if (enableRatioTest)
    {
        // To avoid NaN's when best match has zero distance we will use
            inversed ratio.
        const float minRatio = 1.f / 1.5f;

        // KNN match will return 2 nearest matches for each query descriptor
        m_matcher->knnMatch(queryDescriptors, m_knnMatches, 2);

        for (size_t i=0; i<m_knnMatches.size(); i++)
        {
            const cv::DMatch& bestMatch   = m_knnMatches[i][0];
            const cv::DMatch& betterMatch = m_knnMatches[i][1];

            float distanceRatio = bestMatch.distance / betterMatch.
            distance;

            // Pass only matches where distance ratio between
            // nearest matches is greater than 1.5 (distinct criteria)
            if (distanceRatio < minRatio)
            {
                matches.push_back(bestMatch);
            }
        }
    }
    else
    {
        // Perform regular match
        m_matcher->match(queryDescriptors, matches);
    }
}
```

Figure 6.14 Visualization of matches that were refined using homography matrix estimation using a RANSAC algorithm.

Usually, the ratio test removes all observed deviations. However, it happens that in some cases false-positive matches can be missed. For more matching, improvement can be performed by outlier filtration using the random sample consensus method (RANSAC). If we start from the assumption that the object to be recognized is a plane object–image and we do not expect that image will be deformed, it is possible to find a homographic transformation between the character points in the target image and the characteristic points in the query image. Homography transformations [14] will transfer points from a pattern to the new coordinate system of the query image. cv::findHomography function will find this transformation. This function uses RANSAC to find the best homography matrix by testing each subsets of input points. As a result, depending on the calculated reprojection error, this function will mark each correspondence as inlier or outlier (Figure 6.14).

```
/*Filter out geometrically incorrect matches,
with homography matrix estimation using a RANSAC algorithm */
bool ARPatternDetector::refineMatches_WithHomography
    (
        const std::vector<cv::KeyPoint>& queryKeypoints,
        const std::vector<cv::KeyPoint>& trainKeypoints,
        float reprojectionThreshold,
        std::vector<cv::DMatch>& matches,
        cv::Mat& homography
    )
{
    const int minNumMatchesAllowed = 8;

    if (matches.size() < minNumMatchesAllowed)
```

```
    return false;

// Prepare data for cv::findHomography
std::vector<cv::Point2f> srcPoints(matches.size());
std::vector<cv::Point2f> dstPoints(matches.size());

for (size_t i = 0; i < matches.size(); i++)
{
    srcPoints[i] = trainKeypoints[matches[i].trainIdx].pt;
    dstPoints[i] = queryKeypoints[matches[i].queryIdx].pt;
}

// Find homography matrix and get inliers mask
std::vector<unsigned char> inliersMask(srcPoints.size());
homography = cv::findHomography(srcPoints,
                                dstPoints,
                                cv::FM_RANSAC,
                                reprojectionThreshold,
                                inliersMask);

std::vector<cv::DMatch> inliers;
for (size_t i=0; i<inliersMask.size(); i++)
{
    if (inliersMask[i])
        inliers.push_back(matches[i]);
}

matches.swap(inliers);
return matches.size() > minNumMatchesAllowed;
}
```

This step of the homography searching is important because the obtained transformation is the key to finding the exact location of the pattern in the query image.

After performing homography transformations, we already have all the necessary data to find key points locations in 3D. However, by finding more accurate pattern corners, it is possible to more improve its position. We warp the input image by using estimated homography to obtain a pattern that has been found on the query image. As a result, query image should look very close to the target train image. So, we can see that homography refinement is useful to find even more accurate homography transformations. In this way, we obtain another homography and also another set of inlier features. The product of the first (H1) and second (H2) homography will be precise homography (Figure 6.15).

Figure 6.15 Visualization of matches that were refined using refinement homography matrix.

```cpp
bool ARPatternDetector::findPattern(const cv::Mat& image,
ARPatternTrackingInfo& info)
{
    // Convert input image to gray
    getGray(image, m_grayImg);

    // Extract feature points from input gray image
    extractAllFeatures(m_grayImg, m_queryKeypoints, m_queryDescriptors);

    // Get matches with current pattern
    getMatches(m_queryDescriptors, m_matches);

    // Find homography transformation and detect good matches
    bool homographyFound = refineMatches_WithHomography(
        m_queryKeypoints,
        m_pattern.keypoints,
        homographyReprojectionThreshold,
        m_matches,
        m_roughHomography);

    if (homographyFound)
    {
        // If homography refinement enabled improve found transformation
        if (enableHomographyRefinement)
        {
            // Warp image using found homography
            cv::warpPerspective(m_grayImg, m_warpedImg, m_roughHomography,
                m_pattern.size, cv::WARP_INVERSE_MAP | cv::INTER_CUBIC);
            // Get refined matches:
            std::vector<cv::KeyPoint> warpedKeypoints;
            std::vector<cv::DMatch> refinedMatches;
```

```
            // Detect features on warped image
            extractAllFeatures(m_warpedImg, warpedKeypoints,
            m_queryDescriptors);

            // Match with pattern
            getMatches(m_queryDescriptors, refinedMatches);

            // Estimate new refinement homography
        homographyFound = refineMatches_WithHomography(warpedKeypoints,
                                m_pattern.keypoints,
                                homographyReprojectionThreshold,
                                refinedMatches,
                                m_refinedHomography);
// Get a result homography as result of matrix product of refined and
rough homographies:
            info.homography = m_roughHomography * m_refinedHomography;

            // Transform contour with precise homography
            cv::perspectiveTransform(m_pattern.points2d,
                        info.points2d, info.homography);
        }
        else
        {
            info.homography = m_roughHomography;

            // Transform contour with rough homography
            cv::perspectiveTransform(m_pattern.points2d,
                        info.points2d, m_roughHomography);
        }
    }
    return homographyFound;
}
```

So, when we finish all the outlier removal stages, if we still have the number of matches reasonably large (more than 25% of all features from the target image still have correspondences with the pattern from the query image), then we can be sure that the pattern image is located correctly. In this case, the algorithm will proceed to the next step – estimation of the 3D position of the pattern pose but in this case with regards to the camera positions.

For pose estimation usual, we need 2D–3D correspondences to estimate the camera-extrinsic parameters. Therefore, we assign four (4) 3D points to coordinate with the corners of the unit rectangle. This rectangle lies in the *XY* plane (the *Z*-axis is up), and all 2D points correspond to the corners of the image bitmap.

```cpp
void ARPatternDetector::buildPatternFromImage(const cv::Mat& image,
                                              ARPattern& pattern) const
{
    int numImages = 4;
    float step = sqrtf(2.0f);

    // Store original image in pattern structure
    pattern.size = cv::Size(image.cols, image.rows);
    pattern.frame = image.clone();
    getGray(image, pattern.grayImg);

    // Build 2d and 3d contours (3d contour lie in XY plane since it's
       planar)
    pattern.points2d.resize(4);
    pattern.points3d.resize(4);

    // Image dimensions
    const float w = image.cols;
    const float h = image.rows;

    // Normalized dimensions:
    const float maxSize = std::max(w,h);
    const float unitW = w / maxSize;
    const float unitH = h / maxSize;

    pattern.points2d[0] = cv::Point2f(0,0);
    pattern.points2d[1] = cv::Point2f(w,0);
    pattern.points2d[2] = cv::Point2f(w,h);
    pattern.points2d[3] = cv::Point2f(0,h);

    pattern.points3d[0] = cv::Point3f(-unitW, -unitH, 0);
    pattern.points3d[1] = cv::Point3f( unitW, -unitH, 0);
    pattern.points3d[2] = cv::Point3f( unitW,  unitH, 0);
    pattern.points3d[3] = cv::Point3f(-unitW,  unitH, 0);

    extractAllFeatures(pattern.grayImg, pattern.keypoints, pattern.
    descriptors);
}
```

Based on this configuration of corners, the pattern coordinate system will be placed in the center of the pattern location. It will be lying in the *XY* plane, and the *Z*-axis will have the direction of the camera.

6.4.2   Calculating the Camera-intrinsic Matrix

Calculation of camera-intrinsic parameters can be done by using a sample C++ program from the OpenCV distribution package. We need to find calibration. cpp file in OpenCV installation directory on next location: ..\opencv4.x.x\samples\cpp\. Then, create a project and build app cailbration.exe.

This application will find the internal lens parameters of our camera such as focal length, principal point, and distortion coefficients by using a series of pattern images. A sample of pattern images is presented in Figure 6.16. For successful calibration, more than 15 images from various angles and distances are necessary, as presented in Figure 6.16.

Run calibration exe and input next command-line to perform calibration from images:

```
imagelist_creator imagelist.yaml *.png
calibration -w 9 -h 6 -o camera_intrinsic.yaml imagelist.yaml
```

Figure 6.16   Examples of calibration images.

The first command-line creates an image list of YAML format. The calibration tool expects this list as input from all PNG files in the current directory. Use the exact file names of images, such as img1.png, img2.png, etc. The second command-line passes the file imagelist.yaml to the calibration application. This application can also take images from a regular web camera and perform calibration.

At the end of calibration, the following result will be written in a YAML file:

```
%YAML:1.0
---
calibration_time: "MM DD hh:mm:ss YYYY"
nframes: 16
image_width: 1500
image_height: 1125
board_width: 6
board_height: 9
square_size: 2.5000000372529030e-02
flags: 0
camera_matrix: !!opencv-matrix
   rows: 3
   cols: 3
   dt: d
   data: [ 1.1326352715964249e+03, 0., 7.3745365032476752e+02, 0.,
       1.1264825725175474e+03, 5.6887985674826996e+02, 0., 0., 1. ]
distortion_coefficients: !!opencv-matrix
   rows: 5
   cols: 1
   dt: d
   data: [ 3.2156205271785027e-01, -1.2115509886273756e+00,
       -2.4991189393450369e-03, -7.2770269538228608e-03, 0. ]
avg_reprojection_error: 5.2319256825230775e-01
```

Camera calibration matrix is 3 × 3:

$$
\begin{bmatrix}
f_x & 0 & C_x \\
0 & f_y & C_y \\
0 & 0 & 1
\end{bmatrix}
$$

So, components of interest are: f_x, f_y, C_x, C_y. With these data, instance of the camera-calibration object can be created by using the following code for calibration:

```
CameraCalibration calibration(1132.6352715964249f, 1126.4825725175474f,
737.45365032476752f, 568.87985674826996f )
```

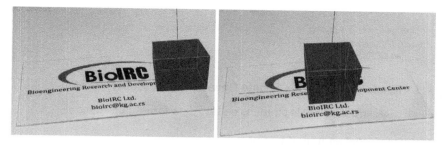

Figure 6.17 AR result with wrong calibration parameters (left); with good calibration parameters (right).

Without correct camera calibration, it is almost impossible to create a natural-looking AR. Depending on the camera lens, the estimated perspective transformation will differ. Wrong calibration parameter will cause the augmented objects to look like they are too close or too far (Figure 6.17).

The pattern location in 3D space will be estimated by using the computePose function:

```
void ARPatternTrackingInfo::computePose(const ARPattern& pattern, const
CameraCalibration& calibration)
{
    cv::Mat RotVec;
    cv::Mat_<float> TransVec;
    cv::Mat raux,taux;

    cv::solvePnP(pattern.points3d, points2d, calibration.getIntrinsic(),
calibration.getDistorsion(),raux,taux);
    raux.convertTo(RotVec,CV_32F);
    taux.convertTo(TransVec ,CV_32F);

    cv::Mat_<float> rotMat(3,3);
    cv::Rodrigues(RotVec, rotMat);

    // Copy to transformation matrix
    for (int col=0; col<3; col++)
    {
        for (int row=0; row<3; row++)
        {
            pose3d.r().mat[row][col] = rotMat(row,col); // Copy rotation
component
        }
        pose3d.t().data[col] = TransVec(col); // Copy translation component
    }
```

```
        /*Since solvePnP finds camera location, w.r.t to marker pose,
        to get marker pose w    .r.t to the camera we invert it.*/
    pose3d = pose3d.getInverted();

}
```

6.4.3 Rendering AR with a Simple OpenGL Object (Cube)

`DrawingAR` structure holds all the necessary data for visualization:

- The image is taken from the camera.
- The camera-calibration (3×3 matrix).
- (if present) the pattern poses in 3D.
- All the internal data related to OpenGL.

```
class DrawingAR
{
public:
    DrawingAR(std::string windowName, cv::Size frameSize, const
CameraCalibration& c);
    ~DrawingAR();

    bool                isPatternPresent;
    Transformation      patternPose;

    // Set the new frame for the background
    void updateBackground(const cv::Mat& frame);
    void updateWindow();

private:
    friend void DrawingAR_DrawCallback(void* param);
    //  Render entire scene in the OpenGl window
    void draw();

    //Draws the background with video
    void drawCameraFrame();
    // Draws the AR
    void drawAugmentedScene();
    // Builds the right projection matrix from the camera calibration for AR
    void buildProjectionMatrix(const CameraCalibration& calibration, int w,
                        int h, Matrix44& result);
    // Draws the coordinate axis
    void drawCoordinateAxis();
    // Draw the cube model
    void drawCubeModel();
}
```

Initialization of the OpenGL window in the constructor of the DrawingAR class:

```
DrawingAR::DrawingAR(std::string windowName, cv::Size frameSize, const
CameraCalibration& c)
    : m_isTextureInitialized(false)
    , m_calibration(c)
    , m_windowName(windowName)
{
    // Create window with OpenGL support
    cv::namedWindow(windowName, cv::WINDOW_OPENGL);
    // Resize it exactly to video size
    cv::resizeWindow(windowName, frameSize.width, frameSize.height);

    // Initialize OpenGL draw callback:
    cv::setOpenGlContext(windowName);
    cv::setOpenGlDrawCallback(windowName, DrawingAR_DrawCallback, this);
}
```

Considering that we have a separate class for storing the visualization state, cv:: setOpenGlDrawCallback will be modified to call and pass an instance of DrawingAR as the parameter. The modified callback function is:

```
void DrawingAR_DrawCallback(void* param)
{
    DrawingAR * ctx = static_cast<DrawingAR*>(param);
    if (ctx)
    {
        ctx->draw();
    }
}
```

DrawingAR calls all the processes for rendering the AR. Each frame rendering starts withdrawing a background with an orthographic projection. Then function renders a 3D model with the correct perspective projection and model transformation.

```
void DrawingAR::draw()
{
  glClear(GL_DEPTH_BUFFER_BIT | GL_COLOR_BUFFER_BIT); // Clear entire
screen:
  drawCameraFrame();                                  // Render background
  drawAugmentedScene();                               // Draw AR
  glFlush();
}
```

To draw an AR, it is necessary to set the correct perspective in OpenGL projection that matches camera calibration.

```
void DrawingAR::drawAugmentedScene()
{
  // Init augmentation projection
  Matrix44 projectionMatrix;
  int w = m_backgroundImage.cols;
  int h = m_backgroundImage.rows;
  buildProjectionMatrix(m_calibration, w, h, projectionMatrix);

  glMatrixMode(GL_PROJECTION);
  glLoadMatrixf(projectionMatrix.data);

  glMatrixMode(GL_MODELVIEW);
  glLoadIdentity();

  if (isPatternPresent)
  {
    // Set the pattern transformation
    Matrix44 glMatrix = patternPose.getMat44();
    glLoadMatrixf(reinterpret_cast<const GLfloat*>(&glMatrix.data[0]));

    // Render model
    drawCoordinateAxis();
    drawCubeModel();
  }
}
```

ArExample03 is example project for demonstration that supports the processing of single image, recorded videos file, and live views from a web camera. For this purpose, two functions were created:

The function processVideo processing the video file and the function processSingleImage is processing a single image:

```
void processVideo(const cv::Mat& patternImage, CameraCalibration&
calibration, cv::VideoCapture& capture);

void processSingleImage(const cv::Mat& patternImage, CameraCalibration&
calibration, const cv::Mat& image);
```

The processFrame function performs full detection routine on camera frame and draws the scene using drawing context:

```cpp
bool processFrame(const cv::Mat& cameraFrame, PipelineAR& pipeline,
DrawingAR& drawingCtx)
{
    // Clone image used for background (we will draw overlay on it)
    cv::Mat img = cameraFrame.clone();

    // Draw information:
    if (pipeline.m_patternDetector.enableHomographyRefinement)
      cv::putText(img, "Pose refinement: On   ('h' to switch off )",
       cv::Point(10,15), cv::FONT_HERSHEY_PLAIN, 1, CV_RGB(0,200,0));
      else
        cv::putText(img, "Pose refinement: Off  ('h' to switch on)",
        cv::Point(10,15), cv::FONT_HERSHEY_PLAIN, 1, CV_RGB(0,200,0));

      cv::putText(img, "RANSAC threshold: " + ToString(pipeline.
m_patternDetector.homographyReprojectionThreshold)
      + "( Use'-'/'+' to adjust)", cv::Point(10, 30), cv::FONT_HERSHEY_PLAIN,
1, CV_RGB(0,200,0));

    // Set a new camera frame:
    drawingCtx.updateBackground(img);

    // Find a pattern and update it's detection status:
    drawingCtx.isPatternPresent = pipeline.processFrame(cameraFrame);

    // Update a pattern pose:
    drawingCtx.patternPose = pipeline.getPatternLocation();

    // Request redraw of the window:
    drawingCtx.updateWindow();

    // Read the keyboard input:
    int keyCode = cv::waitKey(5);

    bool bQuit = false;
    if (keyCode == '+' || keyCode == '=')
    {
      pipeline.m_patternDetector.homographyReprojectionThreshold += 0.2f;
     pipeline.m_patternDetector.homographyReprojectionThreshold =
      (std::min)(10.0f, pipeline.m_patternDetector.
homographyReprojectionThreshold);
    }
    else if (keyCode == '-')
    {
```

```
            pipeline.m_patternDetector.homographyReprojectionThreshold -= 0.2f;
             pipeline.m_patternDetector.homographyReprojectionThreshold =
            (std::max)(0.0f, pipeline.m_patternDetector.
homographyReprojectionThreshold);
        }
        else if (keyCode == 'h')
        {
            pipeline.m_patternDetector.enableHomographyRefinement =!
         pipeline.m_patternDetector.enableHomographyRefinement;
        }
        else if (keyCode == 27 || keyCode == 'q')
        {
            bQuit = true;
        }

        return bQuit;
}
```

The initialization of `PipelineAR` and `DrawingAR` is done in the function `processSingleImage` or `processVideo`:

```
void processSingleImage(const cv::Mat& patternImage, CameraCalibration&
calibration, const cv::Mat& image)
{
    cv::Size frameSize(image.cols, image.rows);
    PipelineAR pipeline(patternImage, calibration);
    DrawingAR drawingCtx("Markerless AR", frameSize, calibration);

    bool bQuit = false;
    do
    {
        bQuit = processFrame(image, pipeline, drawingCtx);
    } while (!bQuit);
}
```

Example application **ArExample03** can be run in the following ways:

- Run on a single image:

```
    ArExample03.exe pattern.png test_image.png
```

- Run on a recorded video file:

```
    ArExample03.exe pattern.png test_video.avi
```

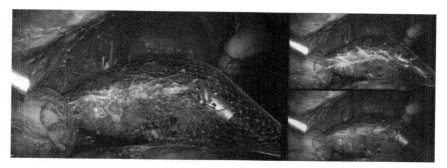

Figure 6.18 Accurate tracking of liver tumors for augmented reality in robotic assisted surgery. *Source*: [15] Haouchine et al., 2014/with permission of IEEE.

- Run using live video from a web camera:

```
ArExample03.exe pattern.png
```

An example of application marker-less AR in biomedical engineering is shown in Figure 6.18.

6.5 Conclusion

In this chapter, the basic framework of biomedical engineering based on AR was presented. The real AR model will be more complex and even more powerful.

In any case, AR technology is a powerful tool for the development of biomedical engineering in the future. Its application will help people to observe, train, and act in a new perspective and method in the field of clinical application in medicine or biomedicine and will give more intuitive results than traditional methods. From the current development of technology, we can conclude that AR technology belongs to the future, but it is also part of the present.

References

1 Filipovic, N., Rosic, M., Tanaskovic, I., Milosevic, Z., Nikolic, D., Zdravkovic, N., Peulic, A., Kojic, M., Fotiadis, D., Parodi, O. (2012). ARTreat project: three-dimensional numerical simulation of plaque formation and development in the arteries. *IEEE Transactions on Information Technology in Biomedicine: A Publication of the IEEE Engineering in Medicine and Biology Society*, 16 (2): 272–278,

2 Parodi, O., Exarchos, T., Marraccini, P., Vozzi, F., Milosevic, Z., Nikolic, D., Sakellarios, A., Siogkas, P., Fotiadis, D., Filipovic, N. (2012). Patient-specific prediction of coronary plaque growth from CTA angiography: a multiscale model for plaque formation and progression. *IEEE Transactions on Information Technology in Biomedicine: A Publication of the IEEE Engineering in Medicine and Biology Society*, 16(5): 952–65

3 Filipovic, N., Nikolic, D., Saveljic, I., Milosevic, Z., Exarchos, T., Pelosi, G., and Parodi, O. (2013). Computer simulation of three dimensional plaque formation and progression in the coronary artery, *Computers and Fluids*, 88: 826–833,

4 Nikolić, D., Radović, M., Aleksandrić, S., Tomasević, M., Filipović, N. (2014). Prediction of coronary plaque location on arteries having myocardial bridge, using finite element models. *Computer Methods and Programs in Biomedicine*, 117 (2):137–144, doi: 10.1016/j.cmpb.2014.07.012

5 Azuma, R. T. (1997). A survey of augmented reality. In *Presence: Teleperators and Virtual Environments*, 6(4), 355–385.

6 Kojic, M., Filipovic, N., Stojanovic, B., Kojic, N. (2008). *Computer Modeling in Bioengineering – Theoretical Background, Examples and Software*. Wiley, ISBN: 978-0-470-06035-3,

7 Quarles J., Fishwick P., Fischler I., Lampotang S., Fischler I., Lok B. (2010). A mixed reality approach for interactively blending dynamic models with corresponding physical phenomena. *ACM Transactions on Modeling and Computer Simulation* 20: 1–23. https://doi.org/10.1145/1842722.1842727

8 Sutherland, I. E. (1968). A head-mounted three dimensional display. *Proceedings of AFIPS* 68: 757–764

9 Tian, Y. (2010). *Research on Occlusion Handling Between Virtual and Real Worlds in Augmented Reality Systems*. Wuhan: Huazhong University of Science and Technology.

10 Zhang, J., Luo S. (2009). Research on medical augmented reality modeling method. *Journal of System Simulation* 21(12): 3658–3661.

11 Botden, S. M. B. I., de Hingh, I. H. J. T., Jakimowicz, J, J. (2009). Suturing training in augmented reality: gaining proficiency in suturing skills faster. *Surgical Endoscopy and Other Interventional Techniques*, 23: 2131–2137.

12 Oostema, J. A., Abdel, M. P., Gould, J. C. (2008). Time-efficient laparoscopic skills assessment using an augmented-reality simulator. *Surgical Endoscopy and Other Interventional Techniques* 22:2621–2624.

13 Garrido-Jurado, S., Muñoz-Salinas, R., Madrid-Cuevas, F. J., and Marín-Jiménez, M. J. (2014). Automatic generation and detection of highly reliable fiducial markers under occlusion. *Pattern Recognition* 47(6): 2280–2292. 10.1016/j.patcog.2014.01.005

14 Monnin, D., Bieber, E., Schmitt, G., Schneider, A. (2010). An effective rigidity constraint for improving RANSAC in homography estimation. In: Blanc-Talon J.,

Bone D., Philips W., Popescu D., Scheunders P. (eds) *Advanced Concepts for Intelligent Vision Systems. ACIVS 2010.* Lecture Notes in Computer Science, vol 6475. Springer Berlin, Heidelberg. https://doi.org/10.1007/978-3-642-17691-3_19

15 Haouchine N., Dequidt J., Peterlik I., et al. (2014). Towards an accurate tracking of liver tumors for augmented reality in robotic assisted surgery. *International Conference on Robotics and Automation (ICRA),* Hong Kong, China (June 2014).

7

Augmented Reality Balance Physiotherapy in HOLOBALANCE Project

Nenad D. Filipovic[1,2] and Zarko Milosevic[3]

[1] *Bioengineering Research and Development Center, BIOIRC Kragujevac, Serbia*
[2] *Faculty of Engineering, University of Kragujevac Kragujevac, Serbia*
[3] *Institute for Information Technology, University of Kragujevac, Kragujevac, Serbia*

7.1   Introduction

Human balance is multifactorial and relies on the complex integration of visual, somatosensory, vestibular information, and musculoskeletal function. Balance disorders due to age-related progressive loss of functioning of sensory information and the inability to control body movements are considered a global epidemic according to the World Health Organization [1]. Balance and gait physiotherapy has been shown to improve postural stability, increase balance confidence, and quality of life according to NICE guidelines [2]. Currently, there is a total lack of personalized coaching solutions for people with balance disorders to engage in balance and gait physiotherapy and increase physical activity (PA).

The overall objective of HOLOBALANCE was to develop and validate a new personalized hologram coach platform for virtual coaching, motivation, and empowerment of the aging population with balance disorders.

The coaching part was realized through new forms of accessible user interaction, (holograms, augmented reality games, and vocal instructions), along with easy-to-use wearable (smart bracelet, smart glasses, and sensorized soles) and ambient sensors (motion capture) that can be customized to implement and coach the user with specific, individualized exercises.

HOLOBALANCE engaged the experts related to the management of people suffering from balance disorders, including physiotherapists, Ear Nose Throat experts (ENTs), neurologists, psychologists, and gerontologists, through an expert panel with visual analytics capabilities, toward developing a multi-stakeholder user-centered coaching ecosystem.

Careers were also involved through easy-to-use apps. Suggestions for specific exercises and tasks and an activity plan were provided by the experts on a daily basis, which were then refined and updated through progressive learning algorithms. Three different types of coaching are provided by the HOLOBALANCE platform:

- short-term balance physiotherapy (BP),
- short-term cognitive training (CT) combined with auditory tasks, and
- long-term multilevel motivation and PA promotion.

The platform consists of a cloud-based big data analytics component, with intelligent data mining algorithms that assess personalized heterogeneous data (motion capture sensor data, wearable sensor data, physiotherapy exercises movement data, and clinical profiles), emotional and behavioral data (through emotional computing and behavioral modeling minimizing attention theft), and provide personalized advice, guidance and follow up for motivation, goal setting, BP, CT, and PA.

The correct performance of the exercises is measured objectively in real time, through the wearable and ambient sensors. Validation of the Virtual Coach (VC) platform was performed through a proof-of-concept study in four clinical sites, in 80 participants.

The overall aim was to deliver a radically new cost-effective VC, consisting of:

- a hologram-based surrogate balance physiotherapist,
- an augmented reality cognitive game, combined with auditory exercises, and
- a PA planner.

The VC will be daily accessible by the user to improve balance, cognition, and PA, by promoting exercise compliance, motivation, functional capacity, self-management, and self-monitoring. The long-term aim is to promote behavioral changes that maintain the users' independence and engagement in physical and social activities.

Multilevel motivation strategies, real time feedback, online connection and competition between users and a continuous communication channel with the healthcare professional, along with a visual analytics expert panel will be integrated to improve the platform's outcome.

HOLOBALANCE key points

- User-centric design (UCD) using Human Computer Interaction methodology.
- Holograms acting as virtual balance physiotherapists.
- New augmented reality games for CT.
- Smart glasses with audio for vocal instructions and cognitive/auditory training.

- Capitalization on FP7 EMBalance project data and knowledge and FI-WARE generic enablers.
- Multilevel motivation through emotional computing, behavioral modeling, and intelligent data analytics.
- Integration into a radically new VC for aging population with balance disorders.
- Exploitation and commercialization potential, as a service and as a product and capitalization based on the IoT framework and business model of C3PO: Continuous Care & Coaching Platform.

7.2 Motivation

Understanding the demographic changes that are likely to unfold over the coming years, as well as the challenges and opportunities that they present in terms of people's health, wellbeing, and quality of life, is the key to the design and implementation of new Information and Communications Technology (ICT) tools for coaching people, as they age. It is of utmost importance that ICT solutions are developed and adopted, so that older people live as independently as possible with dignity, pleasure, and security.

The facts are undeniable: One in three people over the age of 65 fall annually, with the majority of these being caused by balance disorders. Balance disorders in older people have wide-ranging physical and psychological consequences and increase the likelihood of frailty, cognitive decline, sedentary behavior, social exclusion, falls, and injury-related death [3–7].

The National Institute of Clinical Excellence (NICE) UK Guidelines (2013) [8] recommend an early detailed individualized assessment and treatment intervention for older adults with balance disorders at risk of falls, but despite the strength of available evidence, compliance and implementation to date have been poor or nonexistent [9].

Unsurprisingly, falls are a major burden on health and social resources costing the UK National Health Service more than £2 billion per year and $67.7 billion per year in the United States alone by 2020. The cost of falls is estimated at 281 EUR per inhabitant [10], which would mean an estimated 25 billion EUR direct medical cost every year for the entire EU-region [11]. Cost-effective measures are urgently needed to address this highly unmet societal need. Sustaining levels of activity is important for an older person's well-being and can extend years of active independent life for older persons [12], especially for a large percentage of this population, whose impaired mobility leads to various degrees of disability. The effects of such impairment include decreased PA, social life restriction, dependence on others, other acute injuries, as well as feeling of depression and social isolation.

Balance Disorders Key Points

- 80% of older people suffer from vestibular dysfunction-related balance disorders, leading to falls. One out of three people above 65 falls annually.
- The cost of falls in the EU is estimated at 25 billion euros.
- Individualized balance disorders diagnosis and treatment can prevent falls.
- The prevalence of aging population at risk overwhelms the availability of experts to provide this treatment.
- Virtual BP and coaching could address this challenge.

Physical inactivity has been identified as a global epidemic by the WHO, with the greatest levels of inactivity noted in older adults [13, 14]. Customized BP intervention is the gold standard of care for persons with postural deficits who are at risk of falling or have experienced a fall. They are asked to perform individualized exercises daily in a safe environment.

Additionally, multisensory BP can improve static and dynamic balance and gait, reduce symptoms of imbalance, comorbid depression, anxiety, increase PA [15], and ultimately result in an increase of self-confidence and quality of life [16]. It is considered a safe and effective intervention. The joint American and British Geriatric Society guidelines [17] recommend that a suitable balance exercise program is a crucial component for balance rehabilitation.

Currently, multifactorial balance exercise programs incorporate multicomponent strength, balance, aerobic, and gait exercises. However, relevant intervention strategies vary widely and within current healthcare systems, the ability to provide optimal care for optimal results is often challenging.

Best care requires a structured, progressive exercise regime, with minimum exercise dose [18]. However, less than 50% of community-based physiotherapists link their interventions to fall risk factors or refer patients to other healthcare professionals for further management. Furthermore, approximately 50% of those who attended physiotherapy received only a single set of balance exercises and only a subset of them were provided with a home exercise program to maintain or continue progress [19].

However, take up and adherence to these interventions is often poor, with possible reasons for this, including problem denial, disbelief that balance can be improved, or a lack of perceived relevance by older people [20]. The challenges in providing effective BP coaching include lack of compliance, difficulty in proper exercise performance, and limited access to specialized physiotherapists and balance clinics. For this reason, it is important to come up with an alternative approach that can be a surrogate physiotherapist with daily presence in a user's home and provide ongoing support remotely, monitor and assess the activities and the correct execution of balance exercises as well as their progression, and can motivate and empower the user.

This can be achieved by the incorporation of holograms, i.e. visual artifacts super-imposed onto the real world through augmented reality with light diffraction technology. Moreover, with a set of wearable (i.e. sensorized soles, smart glasses, and smart bracelet) and ambient sensors (motion capture sensor), an objective assessment of exercise and task performance can be performed remotely, when the user is at home. This allows for real-time feedback, which may motivate and enhance self-management of users and promote long-term behavior change [21]. For the physiotherapist and other healthcare professionals, the use of this sensor technology allows for ongoing daily activity coaching regarding quality of exercise performance, PA levels, activities of daily living performance and compliance with goals in order to provide feedback, and update of instructions and exercise progression on a real time basis, which may further motivate and enhance self-management of users [22]. Motivation can be further enhanced through augmented reality gamification. Older adults with balance disorders, predominately report difficulty walking, a constant anxiety and fear of falling resulting in limitation of and decreased participation in physical activities and increased social isolation, to the point of social withdrawal [23]. Psychological motivational elements including emotion identification, digital badges, and virtual communities will be introduced to improve exercise compliance, increase PA and social participation.

BP Key Points

- Consists of personalized sets of exercises, defined by a healthcare professional and can be performed daily in a home environment.
- There is evidence from Cochrane reviews, NICE guidelines, and the American Physical Therapy Association that balance and gait physiotherapy is the only effective treatment for balance disorders.
- Although exercises are brief and easy to perform, there is up to 50% loss in follow-up rate in older adults.
- Supervision significantly increases compliance and effectiveness.

The HOLOBALANCE project is based on the following seven objectives:

1) Perform comprehensive analysis of the evidence basis for concurrent balance, CT and PA promotion interventions, and formulate a library of guidelines, modeled in an interpretable form to enable the delivery of HOLOBALANCE services.

2) Design and develop new forms of accessible user interaction, based on holograms and augmented reality games (holograms for BP, augmented reality games for CT, and auditory tasks for combined cognitive and auditory training), with customized avatars and environments, using human–computer interaction techniques, able to simulate the exercises of the library.

3) Develop secure cloud services, harnessing the existing IoT environment of C3PO and generic enablers from the FI-WARE catalogue and a sensor network to monitor data from easy-to-use wearable and motion sensors, enabling the inference of emotional and behavioral data and recording parameters of the execution of exercises, the daily PA plan and user satisfaction through the accessible user interaction environment, minimizing the attention theft. Additionally, to develop an expert panel based on visual analytics and the relevant communication mechanisms for careers and healthcare professionals.

4) To analyze retrospective BP, CT, and PA data of the pilot partners (UCL, UOA, KCL, and UKLFR) to develop and train autonomous learning algorithms for personalizing the virtual coaching and providing appropriate personalized exercises according to the user's profile, symptoms, emotional and behavioral data, satisfaction, and preferences.

5) To formulate the multilevel motivational concept of HOLOBALANCE platform, by integrating the customized holograms (selection of avatar and surrounding environment), augmented reality cognitive games, emotional and behavioral modeling with other forms of psychological motivation, including digital badges and virtual communities' social networks in order to promote exercise compliance, increased PA levels and self-management.

6) To deliver the virtual coaching platform, through a unified ICT environment for personalized advice, guidance, and follow-up of people with balance disorders, and to conduct a proof-of-concept study at the four pilot sites (UCL, UOA, KCL, and UKLFR) for evaluating the specific outcomes (i.e. Objective tests: MiniBEST, Functional Gait Assessment [FGA], cognitive function domains; subjective questionnaires: Rapid Assessment of Physical Activity Scale [RAPA], Falls Efficacy Scale International [FES-I], Euroqol [EQ-5D], and Activities Specific Balance Confidence Scale [ABC], as well as assessing the overall platform usability, user interaction experience and impact). The overall duration of the proof-of-concept study will be 12 months, recruiting in total 80 participants into the intervention group and 80 into the control group.

7) To utilize the outcomes of objective 6, along with currently available BP, CT, and PA guidelines and costs in different healthcare systems and perform a socioeconomic analysis of the HOLOBALANCE platform. Additionally, to deliver an exploitation/commercialization and sustainability plan, taking into account the socioeconomic analysis, legal, ethical, and regulatory requirements, data ownership, the data protection/privacy strategy, potential liabilities, and consumer protection. The user preferences and experience with the HOLOBALANCE platform have been explored, to gain insights on the perceived acceptance, satisfaction, and expectations of both the target group as well as

the other stakeholders (e.g. physiotherapists, psychologists, and healthcare professionals) in the three different European geographic regions (Greece, the United Kingdom, and Germany).

7.3 Holograms-Based Balance Physiotherapy

The hologram-based balance physiotherapist tool contains of software and devices system that gives patients opportunity to receive personalized exercise instructions as well as feedback through VC.

In order to provide a system with optimal usability, it was essential that the development of the VC was driven by a UCD approach. Using this iterative paradigm, the focus was on developing natural, intuitive, and usable techniques for the interaction and communication between patients and the VC. Such an iterative development circle includes the analyses of requirements, different design concepts, prototype implementation, and evaluation of the developed concepts. After each circle, new insights evolved and lead to re-prototyping of user interface concepts until the specified usability goals are achieved. The quantitative measures were defined together with the expert panelists and included specific requirements regarding task time, precision and accuracy of motions, level of user experience, etc.

The UCD approach involves the end users throughout the product development and testing process in order to ensure all the safety and user experience requirements in the final product. The approach is specified as part of ISO 9241 section 210 and focuses on enhancing effectiveness and efficiency, final user satisfaction, accessibility, and counteracts possible adverse effects of use on human health, safety, and performance [1].

7.4 Mock-ups

Three mock-ups were created for displaying a holographic VC. Two of our mock-ups are based on augmented reality head-mounted displays, specifically, Meta 2 and HoloLens. The advantage of such displays is that the user does not need to have any extra space for the device. The third solution is a projection-based holographic display and needs a dedicated space of about 9 m^3. The advantage of the projection based, however, is that the user does not have to wear any display device.

7.4.1 Meta 2

The first mock-ups of the balance physiotherapy hologram (Figures 7.1, 7.2) were developed by UHAM using Meta 2, Kinect 2, and a gaming laptop as the main Hardware components. Unity3D game engine and C# programming language were also used for the development of the software. Kinect Gesture builder was also used to capture the performance of the exercises once and storing them in

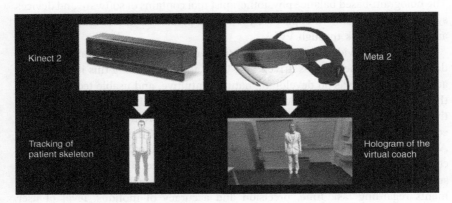

Figure 7.1 Balance physiotherapy mock-up motion capture and display devices.

Figure 7.2 Balance physiotherapy hologram using Meta 2 and Kinect 2 for standing–bending exercise.

Figure 7.3 First versions of avatars used with Meta 2. Realistic Female Virtual Coach, Realistic Male Virtual Coach, Cartoon Female Virtual Coach, and Cartoon Male Virtual Coach.

a database. This database was later used to compare the live performances of the exercises with the correct stored gesture in the database and to provide proper feedback in real time to the user.

Meta 2 was used as the visual as well as audio display by which users could see an augmented holographic VC in the physical environment (e.g. home, office). Moreover, Kinect 2 was used to capture the user's motions. Upon correct detection of the exercise performance, users received an auditory feedback (e.g. "Well done") from the VC. VC provided the instructions and demonstrated the correct performance of the exercises.

In addition, four versions of the VC (Figure 7.3) were designed and implemented using Adobe Fuse CC software.

7.4.2 HoloLens

First mock-up for HoloLens was developed by using Kinect 2. Unity3D game engine and C# programming language were also used for the development of the software. Cinema mock-up 2 Unity3D plugins was used to capture motions from the real person performing exercise without markers required. Recorded motions were applied to the humanoid avatar using Unity Mecanim animation system. Interaction with humanoid avatar, hand and voice gestures, was also implemented allowing the user to interact and initiate or terminate demonstration of the exercise by avatar either by tap or voice gesture (Figure 7.4).

After starting HoloLens, VC will be presented in front of the user. If VC is out of focus, direction indicator in the form of the arrow will guide the user toward the VC (Figure 7.5).

When the user is gazing out of the VC, white pointer will be shown. When gazing at the VC, ring pointer will be shown. Ring pointer is adapted to the shape of the avatar (Figure 7.6).

VC initially is standing still. In order to initiate exercise users have to gaze at the avatar, which is indicated via gazing pointer. Exercise can be initiated via two types of interactions: hand or voice gesture.

Hand gesture can be performed via air tap while gazing at the VC where the user by taping via one of the hands will initiate exercise. VC is performing the exercise until second air tap after which it returns in standing still position (Figure 7.7).

Figure 7.4 HoloLens Virtual Coach (VC).

(a) (b)

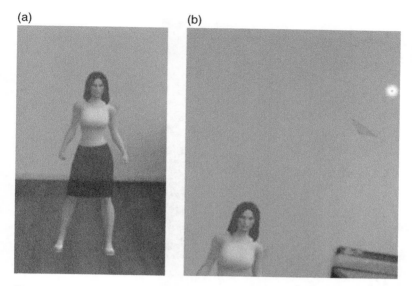

Figure 7.5 Virtual Coach initial standing position (a) and direction indicator (b).

(a) (b)

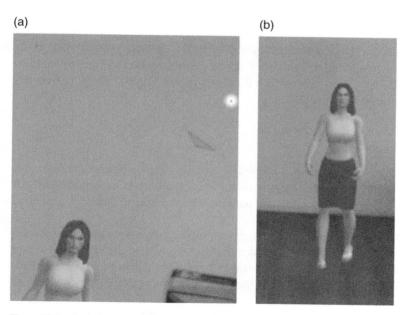

Figure 7.6 Gazing pointers. (a) White pointer when gazing out of VC and (b) ring pointer when gazing at the VC.

Figure 7.7 Hand gesture. Interacting with VC. VC demonstrates exercise after tap gesture.

Voice gesture can be performed by saying "Begin" or "Start" while gazing at the avatar, which will initiate exercise. Exercise demonstration can be terminated by saying "End" or "Stop" after which it returns in standing still position. Gestures can be combined (Figure 7.8).

7.4.3 Holobox

Hologram box was developed by BioIRC by using highly efficient holographic foil and high lumen projector to create best possible 3D experience without using any type of device on the patient side. Schematic setup is presented in Figure 7.9.

Unity3D game engine and C# programming language were also used for the development of the software. Cinema mock-up 2 Unity3D plugins was used to capture motions from the real person performing exercise without markers required. Recorded motions were applied to the humanoid avatar using Unity's Mecanim animation system.

Users have to stand in front of the holographic foil. When animation starts, VC will be presented in front of the user and perform exercise (Figure 7.10).

Figure 7.8   Voice gesture. Interacting with VC. VC demonstrates exercise after voice command.

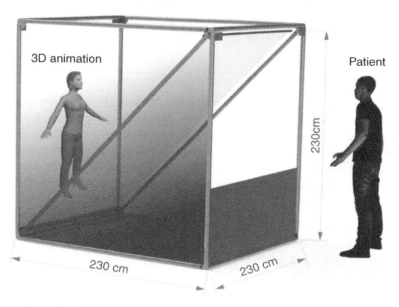

Figure 7.9   HoloBox schematic setup.

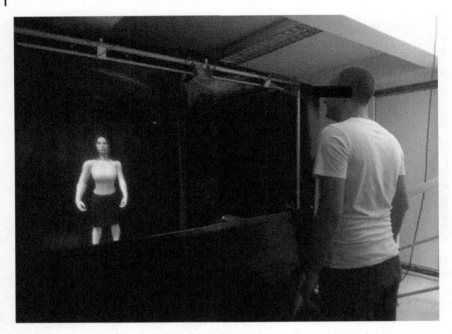

Figure 7.10 Test version – demonstrates exercise.

7.4.4 Modeling of BP in Unity 3D

Augmented reality hologram models were created combining three-dimensional biomechanical models of BP. The fast solvers for rigid body motion, body transform, and muscle setup were used in Unity3D. A linear elasticity for solid domain was used for body biomechanics and mesh motion. These solvers were implemented in WRL format and then converted to 3DS, which can be incorporated into unity3D. The following procedures were necessary for creating animation in unity 3D:

- Setting up the Scene
- Adding Objects
- Creating and Applying Material
- Applying Multiple Blocks
- Creating solution for Visual Studio

Displacement and strain distribution were presented for parametric patient-specific models. All the models are correlated with the same experimental protocols with body and head moving and eye tracking, where it is possible. Unity mixed reality plugins have been used in order to combine biomechanical models of simplified spring-damper-mass model with real environment for BP holograms.

With the help of this unity library, simulation is handled and controlled easily in the unity 3D environment. Generally, there are two engines related to the physics; 3D physics and 2D physics. Both engines are the same except that the 3D dimension of 3D physics is implemented through different modules. For instance, Rigid body 2D is used for the 2D physics and Rigid body module is used for the 3D physics engine.

Physics library of Unity 3D addresses the following four main components:

- Rigid bodies: the core component used to provide the physical behavior of the Unity 3D object.
- Colliders: the component that describes the shape of the Unity 3D object.
- Joints: used to attach two rigid body objects together or one rigid body object with a fixed point.
- Character controller: used to provide the character a capsule shaped, simple collider.

The Body Transform is the mass center of the character. It is used in Mecanim's retargeting engine and provides the most stable displacement model. Body Orientation is an average of lower and upper body orientation relative to the Avatar T-Pose. Body Transform and Orientation are stored in the Animation Clip (using the Muscle definitions set up in the Avatar). They are the only world-space curves stored in the Animation Clip. Everything else are stored relative to the body transform.

7.5 Final Version

Final version of the HOLOBALANCE platform differs from mock-ups described in the previous section in a manner that Meta 2 or Microsoft HoloLens has been replaced with head-mounted device and smartphone with capability of world tracking. Main reasons for that significant shift is:

- Meta 2 is not wireless and patients have to be connected with cable to additional PC computer while performing the exercise, which was bringing negative user experience.
- HoloLens on the other hand is totally wireless and standalone device without any dependency on additional hardware but the main drawback was that field of view is very narrow (only 35°) which makes it incapable for presenting whole holograms. Holograms have been cut out unless places far away from the user which makes it unusable.

Smartphone with head-mounted adapter overcomes both these drawbacks. It is totally wireless standalone solution providing field of view of 76°.

HOLOBALANCE modules are composed of the hardware and the software modules. Hardware modules contains:

- Balance Physiotherapist Hologram (BPH) hardware devices.
- Motion Capture and Wearable Sensors (MCWS). Those modules capture the motion of the patient in the space and provide the hologram of the VC.

The software modules of HOLOBALANCE are:

- BPH application
- Cognitive Training Game (CTG)
- Auditory Training Tool
- Cloud Backend Infrastructure (CBI)
- Emotional Computing module (ECM)
- Behavioral Modeling module (BM)
- Progressive Learning module (PL)
- Expert Panel (EP)
- Edge Computer Data Capture (ECDC) module
- Activity Planning Application (APA) and Wearable Devices for Motion Capture (WDMC)

Hardware devices and components of HOLOBALANCE solution

Head Mounted Adapter	Material: Mixed Reality Headset
	Development kit is required: NO
	Projection type: low-cost mirrors and lenses
	Gesture control compatibility: Bluetooth controller or gesture control
	Diagonal Field of View: 76° FOV
	EC declaration of conformity is required: NO
Smartphone	Connectivity: Bluetooth Low Energy, WiFi, 4G
	Supported devices
	iOS: iPhone 6S, iPhone 6S+, iPhone 7, iPhone 7 Plus, iPhone 8, iPhone 8 Plus, iPhone X
	Android: Samsung Galaxy S8, Google Pixel/Google Pixel XL
	Tango Phone: Asus Zenfone AR
	CE marking is required: YES
Edge Computer	Connectivity: Bluetooth Low Energy
	Desktop or low factor computer
	CE marking is required: YES
Motion Sensors	Connectivity: Bluetooth Low Energy
	BLE Beacon support is required: YES
	Development kit is required: YES
	Minimum number of devices synchronized and used simultaneously: 5
	Range: more than 15 m

Hardware devices and components of HOLOBALANCE solution

Minimum set of sensors: A 3-axis gyroscope, a 3-axis accelerometer, a 3-axis magnetometer, and one 16-bit ADC (Analog to Digital Converter) for each sensor
Accelerometer specs:

- Range: configurable, up to +-16 g
- Sensitivity: $\geq$ 1000 LSB/g at +-2 g
- Numeric resolution: 16 bit
- RMS Noise: <0.01 m/s^2

Gyroscope specs:

- Range: configurable, up to +- 2000 dps
- Sensitivity: 131 LSB/dps at +-250
- Numeric resolution: 16 bit
- RMS Noise: <0.1 dps

Magnetometer specs:

- Range: configurable, up to +-8,1 Ga
- Sensitivity: $\geq$ 1000 LSB/g at +-1.3 g
- Numeric resolution: 16 bit
- RMS Noise: <0.01 m/s^2

Interface for flash programming is required: YES
Minimum onboard 1-MB serial flash memory
Upgradable firmware
Weight: less than 100 g
EC declaration of conformity is required: YES

Depth Camera   Connectivity: USB 3.0
Depth Stream Output Resolution (active pixels): at least 1280×720
Depth Stream Output Frame Rate: up to 90 fps
Depth technology: Active Infrared light
Format (depth): 10-bit RAW
Sensor Aspect Ratio: 16 : 9
F Number: $f/2.0$ and above
Focal Length: at least 1.88 mm
Focus: Fixed
Shutter Type: Rolling Shutter
Horizontal Field of View: more than 65
Vertical Field of View: more than 40
Diagonal Field of View: more than 70
Distortion: $\leq 1.5\%$
Development kit is required: YES
CE marking is required: YES

Pressure Mat   Connectivity: USB 2.0 or 3.0 and/or Bluetooth Low Energy
Individual sensor area 10×15 mm

(Continued)

Hardware devices and components of HOLOBALANCE solution

Wearable devices	Maximum distance between sensors: 3 cm
	Overall area dimensions: 1×2 m
	Sensing area dimensions: 0.90×1.80 m
	Durable floor mat
	Sensor response time latency < 3%
	Accuracy > 90%
	Drift < 10%
	Repeatability < 5%
	Measure range > 250 mmHg
	Hysteresis: ±5%
	Development kit is required: YES
	Measurement configuration is required: YES
	Capability of raw data gathering at real time is required: YES
	EC declaration of conformity is required: YES
	Sensors: MEMS 3-axis accelerometer, optical heart rate PPG sensor
	Connection: Bluetooth 4.0
	EC declaration of conformity is required: YES
Headphones	Type: Over the ear
	Frequencies: 15–25 000 Hz
	Connection: Wired
	Sensitivity: >104 dB SPL/V
	EC declaration of conformity is required: YES

Final version of the platform uses a set of sensors and image capture devices to assess the execution of balance training exercises and provide real-time feedback to the user. Apart from the real-time requirements, the processed data that result from the execution of the exercises, the cognitive games, and the auditory training sessions, along with data of the PA of the user are analyzed on the cloud, to provide detailed insights and be used to motivate the user.

In that sense, there are two nodes of the system:

- one placed near the user (in the home environment) and
- one on the cloud infrastructure.

The home node or MCWS computer is based on a commercial desktop or low-factor computer, capable enough to run real-time evaluations on the sensor-captured parameters, so to give feedback to the VC if an exercise is executed in the proper way or if the user should improve it. MCWS computer represents dependency of the system with which continuous and synchronized connection have to be established with the devices and sensors, and have enough processing power to evaluate the input and provide feedback in real time. Low latency is crucial for this task.

The MCWS will interface with two modules. Namely, the MCWS will feed data to

- Virtual coaching module
- Expert Panel

As presented in Figure 7.11, the MCWS subsystem will

- Gather the sensors data
- Calculate the relative parameters, depending on the performed exercise
- Inform the aforementioned modules

For the Virtual coaching module, the communication will be performed in a "semi-real time" fashion, while the Expert panel will be updated after the end of each training session.

The information that the MCWS module will send to the Virtual coaching module will be an abstract reasoning representation of the calculated parameters and will concern the performance of the end user on a specific exercise. Indicative examples are "You did well" or "You have to perform the exercise again." The Expert panel will receive on a structured form all the calculated parameters of the performed exercises.

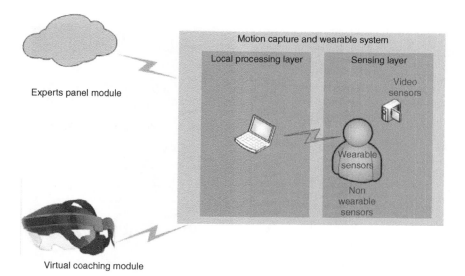

Figure 7.11 Conceptual architecture of the interfacing modules with the MCWS.

7.5.1 Balance Physiotherapy Hologram (BPH)

Final version of BPH interacts with a patient by providing personalized exercises instructions as well as feedback for the performance through the VC. BPH module provides different versions of avatars with various verbal capabilities in English, German, and Greek that are controlled and triggered by the MCWS module based on the BPH design as RESTful service. That approach allows BPH to be remotely guided and controlled in an extensive number of use cases. Control is performed in two-way communication between MCWS and BPH module giving the possibilities to achieve very complex interactions and behavior.

One version of the software is hosted on the smartphone where the user wears the head-mounted adapter to keep a smartphone at a set location on the head. The head-mounted adapter allows the user to see images generated from the smartphone screen as holograms in the space in front of the user.

Two new avatars have been designed in order to extend possibilities to the user and give additional user experience. Initial version of female avatar has been replaced with a new one and additional male avatar has been added (Figure 7.12).

Both avatars, shown in Figure 7.12, have been re-engineered to be lighter and give better application performance according to results collected via UCA (User Centric Approach). Their movement and lip sync have been improved giving a more realistic demonstration. Avatars have been created without doctor coat, as it was in initial version, in order to align with feedback collected during testing the

Figure 7.12 Avatars.

whole system with patients' groups explaining that more casual avatars are more appropriate and leads to the better user experience.

Avatars and languages can be selected from starting menu by pressing one of the available buttons. Saved settings will be persisted and loaded during the next application initiation (Figure 7.13).

Instructions with progressions of all exercises required have been integrated in the BPH module for both, male and female, avatars (Figures 7.13–7.23).

Interruptions and instructions have also been integrated for both avatars, which are available for triggering all the time during lifetime of BPH module.

Eye control has been added, so that the avatar follows the finger in sitting exercise. With this modification, the avatar is gazing at the index finger while moving the head (Figure 7.24).

Eye blinking has been added, so that the avatar opens and closes the eyes when needed to provide a more natural behavior (Figures 7.25 and 7.26).

7.5.2 BPH–MCWS Communication

Communication between MCWS and BPH is performed via REST protocol with simplified messaging system leading to more plane messages and significantly simpler structure.

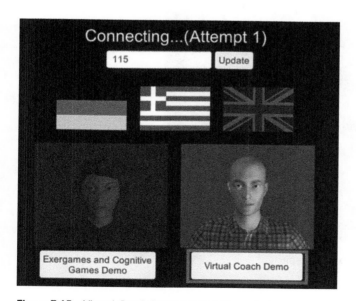

Figure 7.13 Virtual Coach Demo mode with controls.

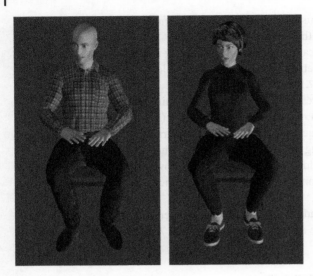

Figure 7.14 Sitting exercise – yaw.

Figure 7.15 Sitting exercise – pitch.

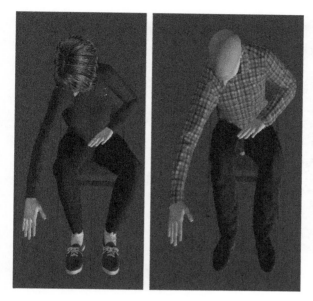

Figure 7.16   Sitting exercise – bend over.

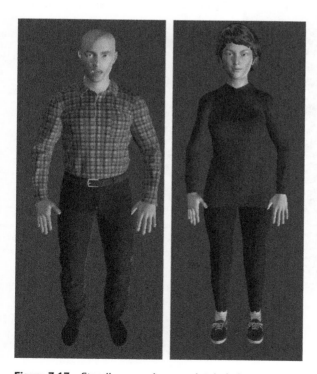

Figure 7.17   Standing exercise – maintain balance.

Figure 7.18 Standing exercise – bend over.

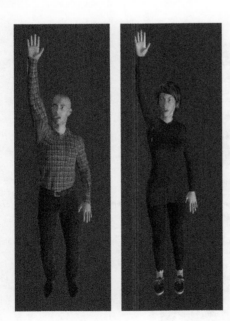

Figure 7.19 Standing exercise – reach up.

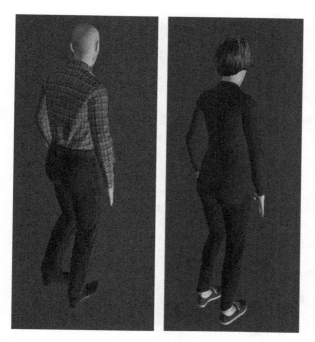

Figure 7.20 Standing exercise – turn 180°.

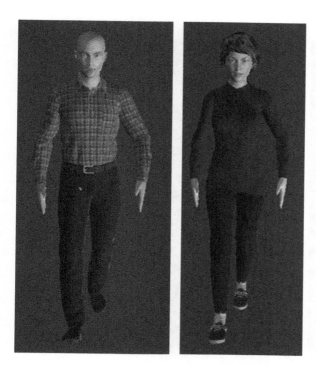

Figure 7.21 Walking exercise – looking at the horizon.

Figure 7.22 Walking exercise – yaw.

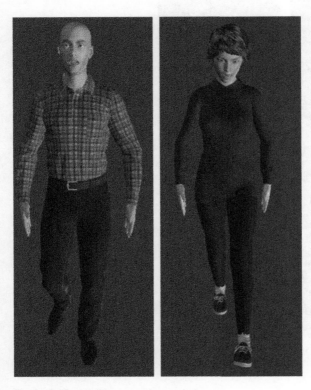

Figure 7.23 Walking exercise – pitch.

Figure 7.24   Avatar following the index finger with the eyes.

Figure 7.25   Opened eyes.

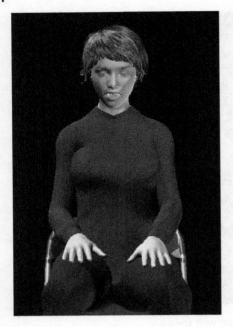

Figure 7.26 Closed eyes.

As a general principle, the Representational State Transfer (REST) design has been adopted as the protocol for the communication between the components. REST or RESTful API design is designed to take advantage of existing protocols. REST can be used over nearly any protocol; it usually takes advantage of HTTP when used for Web APIs. This means that developers do not need to install libraries or additional software in order to take advantage of a REST API design. Data are not tied to methods and resources, so in that sense, REST has the ability to handle multiple types of calls, return different data formats. It is also implementation agnostic, so each module may be developed using different technologies, but being able to communicate with the other.

Every message sent to the BPH from MCWS consists of two attributes: action and value. Four actions have been identified as important (Table 7.1).

7.5.3   Speech Recognition

In order to provide advanced interaction with the user, speech recognition has been integrated in the BPH module allowing collecting of user feedbacks after avatar ask questions. That feature gives better estimation of the patient condition.

Table 7.1   Presents the list of phrases that the BPH uses and that provide the necessary human–machine interaction.

Action	Description	Values
LoadScene	Load Virtual Coach or Cognitive Games	VirtualCoach CognitiveGames
ShowExercise	Triggers avatar to show some of the available exercises with progression. Format of the value is "VC {id_of_execise} P {progression number [0–3]}"	List of available values is shown in the table
ShowInstructions	Avatar interrupts the patient and explains the reason. Format of value "bph {id_of_instruction}"	List of available values is shown in the table.
Exit	Terminate application	No values

Having this functionality present raises the whole system value capturing the scoring of the patient, which results with better clinical decisions and system responses to the patient's answers.

Speech recognition has been achieved by using Google speech to text cloud REST services. After every question that avatar asks to the patient, voice recording is started and when the silence is detected, recording stops. Recorded audio is sent via REST API to the Google cloud platform and as immediate response, recognized text is returned.

The returned text is compared with allowed answers and the matching result is sent back to the MCWS module, which is based on the received data and initiates further actions. Speech recognition workflow is shown in Figure 7.27.

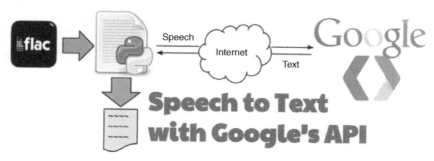

Figure 7.27   Speech recognition workflow. *Source*: Josh Coalson. Alphabet Inc.

7.5.4 Localization

All exercise instructions and interruptions reported in a workflow document have been translated in German, Greek, and English language using Google's text to speech REST API. All recognized unique sentences have been translated by native German and Greek speakers in order to have more reliable results and user experience.

Sentences are sent using Google API and in immediate response, audio files have been received. Conversion has been performed for female and male avatar. Google speech to text does not support male voice for Greek language, so that kind of voices are generated converting female voices to male using third party software.

All sound files have been further used to precompute phonemes (lip syncs) in the form of animations store in application storage. This approach is more preferable against real-time lip sync computation in order to reduce CPU consumption and save battery of devices, which lead to much better application performance.

Speech recognition has also been adapted to support recognition of patients' answers in German, Greek, and English and in native text format. Recognized speech has been sent back to MCWS together with result of recognition in the form of Boolean flag.

7.5.5 Motion Capturing

Motion capture was invented in order to speed up the animation process of complex animations such as human walking. By motion capture, we mean the capturing of the objects' movements in the real world and inserting the data in a three-dimensional model of the world in a virtual environment. The main reason for the usage of this technique is that it is easy to achieve realistic results and can be used as an alternative or a complementary method to video recordings.

This technique can be used to collect:

- movement trajectories
- posture data
- speed profiles
- other performance indices

Advantages of this technique:

- high precision of position
- orientation information regarding specific points of reference (can be chosen by the experimenter)

Disadvantages of this technique:

- employment of obtrusive technology that has to be applied to the body of the target (such as reflective markers),
- data which describes rather the postures and movements of a stick-like figure than that of a body with certain masses, and
- the costs of the installation.

Motion capture systems can be divided into:

- Marker-less Motion Capture
- Marker-based Motion Capture

7.5.6 Marker-less Motion Capture

As the name suggests, these tracking systems do not require artificial enhancements of the environment or the target (Figure 7.28). They just need sensor devices. Marker-less systems compute motion parameters from extracted silhouettes or other features (e.g. edges).

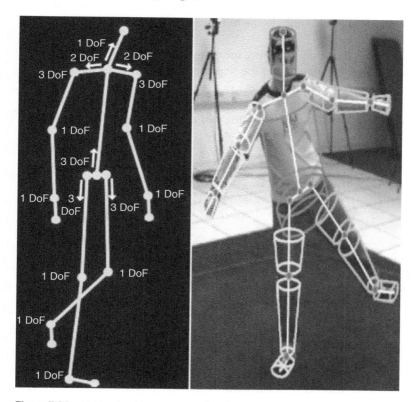

Figure 7.28 Motion tracking – marker-less based system. *Source:* [24] Saini et al., 2015/CC BY 4.0 License.

The movement is recorded in multiple video streams and computer vision algorithms analyze these streams to identify human forms, decomposing these into single, isolated parts used for tracking. The motion capture process is completely done via software, removing all the physical limitations while introducing computational constraints.

The main disadvantage of these systems is that they only measure relative movements and are subject to drift errors, which need to be accounted for.

Mock-ups movement has been addressed using this technique resulting in unnatural movements of avatars, which was inappropriate for usage with patients.

7.5.7 Marker-based Motion Capture

Marker-based tracking systems are the most common type (Figure 7.29). These systems depend on the detection of specific markers.

A marker configuration should cover all the major joints. An optical motion capture system records translational data for individual points corresponding to markers. The marker setup should be able to capture internal rotations of the main joints.

The systems with motion-capturing suites operate in the so-called outside-in mode: the markers are attached to the object of interest and the motion of the markers is observed by appropriate devices, such as high-speed infrared cameras. The markers have to be carefully arranged on the target, so that all relevant movements are captured and the structure of the body can be reconstructed from the data.

Figure 7.29 Motion tracking – marker-based system.

Some popular methods of marker-based motion capture are:

- Acoustical
- Magnetic
- Mechanic
- Optical

7.5.8 Optical Systems

Optical systems are based on photogrammetric methods. With this kind of system, the person wears an especially designed suit, covered with reflectors that are placed in their main articulations. Then, high-resolution cameras are strategically positioned to track those reflectors during the person's movement. Each camera generates the 2D coordinates for each reflector, obtained via a segmentation step. Proprietary software is then used to analyze the data captured by all of the cameras to compute the 3D coordinates of the reflectors. These systems are the most expensive ones in the market due to their cutting-end technological nature.

The advantages of using these systems:

- the very high sampling rate (enables the capture of fast movements),
- freedom offered by these systems, since there are no cables or limited workspace and the reflectors pose no restrain or cumbersome effect.

The disadvantages of this method are the:

- occlusion of some transmitters, especially in small objects such as hands or closely interacting objects,
- the difficulty in identifying reflectors that are too close to one another the lack of interactivity, since the data gathered must be before it is usable.

In order to provide most natural presentation of exercise, avatar movements and exercises have been recorded in Take One studio, whose capture studio is among the biggest in Europe.

With surface area of 200 m^2 and with 54 Vicon cameras, we were able to achieve the highest standard and even the most demanding animation requirements (Figure 7.30).

7.5.9 World Tracking

In this section, the improvements to the context awareness of the hologram with the use of the carpet/paper with the HOLOBALANCE logo as a reference point are described. The improvements described in this section should be applied by BioIRC to the Balance Physiotherapy Hologram.

Figure 7.30 Motion capture professional studio.

The presentation of the VC in the 3D space is facilitated by ARCore library for Android and ARKit library for iOS. Also, the stereoscopic rendering of the VC is done using the Holokit SDK which includes ARCore and ARKit as a basis for the presentation of the virtual objects in the real world. Both ARCore and ARKit libraries detect the features in the real world and search for a plane in the 3D space to place the virtual object on top of it. The problem with this method is that it could result in a random plane in the space with no particular control of its orientation and location for the user. Also, in case the detected plane is lost, the algorithm searches for new planes in the camera view which could again result in the placement of the virtual object in a different location. For BPH application, this led to a "jumping" or "sliding" of the VC to a newly detected plane. This effect is observed more frequently as long as the user performs walking or other dynamic exercises.

One of our goals was to decrease the number of confusing misplacements or "jumps" of the VC. In addition to that and instead of making the VC appear on a random plane, our approach aimed to fix the position of the VC on a desired location in the real world (i.e. older adults' homes).

This could be achieved using an image tracking algorithm. To do so, we explored several techniques such as Vuforia engine. Although Vuforia did not give us the best results, it helped us to optimize our trackable image target. We designed several patterns with HOLOBALANCE logo and evaluated their detectable features

Figure 7.31 HOLOBALANCE target image. *Source*: HOLOBALANCE.

for the camera using the Vuforia target rating system. Figure 7.31 shows our final design which was rated with five stars meaning that it is rich in details and can result in the best target detection. To be able to detect the image from a 2-m distance (HOLOBALANCE min distance required for performing exercises at home), we printed the image on an 84 × 84 mm square (a square on an A0 paper). Since folding and unfolding the image target can harm the detection performance, we printed the image on a foldable carpet with the same size.

In the next step, we used the image tracking feature of the ARKit v1.5 which was available inside the Holokit SDK. The results were acceptable but it was only available for iOS. Since the HOLOBALANCE technical team decided on Android as the main development operating system, we searched further for a solution. Luckily, Unity3D released its AR Foundation library which supports AR implementations including image tracking for both Android and iOS. This way we could make the project cross-platform and available for iOS and Android. The stereoscopic rendering is still implemented with the Holokit SDK.

Implementation in unity is done by the following steps:

- AR Session (component provided by the AR Foundation) should be created.
- The AR Tracked Image Manager should be then attached to the AR Session Origin.
- The ReferenceImageLibrary component includes HOLOBALANCE target image (Figure 7.7).
- The TrackedImagePrefab, which appears once the image from the ReferenceImageLibrary is detected, is the VC prefab.
- To make the VC appear, the HOLOBALANCE target image on the carpet has to be fully and clearly perceived by the mobile phone camera.
- The AR Foundation uses visual odometry and the IMU data of the mobile phone to make the VC stay on the carpet even when the carpet is no longer visible in the camera view.

Although the image tracking algorithm places the VC more precisely than the plane detection algorithm, some "sliding" of the VC could be still observed.

In order to improve the tracking stability, we used additionally the AR Foundation point clouds and the AR Foundation reference points. To do so, we attached the ARPointCloudManager to the AR Session Origin to increase the number of feature points used for the orientation of the mobile device in its environment.

Furthermore, we used the AR Foundation reference points to create a reference point at the position of the carpet once it was detected. To do so, we attached the script called "UseReferencePoint" to the AR Session Origin. It is subscribed to the events evoked by the AR TrackedImageManager. Once an event is evoked (the image detected), we set a reference point at the image location. By checking the tracking state of the tracked image, we know at which time the carpet is visible in the camera view. In case the carpet is visible, the reference point is updated. Unfortunately, when the phone was moved very quickly from side to side, the VC was still sometimes misplaced. This was caused by the blurry camera picture and the resulting loss of all feature points that were used for the orientation of the device. When the camera picture becomes clear again, AR Foundation uses new feature points to orientate the mobile phone. The image tracking algorithm would place the VC based on the last location saved when it was able to locate the position of the phone in its environment. Therefore, to improve the user experience, we decided to make the VC disappear instead of having it placed in the wrong place in the environment. To do so, we included the disabling of the TrackedImagePrefab in the "UseReferencePoint" script after the ARSessionState is no longer trackable (i.e. "SessionTracking"). After the complete loss of all feature points, the participant would have to look at the carpet to enable the VC again.

Although the image tracking with our adjustments makes the user experience smoother, there are still some extreme cases where the VC can no longer be placed correctly and disappears. Although this can easily be solved by looking at the carpet again, there is room for improvement since a variety of feature recognition algorithms exist and could be integrated into the project. However, because of the limited processing power of the GPU in mobile devices, not all of these algorithms are suitable. In future, a version of the Scale-invariant feature transform (SIFT) algorithm or Simultaneous Localization and Mapping (SLAM) algorithm may be used to recognize features again and correctly position the VC on the carpet. Attributes of recognized feature points in the environment would have to be saved to recognize them again.

Another option to save information about the device's environment is to create a map of its surroundings. The map implementation for AR Foundation is called ARWorldMaps and is currently only supported by ARKit (i.e. only for iOS). If implemented, the physical room has to be scanned before starting an AR experience and actual interaction with the VC. This could eventually lead to an improvement of the user AR experience.

7.6 Biomechanical Model of Avatar Based on the Muscle Modeling

Avatar animation is the art of bringing virtual characters to life through the design of solutions that allow the reproduction and/or synthesis of new motions. In this section, biomechanical model of avatar used for BPH has been presented. Once the avatar has been properly configured, the animation system allows to start using the Muscles & Settings tab of the Avatar's Inspector from Unity3D. In the Additional Settings, Translate DoF option was used to enable translation animations for the humanoid. Translation DoF is available for Chest, UpperChest, Neck, LeftUpperLeg, RightUpperLeg, LeftShoulder, and RightShoulder muscles. The "Avatar" system is how Unity identifies that a particular animated model is humanoid in layout, and which parts of the model correspond to the legs, arms, head, and body (Figure 7.32).

Figure 7.32 Unity's Avatar structure.

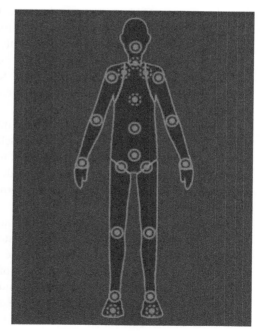

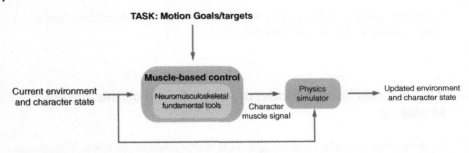

Figure 7.33 Animation using a muscle-based control framework. *Source*: Adapted from Ruiz et al. [28].

Throughout the years, several solutions have been proposed to mimic how humans and animals control their motions in order to animate virtual characters [25–27]. These solutions are split into two main approaches: kinematic-based methods and physics-based methods. The former involves animating characters by specifying limb or joint trajectories. The latter involves animating characters by specifying actuator trajectories, such as joint torques or muscle signals.

A physics-based animation is based on the development of physics simulators, which aim at replicating real environments by modeling the physical laws and conditions that define it. Muscle-based control assumes that all the joints of the character are actuated by muscles and produces muscle signals only.

Figure 7.33 depicts how muscle-based control is integrated in a physics-based framework. In this diagram, the physics simulator updates the state of the character and the environment as a result of the external forces in the environment and internal forces in the character. The external forces in the environment include gravity, force perturbations, and forces due to obstacles, slopes, and changes in the friction coefficient.

The internal forces in the character are the muscle, ligament, and joint-contact forces, which are generated thanks to the muscle signals produced by the muscle-based control method. To design these control methods, knowledge in neuromusculoskeletal simulation is vital since it provides a basis on which to: model the characters, automate the motion generation process, and solve the motion redundancy problem.

Physical-based modeling considers a system of the segments, joints, masses, inertias, and actuation capacities of these systems. Musculoskeletal modeling is common to different types of virtual characters, such as animals, humans, and even imaginary creatures. The core problem of musculoskeletal modeling is the definition of anatomically realistic segments and joints. In biomechanics, the models are often anatomically based [29, 30]. One of the main issues in musculoskeletal modeling is the scaling of the model. As motion is generally recorded to be

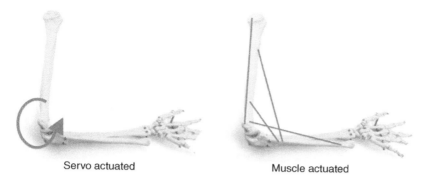

Servo actuated   Muscle actuated

Figure 7.34 A biomechanical upper limb model. On the left, the elbow is actuated by a virtual servo. On the right, the elbow is actuated by muscles.

analyzed, the lengths of the segments are easily computed from marker positions [31].

Muscles were incorporated in the simulation models as non-direct actuators of the joints (Figure 7.34). A more straightforward inclusion of muscles began with the use of mass-spring systems. Contrary to the usual angular-spring dampers and PD-controllers, these systems (like muscles) are non-direct actuators of the joints. This means that they produce first forces, not torques, which interact with the skeletal system to produce motion. Like muscles, these systems use force action lines, determined by their insertion or attachment site to the skeletal structure.

In order to simulate a bio-inspired natural motion, a motor control model with the minimization of a cost function ($f(X)$) could be applied. Formulation of this model consists in a nonlinear constrained optimization problem:

Find X which minimizes $f(X)$ subject to,

$$gj(X) = 0; j = 1, 2, ..., m$$
$$hk(X) = 0; k = 1, 2, ..., p$$

where the constraints enforce that:

- the computed forces solve the dynamic equations;
- the muscles are only pulling and they have physiological based force limits;
- the computed forces may respect any additional set of unilateral (g) or bilateral constraints (h).

The optimization can be a static or dynamic one. A static optimization refers to the process of minimizing or maximizing an objective function at a time instant, while a dynamic optimization refers to the process of minimizing or maximizing an objective function over an interval of time of nonzero duration.

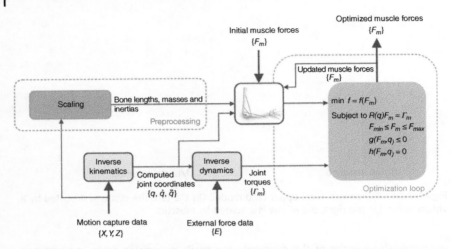

Figure 7.35 Inverse dynamics-based optimization. The optimization problem iterates until the cost function is minimized and the constraints are satisfied. *Source*: Adapted from Ruiz et al. [28].

The model can be used both in the inverse dynamics and forward dynamics framework. The schemes of these frameworks are described in Figures 7.35 and 7.36. In the inverse dynamics-based optimization (Figure 7.35), motion and external forces are applied to a musculoskeletal model. The joint-generated torques can be used in an optimization procedure that computes the muscle forces that satisfy the task, motion objectives, and constraints. The purpose of using inverse dynamics-based optimization for animation is to provide characters with a certain degree of adaptability to perturbations. Computing muscle forces from motion capture data allows the simulation of new motions (e.g. fatigued or injured motions), by replaying the computed muscle forces while altering muscle parameters [32–34]. In the forward dynamics-based optimization problem (Figure 7.36), only initial muscle excitations and external forces are applied to the model.

7.6.1 Muscle Modeling

The most important relation in muscle mechanics is Hill's equation. It refers to the mechanical behavior of skeletal muscle in the tetanized condition. This equation is given as

$$(v + b)(S + a) = b(S_0 + a) \tag{7.1}$$

where S represents tension (tensional stress) in muscle; v is the velocity of the contraction; a, b, and S_0 are constants [35]. The constant S_0 is the maximum tension that can be produced under isometric tetanic contraction. The Hill equation can be rewritten in dimensionless form as

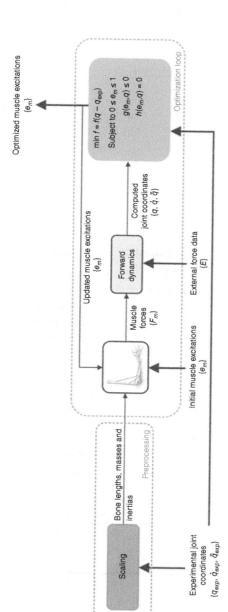

Figure 7.36 Forward dynamics-based optimization. The optimization problem iterates until the cost function is minimized and the constraints are satisfied. *Source:* Adapted from Ruiz et al. [28].

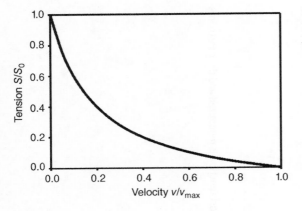

Figure 7.37 Tension–velocity curve corresponding to a muscle in the tetanized condition.

$$\frac{S}{S_0} = \frac{1 - (v/v_0)}{1 + c(v/v_0)} \tag{7.2}$$

in which the maximum velocity is $v_0 = \dfrac{bS_0}{a}$, and the constant $c = \dfrac{S_0}{a}$. A graphical representation of the equation is given in Figure 7.37.

The maximum tension in the tetanized condition is strongly dependent on the muscle stretch ratio. The tension–sarcomere length (or stretch) relationship, shown in Figure 7.38, is due to Gordon [36] and corresponds to the intact skeletal muscle. It can be noticed that the maximum tension corresponds to the slack

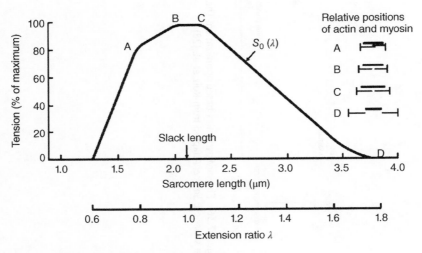

Figure 7.38 Isometric tension–length curve. *Source*: Gordon [36].

Figure 7.39 Hill's functional model of a muscle. CE is the contractile element, SEE is the series elastic element, and PEE is the parallel elastic element; σ is the stress in the muscle fiber direction. *Source*: Fung [35].

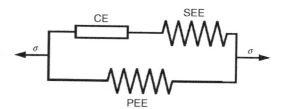

muscle length, i.e. to the extension ratio, or stretch, equal to 1.0. A decrease in stress under shortening and extension can be explained by a change of the number of cross-bridges between myosin and actin fibers under non-slack conditions.

A simple model that reflects the mechanical behavior of muscle described in Figures 7.37 and 7.38 is Hill's three-component model. A graphical representation of this model is given in Figure 7.39, where CE is the contractile element, SEE is the series elastic element, and PEE is the parallel elastic element [35].

The CE has the characteristic that is described by Hill's equation and Gordon's curve, and will be adequately included in the model. The tension–stretch relationship for the nonlinear elastic element (SEE) is given by

$$S = (S^* + \beta)e^{\alpha(\lambda - \lambda^*)} - \beta \tag{7.3}$$

in which S^* represents the tension corresponding to a stretch λ^*, while α and β are material constants. The surrounding connective tissue within a muscle is represented by a linear elastic element that is coupled in parallel to the series of contractile and nonlinear elastic elements.

These models of muscle and dynamic model of motion have been incorporated in the avatar model for BPH.

References

1 World Health Organization 2014. *Health Topics Physical Activity*. https://www.who.int/news-room/fact-sheets/detail/physical-activity.

2 https://www.nice.org.uk/guidance/cg161/chapter/1-Recommendations#preventing-falls-in-older-people-2.

3 Scuffham, P., Chaplin, S., and Legood, R. (2003). Incidence and costs of unintentional falls in older people in the United Kingdom. *J. Epidem. Common Health* **57** (9): 740–744.

4 British Orthopaedic Association HQIP (2012), British Geriatrics Society The National Hip Fracture Database National Report – National Summary 2012.

5 Bridenbaugh, S.A. and Kressig, R.W. (2015). Motor cognitive dual tasking: early detection of gait impairment, fall risk and cognitive decline. *Z. Gerontol. Geriatr.* **48**: 15–21.

6 Frewen, J., King-Kallimanis, B., Boyle, G., and Kenny, R.A. (2015). Recent syncope and unexplained falls are associated with poor cognitive performance. *Age Ageing* **44**: 282–286.

7 Deshpande, N., Metter, E.J., Lauretani, F. et al. (2008). Activity restriction induced by fear of falling and objective and subjective measures of physical function: a prospective cohort study. *J. Am. Geriatr. Soc.* **56**: 615–620.

8 NICE (2013). *Clinical Guideline 161: The Assessment and Prevention of Falls in Older people, ed. N.I.f.C. Excellence*. Royal College of Nursing.

9 Royal College of Physicians (2010). Falling standards, broken promises. In: *Report of the National Audit of Falls and Bone Health in Older People*. London: RCP.

10 Hartholt, K. (2011). *Falls and Drugs in Older Population: Medical and Societal Consequences*. Rotterdam: Erasmus University.

11 http://www.eurosafe.eu.com/uploads/inline-files/Joint%20Declaration_Sept%202015.pdf.

12 National Blueprint (2001). Increasing physical activity among adults age 50 and older. *J. Aging Phys. Act.* **9** (suppl): 1–96.

13 Roth, M. (2009. UK Data Archive Study Number 6397). Self-reported physical activity in adults. In: *Health Survey For England 2008. Physical Activity and Fitness*, vol. **1** (eds. J. Mindell and V. Hirani), 21–88. Health Survey for England.

14 Bauman, A.E. (2004). Updating the evidence that physical activity is good for health: an epidemiological review 2000–2003. *J. Sci. Med. Sport* **7** (1 Suppl): 6–19.

15 Iliffe, S., Kendrick, D., Morris, R. et al. (2015). ProAct65+ research team. Promoting physical activity in older people in general practice: ProAct65+ cluster randomised controlled trial. *Br. J. Gen. Pract.* **65** (640): e731–e738.

16 Ricci AN. Mayra C Aratani, Flávia Doná et al. A systematic review about the effects of the vestibular rehabilitation in middle-age and older adults. *Rev. Bras. Fis.* 2010;**14**(5):361–371.

17 Kenny, R.A., Rubenstein, L.Z., Tinetti, M.E. et al. (2011). Summary of the Updated American Geriatrics Society/British Geriatrics Society clinical practice guideline for prevention of falls in older persons. Panel on Prevention of Falls in Older Persons, American Geriatrics Society and British Geriatrics Society. *J. Am. Geriatr. Soc.* **59** (1): 148–157.

18 Shubert, T.E. (2011). Evidence-based exercise prescription for balance and falls prevention: a current review of the literature. *J. Geriatr. Phys. Ther.* **34** (3): 100–108.

19 Peel, C., Brown, C.J., Lane, A. et al. (2008). A survey of fall prevention knowledge and practice patterns in home health physical therapists. *J. Geriatr. Phys. Ther.* **31**: 64–70.

20 http://www.eurosafe.eu.com/uploads/inline-files/Joint%20Declaration_Sept%202015.pdf.

21 Department of Health (2012). *Long Term Compendium of Information*, 3e. [online]. (viewed 03 November 2014). http://tinyurl.com/m9zyqe8. https://assets.publishing.service.gov.uk/government/uploads/system/uploads/attachment_data/file/216528/dh_134486.pdf.

22 Dobkin, B.H. and Dorsch, A. (2011). The promise of mhealth: daily activity monitoring and outcome assessment. *Neurorehabil. Neural Repair* **25** (788): 1–12.

23 Harun, A., Li, C., Bridges, J.F., and Agrawal, Y. (2016). Understanding the experience of age-related vestibular loss in older individuals: a qualitative study. *Patient* **9**: 303–309.

24 Saini, S., Zakaria, N., Rambli, D.R.A., and Sulaiman, S. (2015). Markerless Human Motion Tracking Using Hierarchical Multi-Swarm Cooperative Particle Swarm Optimization, 15 May 2015. https://doi.org/10.1371/journal.pone.0127833

25 Pavlou, M., Acheson, J., Nicolaou, D. et al. (2015). Effect of developmental binocular vision abnormalities on visual vertigo symptoms and treatment outcome. *J. Neurol. Phys. Ther.* **39** (4): 215–224. https://doi.org/10.1097/NPT.0000000000000105.

26 Liston, M.B., Bamiou, D.E., Martin, F. et al. (2014). Peripheral vestibular dysfunction is prevalent in older adults experiencing multiple non-syncopal falls versus age-matched non-fallers: a pilot study. *Age Ageing* **43** (1): 38–43. https://doi.org/10.1093/ageing/aft129.

27 Wilhelms, J. (1987). Toward automatic motion control. *Comput. Graph. Appl. IEEE* **7** (4): 11–22.

28 Ruiz, A.L.C., Pontonnier, C. et al. (2016). Muscle-Based control for character animation. *Comput. Graph. Forum* **36**: 1–27.

29 Wu, G., Siegler, S., Allard, P. et al. (2002). ISB recommendation on definitions of joint coordinate system of various joints for the reporting of human joint motion part I: ankle, hip, and spine. *J. Biomech.* **35** (4): 543–548.

30 Wu, G., Van Der Helm, F.C.T., Veeger, H.E.J.D. et al. (2005). ISB recommendation on definitions of joint coordinate systems of various joints for the reporting of human joint motion—part II: shoulder, elbow, wrist and hand. *J. Biomech.* **38** (5): 981–992.

31 Lund, M.E., Andersen, M.S., De Zee, M., and Rasmussen, J. (2011). Functional Sscaling of musculoskeletal models. In: *Congress of the International Society of Biomechanics*. ISB.

32 Komura, T., Shinagawa, Y., and Kunii, T.L. (2000). Creating and retargetting motion by the musculoskeletal human body model. *Vis. Comput.* **16** (5): 254–270.

33 Lee, Y., Park, M.S., Kwon, T., and Lee, J. (2014). Locomotion control for many-muscle humanoids. *ACM Trans. Graph.* **33** (6): 218:1–218:11.

34 Wang, J.M., Hamner, S.R., Delp, S.L., and Koltun, V. (2012). Optimizing locomotion controllers using biologically-based actuators and objectives. *ACM Trans. Graph.* **31** (4): 25:1–25:11.

35 Fung, Y.C. (1993). *Biomechanics: Mechanical Properties of Living Tissues*, 2e. New York: Springer-Verlag.

36 Gordon, A.M., Huxley, A.F., and Julian, F.J. (1966). The variation in isometric tension with sarcomere length in vertebrate muscle fibers. *J. Physiol.* **269**: 441–515.

8

Modeling of the Human Heart – Ventricular Activation Sequence and ECG Measurement

Nenad D. Filipovic[1,2]

[1]*Faculty of Engineering, University of Kragujevac, Kragujevac, Serbia*
[2]*Bioengineering Research and Development Center (BioIRC), Kragujevac, Serbia*

In this chapter, a detailed heart and torso model of electrical field is presented. Heart model geometry includes seven different regions. The monodomain model of the modified FitzHugh–Nagumo model of the cardiac cell is used. Six electrodes are positioned at the chest to model the precordial leads and the results are compared with real clinical measurements. The inverse ECG method is used to optimize the potential on the heart's surface. Whole heart electrical activity in the torso-embedded environment, with a spontaneous initiation of activation in the sinoatrial node, incorporating a specialized conduction system with heterogeneous action potential morphologies throughout the heart, is presented. Body surface potential maps in a healthy subject during the progression of ventricular activation in nine sequences are included. The electro model coupled with the mechanical model using the linear elastic and orthotropic material model based on Holzapfel experiments are also presented. Future research will go more deeply into silico clinical trials where we will compare some clinical pathology and findings on the body surface with standard 12 ECG electrode measurements.

8.1 Introduction

One of the first heart models was presented in Hodgkin and Huxley [1]. In the meantime, numerous mathematical models have been developed [2] and many of them proposed an electromechanical model that simulated the left ventricle or the whole heart. For example, Nash and Panfilov [3] presented a computational framework to couple a three-variable FitzHugh–Nagumo (FHN)-type [4] excitation–tension model to governing equations of nonlinear stress equilibrium

Computational Modeling and Simulation Examples in Bioengineering, First Edition. Edited by Nenad D. Filipovic. © 2022 The Institute of Electrical and Electronics Engineers, Inc. Published 2022 by John Wiley & Sons, Inc.

employing the electromechanical and mechanoelectric feedback. Niederer et al. [5] made a model for binding of Ca^{2+} to TnC, the kinetics of tropomyosin, the availability of binding sites, and the kinetics of crossbridge binding after perturbations in sarcomere length. Gurev et al. [6] derived methods to construct finite element electromechanical models of the heart and develop anatomically accurate ventricular mesh based on magnetic resonance and diffusion tensor magnetic resonance imaging of the heart. There is also a construction of the ventricular meshes where authors did not consider the influence of mechanical contraction on the cardiac electrophysiology [6]. Göktepe and Kuhl [7] proposed an implicit and entirely finite element-based approach to the two-way coupled excitation–contraction problem. The electrophysiology was described by a FHN-type model in [7]. Doyle et al. [8] applied parallel computing to the simulation of heart mechanics. Lafortune et al. [9] developed a parallel electromechanical model of the heart. Lafortune et al. described the electrophysiology by the simple three-variable FHN-type model [4] or the three-variable Fenton–Karma (FK) model [10] and employed a "one-way coupling" in which the displacements do not affect the electrophysiology. Moreover, the influence of the heart's mechanical behavior on its electrical behavior, which is termed as "mechanoelectric feedback," has drawn researchers' attention [10–13].

It is very important to use a detailed, complex, and high-resolution anatomically accurate model of the whole heart electrical activity, which requires extensive computation times, dedicated software, and even the use of supercomputers [14–17].

We have recently developed a methodology for developing a real 3D heart model, using linear elastic and orthotropic material model based on Holzapfel experiments. Using this methodology, we can accurately predict the transport of electrical signals and displacement field within heart tissue. [17].

Muscles in the body are activated by electrical signals that are transmitted from the nervous system to muscle cells, affecting the change of the cell membranes' potentials. Additionally, calcium current and concentration inside the cells are the main cause of generating active stress within muscle fibers.

Simulations of an electrical model of the heart require an adequate model of the transfer between cardiac electrical activity and ECG signals measured on the torso surface. In addition, the process requires solving a mathematical inverse problem for which no unique solution exists. Clinical validation in humans is very limited since simultaneous whole heart electrical distribution recordings are inaccessible for both practical and ethical reasons [16].

In this chapter, we present the methodology and concepts used to generate an electromechanical model of the heart and provide the numerical simulation results.

8.2   Materials and Methods

Cardiac cells are filled with and surrounded by an ionic solution, mostly sodium $Na+$, potassium $K+$, and calcium $Ca2+$. These charged atoms move between the inside and the outside of the cell through proteins called ion channels. Cells are connected through gap junctions, which form channels that allow ions to flow from one cell to another.

An accurate numerical model is needed for better understanding of heart behavior in cardiomyopathy, heart failure, cardiac arrhythmia, and other heart diseases. These numerical models usually include drug transport, electrophysiology, and muscle mechanics [18].

We have presented heart geometry and seven different regions of the model where we included: (1) Sinoatrial node; (2) Atria; (3) Atrioventricular node; (4) His bundle; (5) Bundle fibers; (6) Purkinje fibers; (7) Ventricular myocardium (Figure 8.1).

In this study, we used the monodomain model of the modified FHN model of the cardiac cell [19, 20]

$$\frac{dV}{dt} = -kc_1(V_m - B)\left(-\frac{(V_m - B)}{A} + a\right)\left(-\frac{(V_m - B)}{A} + 1\right) - kc_2R(V_m - B)$$

$$\frac{dR}{dt} = -ke\left(\frac{(V_m - B)}{A} - R\right) \tag{8.1}$$

where Vm is the membrane potential, R is the recovery variable, a is relating to the excitation threshold, e is relating to the excitability, A is the action potential

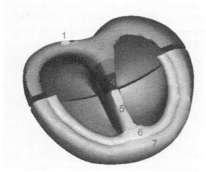

1. Sinoatrial node
2. Atria
3. Atrioventricular node
4. His bundle
5. Bundle branches
6. Purkinje fibers
7. Ventricular myocardium

Figure 8.1   Heart geometry and seven different regions of the model: (1) Sinoatrial node; (2) Atria; (3) Atrioventricular node; (4) His bundle; (5) Bundle fibers; (6) Purkinje fibers; (7) Ventricular myocardium.

amplitude, B is the resting membrane potential, and c_1, c_2, and k are membrane-specific parameters.

The monodomain model [20, 21] with incorporated modified FHN equations is

$$\frac{\partial V_m}{\partial t} = \frac{1}{\beta C_m}\left(\nabla \bullet (\sigma \nabla V_m) - \beta(I_{ion} - I_s)\right), \quad D = \frac{\sigma}{\beta C_m} \tag{8.2}$$

where β is the membrane surface-to-volume ratio, C_m is the membrane capacitance per unit area, σ is the tissue conductivity, I_{ion} is the ionic transmembrane current density per unit area, and I_s is the stimulation current density per unit area.

Parameters for the monodomain model with the modified FHN equations are presented in Table 8.1.

The ECG is a tracing of projections of cardiac electrical potentials, called leads, on specific axes, depending on probes' placement. Those leads represent a view of electrical activity of the heart from a particular angle across the body. The 12-Lead ECG has become a standard system used in clinical practice since the American Heart Association published the recommendation in 1954. It records signals from 10 electrodes respecting the following placement (Figure 8.2):

- V1: 4th intercostal space1 to the right of the sternum;
- V2: 4th intercostal space to the left of the sternum;
- V3: midway between V2 and V4;
- V4: 5th intercostal space at the midclavicular line;
- V5: anterior axillary line at the same level as V4;
- V6: midaxillary line at the same level as V4 and V5.

Table 8.1 Parameters for the monodomain model with modified FitzHugh–Nagumo equations

Parameter	SAN	Atria	AVN	His	BNL	Purkinje	Ventricles
A	−0.60	0.13	0.13	0.13	0.13	0.13	0.13
b	−0.30	0	0	0	0	0	0
$c_1 (\text{AsV}^{-1}\,\text{m}^{-3})$	1000	2.6	2.6	2.6	2.6	2.6	2.6
$c_2 (\text{AsV}^{-1}\,\text{m}^{-3})$	1.0	1.0	1.0	1.0	1.0	1.0	1.0
D	0	1	1	1	1	1	1
e	0.066	0.0132	0.0132	0.005	0.0022	0.0047	0.006
A (mV)	33	140	140	140	140	140	140
B (mV)	−22	−85	−85	−85	−85	−85	−85
k	1000	1000	1000	1000	1000	1000	1000
σ (mS·m^{-1})	0.5	8	0.5	10	15	35	8

Figure 8.2 Six electrodes (V1–V6) positioned at the chest to model the precordial leads.

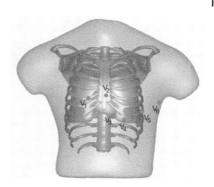

Computer simulations were conducted using the fully coupled heart torso monodomain equations including a detailed description of human ventricular cellular electrophysiology. Myocardial and torso conductivities were based on the literature data, as presented in Table 8.1.

Boundary conditions on all interior boundaries in contact with the torso, lungs, and cardiac cavities are zero flux for Vm; therefore, $-\mathbf{n} \cdot \mathbf{\Gamma} = 0$ where $\mathbf{n}$ is the unit outward normal vector on the boundary and $\mathbf{\Gamma}$ is the flux vector through that boundary for the intracellular voltage, equal to $\mathbf{\Gamma} = -\sigma \cdot \partial Vm/\partial \mathbf{n}$. For the variable Vm, the inward flux on these boundaries is equal to the outward current density $\mathbf{J}$ from the torso/chamber volume conductor; therefore, $-\sigma \cdot \partial Vm /\partial \mathbf{n} = \mathbf{n} \cdot \mathbf{J}$.

In the second part, we implement classical approaches for solving the ECG inverse problem using the epicardial potential formulation [22]. The studied methods are the family of Tikhonov methods and L regularization-based methods [23–25].

ECG measurement was performed in the Clinical Center Kragujevac, University of Kragujevac, on a healthy volunteer.

8.2.1 Material Model Based on Holzapfel Experiments

Figure 8.3 shows the basic definitions of local axes lying in the myocardial sheets (3–4 cell wide) f_0 –fiber direction, s_0-sheet direction; and n_0 normal to the myocardium sheet.

8.2.2 Biaxial Loading: Experimental Curves

Experimental curves for stress vs. stretch are shown in Figure 8.4. We will further use the stretch ratio of sheet stretch λ_2 vs. fiber stretch λ_1, which are shown in the figure.

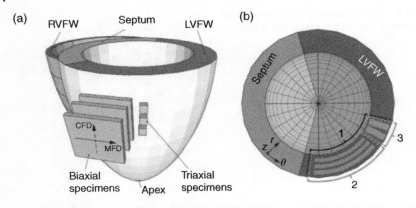

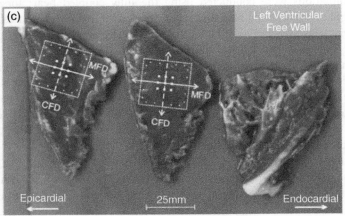

Figure 8.3 Material local coordinate system according to [14]. Biaxial loading of specimens is carried out in the sheet plane (a), while shear tests are performed in all local coordinate planes (b), (c) endocardial and epicardial free wall sample tissue of left ventricular. *Source:* Gerhard Sommer et al., 2015, p.172–192/with permission of Elsevier.

If hysteresis is neglected, then the stress–stretch curves are as shown in Figure 8.5.

8.3 Determination of Stretches in the Material Local Coordinate System

The constitutive curves are experimentally determined in the material local system f_0, s_0, n_0, Figure 8.5. Cauchy (true) stresses are expressed in terms of stretches. We

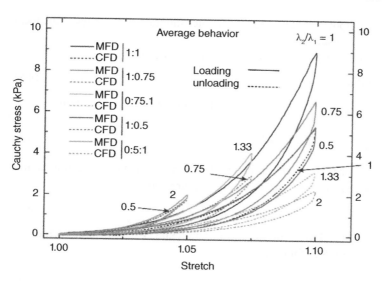

Figure 8.4 Experimental curves with hysteresis for biaxial loading of myocardium tissue [14], MFD-fiber direction f_0, CFD sheet direction s_0. Oblique numbers – unloading.

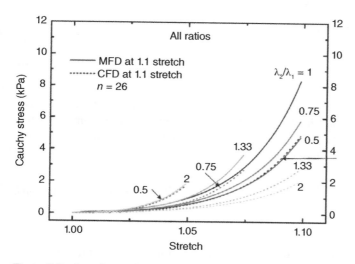

Figure 8.5 Experimental curves without hysteresis for biaxial loading of myocardium tissue [14], MFD-fiber direction f_0, CFD sheet direction s_0. Oblique numbers – unloading.

here present a computational procedure for determination of stretches in the local system in terms of displacements.

Following [16], change of the elementary material element is

$$d^t\mathbf{s} = {}_0^t\mathbf{F}d^0\mathbf{s} \tag{8.3}$$

If we divide this equation by the initial element length d^0s, we obtain

$$\frac{d^t\mathbf{s}}{d^0s} = \frac{d^t\mathbf{s}}{d^t s}\frac{d^t s}{d^0s} = {}_0^t\lambda\,{}^t\mathbf{n} = {}_0^t\mathbf{F}^0\mathbf{n} \tag{8.4}$$

If we take the norm of this equation, we obtain that stretch in a direction of unit vector $^t\mathbf{n}$ in the current configuration can be expressed as

$${}_0^t\lambda = \left|{}_0^t\mathbf{F}^0\mathbf{n}\right| \tag{8.5}$$

where $^0\mathbf{n}$ is unit normal in the initial configuration and ${}_0^t\mathbf{F}$ is deformation gradient. From Eqs. (8.3) and (8.5) follows:

$${}^t\mathbf{n} = \frac{1}{{}_0^t\lambda}{}_0^t\mathbf{F}^0\mathbf{n} \tag{8.6}$$

deformation gradient is defined as

$${}_0^t\mathbf{F} = \frac{\partial^t\mathbf{x}}{\partial^0\mathbf{x}} = \mathbf{I} + \frac{\partial^t\mathbf{u}}{\partial^0\mathbf{x}} \tag{8.7a}$$

or, in a component form

$${}_0^tF_{ij} = \delta_{ij} + \frac{\partial^t u_i}{\partial^0 x_j} \tag{8.7b}$$

where $^0\mathbf{x}$ is position vector of a material point in initial configuration, $^t\mathbf{x}$ is vector of the same material point in the current configuration; $^t\mathbf{u}$ is displacement vector, and δ_{ij} is the Kronecker delta symbol. At a Gauss integration point of a finite element, deformation gradient can be calculated as follows:

$${}_0^tF_{ij} = \delta_{ij} + \frac{\partial^t u_i}{\partial r_k}\frac{\partial r_k}{\partial^0 x_j} = \delta_{ij} + \frac{\partial^t u_i}{\partial r_k}{}^0J_{kj}^{-1} = \frac{\partial^t x_i}{\partial r_k}{}^0J_{kj}^{-1} = {}^tJ_{ik}{}^0J_{kj}^{-1}, \quad \text{sum on } k \tag{8.8}$$

where $^0J_{kj}^{-1}$ is the inverse Jacobian in the initial configuration, and r_k are natural (isoparametric) coordinates of the material point. The Jacobians can be calculated as

$$^0J_{kj} = \frac{\partial^0 x_k}{\partial r_j} = \frac{\partial N_I}{\partial r_j}\,^0X_{Ik}, \quad ^tJ_{kj} = \frac{\partial^t x_k}{\partial r_j} = \frac{\partial N_I}{\partial r_j}\,^tX_{Ik} \quad \text{sum on } I \tag{8.9}$$

where N_I are interpolation functions and $^0X_{Ik}$ and $^tX_{Ik}$ are coordinates of nodal points. Note that

$$\frac{\partial^t x_i}{\partial r_k} = \,^tJ_{ik} = \frac{\partial N_I}{\partial r_k}\,^tX_{Ii}, \quad \text{sum on } I \tag{8.10}$$

Therefore, we can determine deformation gradient from initial coordinates and from either current coordinates or current displacements.

8.4 Determination of Normal Stresses from Current Stretches

We use experimental curves given in Figure 8.5. The concept of determination of normal stresses is displayed in Figure 8.6. For the current iteration (here is omitted iteration counter i), we calculate the stretches λ_1 and λ_2 from Eq. (8.7b) for the fiber direction f_0 and sheet direction s_0, respectively.

Therefore, we know the ratio $\lambda_2/\lambda_1 = c$. Now, we need to calculate the stresses $\bar{\sigma}_i$ in the local system. If the stretch in the local (material) coordinate direction is λ_i, we find the two closest curves – for that direction and the loading/unloading state – among which falls the current ratio c, i.e.

$$c \geq a, \quad c \leq b \tag{8.11}$$

$$\bar{\sigma}_i = \frac{c-a}{b-a}\left(\sigma_i^2 - \sigma_i^1\right) \tag{8.12}$$

where σ_i^1 and σ_i^2 are stresses at the lower and upper curves, respectively. For the 3D model, we use characteristics for the normal direction to be the same as for the sheet direction.

Case when $\lambda_2/\lambda_1 < a_{\min}$ for fiber direction:

$$\bar{\sigma}_B = \frac{c}{a_{\min}}\bar{\sigma}^1 \tag{8.13a}$$

Case when $\lambda_2/\lambda_1 > a_{\max}$ for fiber direction:

$$\Delta\bar{\sigma} = \frac{\Delta\lambda_{ratio}}{a_{\max}}\sigma_{\max} = \frac{c-a_{\max}}{a_{\max}}\sigma_{\max}, \bar{\sigma} = \bar{\sigma}^2 + \left(\frac{c}{a_{\max}} - 1\right)\bar{\sigma}^2 = \frac{c}{a_{\max}}\bar{\sigma}^2 \tag{8.13b}$$

(a)   (b)

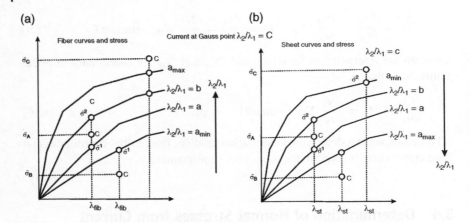

Figure 8.6 Interpolation of normal stresses in the local (material) coordinate system. (a) Fiber curves and stress, (b) Sheet curves and stress.

Case when $\lambda_2/\lambda_1 < a_{min}$ for sheet direction:

$$\overline{\sigma}_C = \overline{\sigma}^2 + \frac{a_{min} - c}{a_{min}}\overline{\sigma}^2 = \left(2 + \frac{c}{a_{min}}\right)\overline{\sigma}^2 \tag{8.13c}$$

Case when $\lambda_2/\lambda_1 > a_{max}$ for sheet direction:

$$\overline{\sigma}_C = \frac{a_{max}}{c}\overline{\sigma}^1 \tag{8.13d}$$

Tangent modulus is related to stress increment with respect to the increment of displacement in the direction of the fiber or in direction of sheet, hence, we obtain

$$\overline{E}_{Ti} = \lambda_i \frac{\partial \overline{\sigma}_i}{\partial \lambda_i}, \quad \text{no sum on } i \tag{8.13e}$$

In the case of a 3D model, we need to calculate the stress in the direction of normal. We assume the same curves as for the sheet direction. Hence, we calculate the stretch λ_n and calculate the normal stress σ_n and transform it to the global system as for the other two directions.

8.4.1 Determination of Shear Stresses from Current Shear Strains

Shear stresses acting on a material element are shown in Figure 8.7. The size of the specimens used in experiments was $4 \times 4 \times 4$ mm.

The stress–strain curves are shown in Figure 8.8. They are given for the three modes according to notation in Figure 8.9.

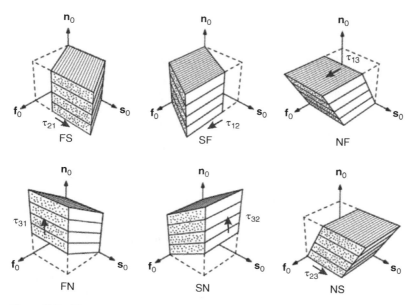

Figure 8.7 Shear stresses acting on material element.

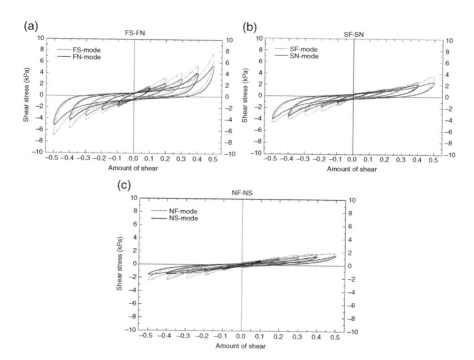

Figure 8.8 Shear stresses in terms of shear deformation for three modes. (a) FS-FN mode, (b) SF-SN mode, (c) NF-NS mode.

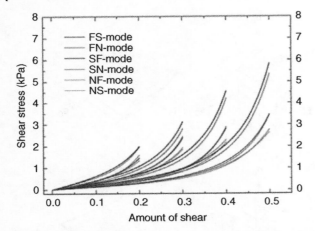

Figure 8.9 Shear stress constitutive curves, no hysteresis (Figure 8.13 [14]).

The curves are obtained by cycling loading and Figure 8.8 [14] corresponds to a different level of ultimate shear strain. Maximum shear strains are at level 0.5 [26]. If hysteresis is neglected, then the constitutive curves are as shown in Figure 8.9.

8.5 Results and Discussion

We firstly presented the whole heart electrical activity in the torso-embedded environment, with a spontaneous initiation of activation in the sinoatrial node, incorporating a specialized conduction system with heterogeneous action potential morphologies throughout the heart (Figure 8.10).

Figure 8.11 illustrates the simulated and measured ECG for six electrodes (V1–V6) positioned at the chest to model the precordial leads for the normal heart volunteer. The normal ECG is in agreement with standard clinical findings [27].

Body surface potential maps in a healthy subject during progression of ventricular activation in nine sequences corresponding to measurement ECG signal are presented in Figure 8.12.

Boundary conditions of the mechanical model are shown in Figure 8.13d. Field of displacement for two different time moments of simulation, for the electromechanical model of the human heart, are shown in Figure 8.13.

The methodology is tested with respect to the real 3D heart model, using the linear elastic and orthotropic material model based on Holzapfel experiments. Using this methodology, we can accurately predict the transport of electrical signals and displacement field within heart tissue. This approach has the potential to

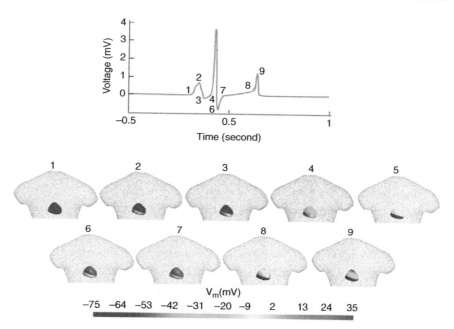

Figure 8.10 Whole heart activation simulation from lead II ECG signal at various time points on the ECG signal. There are 1–9 activation sequences corresponding to ECG signal above. The color bar denotes mV of the transmembrane potential.

be used in a coupled solid–fluid simulation of the whole heart so that an accurate prediction of heart beat for different heart conditions can be obtained.

8.6 Conclusion

In this chapter, heart geometry and seven different regions of the model were presented. It includes: (i) Sinoatrial node; (ii) Atria; (iii) Atrioventricular node; (iv) His bundle; (v) Bundle fibers; (vi) Purkinje fibers; (vii) Ventricular myocardium. The monodomain model of the modified FHN model of the cardiac cell was used. Six electrodes (V1–V6) were positioned at the chest to model the precordial leads and the results were compared with real clinical measurements. The inverse ECG method was used to optimize potential on the heart's surface.

Whole heart electrical activity in the torso-embedded environment was presented, with spontaneous initiation of activation in the sinoatrial node,

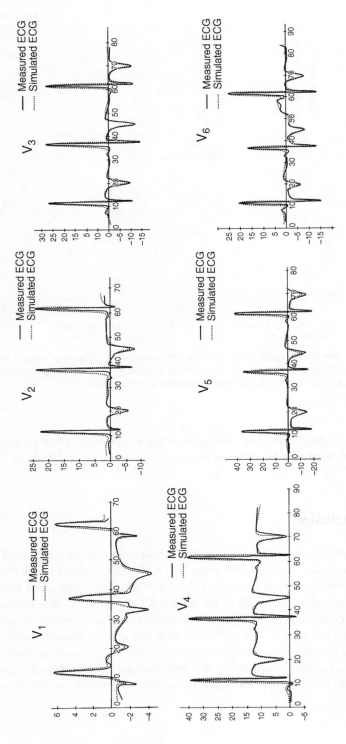

Figure 8.11 Simulated and measured ECG for six electrodes (V1–V6).

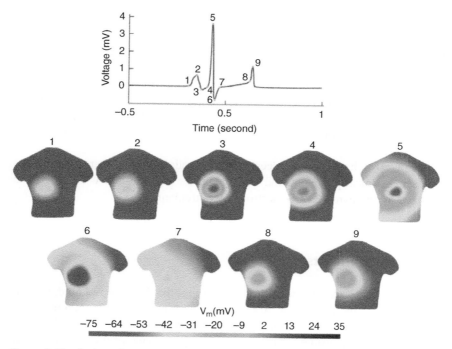

Figure 8.12 Body surface potential maps in a healthy subject during progression of ventricular activation in nine sequences corresponding to the ECG signal above. The color bar denotes mV from the heart activity range.

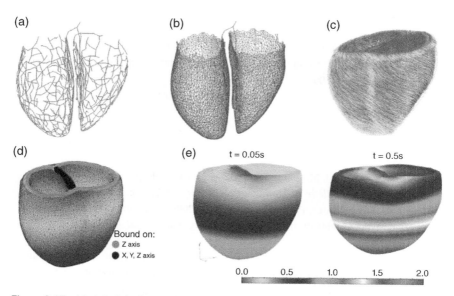

Figure 8.13 Model of the human heart: (a) Purkinje network; (b) uniform Purkinje 3D layer; (c) directions of muscle fibers; (d) boundary conditions in the model; (e) displacement field for two time moments.

incorporating a specialized conduction system with heterogeneous action potential morphologies throughout the heart.

Body surface potential maps in a healthy subject during progression of ventricular activation in nine sequences corresponding to the ECG signal were shown.

The coupling of the electro model with the mechanical model, using the linear elastic and orthotropic material model based on Holzapfel experiments was presented.

Future research will go more deeply into silico clinical trials where we will compare some clinical pathology and findings on the body surface with the results obtained by standard 12 ECG electrode measurements.

Acknowledgements

This paper is supported by the SILICOFCM project that has received funding from the European Union's Horizon 2020 research and innovation programme under grant agreement No 777204. This article reflects only the author's view. The Commission is not responsible for any use that may be made of the information it contains.

References

1 Hodgkin, A.L. and Huxley, A.F. (1952). A quantitative description of membrane current and its application to conduction and excitation in nerve. *The Journal of Physiology* **117** (4): 500–544.

2 Fenton, F.H. and Cherry, E.M. (2008). Models of cardiac cell. *Scholarpedia* **3** (8): 1868.

3 Nash, M.P. and Panfilov, A.V. (2004). Electromechanical model of excitable tissue to study reentrant cardiac arrhythmias. *Progress in Biophysics and Molecular Biology* **85** (2-3): 501–522.

4 FitzHugh, R. (1955). Mathematical models of threshold phenomena in the nerve membrane. *The Bulletin of Mathematical Biophysics* **17** (4): 257–278.

5 Niederer, S.A., Hunter, P.J., and Smith, N.P. (2006). A quantitative analysis of cardiac myocyte relaxation: a simulation study. *Biophysical Journal* **90** (5): 1697–1722.

6 Gurev, V., Lee, T., Constantino, J. et al. (2011). Models of cardiac electromechanics based on individual hearts imaging data: image-based electromechanical models of the heart. *Biomechanics and Modeling in Mechanobiology* **10** (3): 295–306.

7 Göktepe, S. and Kuhl, E. (2010). Electromechanics of the heart: a unified approach to the strongly coupled excitation-contraction problem. *Computational Mechanics* **45** (2-3): 227–243.

8 Doyle, M.G., Tavoularis, S., and Bourgault, Y. (2010). Application of parallel processing to the simulation of heart mechanics. *Lecture Notes in Computer Science* **5976**: 30–47.

9 Lafortune, P., Arís, R., Vázquez, M., and Houzeaux, G. (2012). Coupled electromechanical model of the heart: parallel finite element formulation. *International Journal For Numerical Methods in Biomedical Engineering* **28**: 72–86.

10 Clayton, R.H., Bernus, O., Cherry, E.M. et al. (2011). Models of cardiac tissue electrophysiology: progress, challenges and open questions. *Progress in Biophysics and Molecular Biology* **104** (1-3): 22–48.

11 Kohl, P. and Sachs, F. (2001). Mechanoelectric feedback in cardiac cells. *Philosophical Transactions of the Royal Society A* **359** (1783): 1173–1185.

12 Cherubini, C., Filippi, S., Nardinocchi, P., and Teresi, L. (2008). An electromechanical model of cardiac tissue: constitutive issues and electrophysiological effects. *Progress in Biophysics and Molecular Biology* **97** (2-3): 562–573.

13 Kuijpers, N.H.L., Ten Eikelder, H.M.M., Bovendeerd, P.H.M. et al. (2007). Mechanoelectric feedback leads to conduction slowing and block in acutely dilated atria: a modeling study of cardiac electromechanics. *American Journal of Physiology* **292** (6): H2832–H2853.

14 Gibbons Kroeker, C.A., Adeeb, S., Tyberg, J.V., and Shrive, N.G. (2006). A 2D FE model of the heart demonstrates the role of the pericardium in ventricular deformation. *American Journal of Physiology* **291** (5): H2229–H2236.

15 Pullan, A.J., Buist, M.L., and Cheng, L.K. (2005). *Mathematically Modelling the Electrical Activity of theHeart—FromCell To Body Surface and Back Again*. World Scientific.

16 Trudel, M.-C., Dub'e, B., Potse, M. et al. (2004). Simulation of QRST integral maps with a membrane based computer heart model employing parallel processing. *IEEE Transactions on Biomedical Engineering* **51** (8): 1319–1329.

17 Kojic, M., Milosevic, M., Simic, V. et al. (2019). Smeared Multiscale finite element models for mass transport and electrophysiology coupled to muscle mechanics. *Frontiers in Bioengineering and Biotechnology*, ISSN 2296-4185 **7** (381): 1–16, 2296-4185.

18 Wilhelms, M., Dossel, O., and Seemann, G. (2011). In silico investigation of electrically silent acute cardiac ischemia in the human ventricles. *IEEE Trans Biomed Eng* **58**: 2961–2964.

19 Fitzhugh, R. (1961). Impulses and physiological states in theoretical models of nerve membrane. *Biophysical Journal* **1**: 445–466.

20 Nagumo, J., Arimoto, S., and Yoshizawa, S. (1962). An active pulse transmission line simulating nerve axon. *Proceedings of the IRE* **50**: 2061–2070.

21 Sovilj, S., Magjarević, R., Lovell, N., and Dokos, S. (2013). A simplified 3D model of whole heart electrical activity and 12-lead ECG generation. *Computational and Mathematical Methods in Medicine* https://doi.org/10.1155/2013/134208.

22 Van Oosterom, A. (1999). The use of the spatial covariance in computing pericardial potentials. *IEEE Transactions on biomedical engineering* **46** (7): 778–787.

23 Van Oosterom, A. (2019). The spatial covariance used in computing the pericardial potential distribution. *Front Physiology* **10**: 58.

24 Van Oosterom, A. (2003). Source models in inverse electrocardiography. *International Journal of Bioelectromagn* **5**: 211–214.

25 Van Oosterom, A. (2010). The equivalent double layer: source models for repolarization. In: *Comprehensive Electrocardiology* (eds. P.W. Macfarlane, A. van Oosterom, O. Pahlm, et al.), 227–246. London: Springer https://doi.org/10.1007/978-1-84882-046-3_7.

26 Fenton, F. and Karma, A. (1998). Vortex dynamics in three-dimensional continuous myocardium with fiber rotation: filament instability and fibrillation. *Chaos* **8**: 20–47.

27 Wang, Y. and Rudy, Y. (2006). Application of the method of fundamental solutions to potential-based inverse electrocardiography. *Annals of Biomedical Engineering* **34** (8): 1272–1288.

9

Implementation of Medical Image Processing Algorithms on FPGA Using Xilinx System Generator

Tijana I. Šušteršič[1,2] and Nenad D. Filipović[1,2]

[1] *Faculty of Engineering, University of Kragujevac (FINK), Kragujevac, Serbia*
[2] *Bioengineering Research and Development Center (BioIRC), Kragujevac, Serbia*

9.1 Brief Introduction to FPGA

A field programmable gate array (FPGA) is a general-purpose integrated circuit that can be "programmed" by the designer, which means that it is not preprogrammed by device manufacturer. In comparison to the application-specific integrated circuit (ASIC), an FPGA can be programmed and reprogrammed, adapted to a specific problem, but later updated, even after it has been deployed into a system [1].

After the program design, it is converted into a bitstream configuration program and sent to static on-chip random-access memory. This bitstream represents the product of compilation tools that translate the high-level abstractions created by a designer into low-level and executable form. A FPGA offers a two-dimensional array of configurable resources that can execute a set of arithmetically and logically related features. These include DSP blocks, dual-port memories, multipliers, lookup tables (LUTs), buffers, registers, multiplexers, and digital clock managers. Xilinx FPGAs also feature specialized I/O structures that satisfy different voltage and bandwidth requirements [1].

For the purposes such as digital signal processing (DSP), it is of utmost importance to enable short delivery time with the possibility to modify the device function in the lab or the working site where the device is installed to suit the purpose [2]. Mostly due to the ever-decreasing cost and re-configurability, FPGAs have also found its place in DSP [3]. A lot of engineers give advantage to FPGAs over traditional digital signal processors (DSPs) because of the massive parallel processing capabilities and shorter time to market. Since FPGAs can be configured in hardware, they offer a complete hardware customization while implementing various

Computational Modeling and Simulation Examples in Bioengineering, First Edition. Edited by Nenad D. Filipovic. © 2022 The Institute of Electrical and Electronics Engineers, Inc. Published 2022 by John Wiley & Sons, Inc.

DSP applications [3, 4]. DSP performance comes from the ability of FPGA to create incredibly parallel data processing architectures. Unlike the DSP processor or a microprocessor, where the performance is linked to the processor clock rate, FPGA performance is related to the amount of parallel that can be applied to the algorithms that constitute a signal processing system [1]. Therefore, it is suitable for implementation of many signal processing algorithms in the area of artificial neural networks [5], classification in medical diagnosis [6, 7] etc. The system designer is able to use parallel DSP with a combination of increasingly high clock rates (current system frequencies between 100 and 200 MHz are common today), along with the highly distributed memory architecture. For example, hundreds of terabytes per second would be the raw memory bandwidth of a massive FPGA with a clock rate of 150 MHz [1].

FPGAs have many advantages in applications that involve acquisition of digital data but also large-scale data processing, especially in real time [2]. Therefore, FPGAs are widely used to design applications that require high-speed parallel data processing, such as image processing [3]. Some of the examples of image processing purposes involve filtering [8], image compression [9], wireless communication [10], medical imaging [11], face recognition [12–14], target recognition, robotics application, video surveillance, etc., where high-speed parallel data processing is the main request. This is primarily due to the fact that FPGAs provide the parallelism that allows image processing in real time. The key downside of FPGAs is that it takes time to create a finished output and time to redesign the system if there are requests for that [3].

Another advantage of using the FPGA includes significantly lower nonrecurring engineering costs compared with the custom IC costs (FPGAs are off-shelf commercial equipment), shorter time on the market, and a more configurable FPGA, which makes it possible to modify the design in the final application, even after deployment [1]. As FPGAs continue to increase their chip density according to Moore's law, their potential applications have also expanded [3]. Additionally, FPGA boards are equipped with many I/O ports, AD/DA converters as well as a soft/hard core microprocessor, which allows these boards to behave as a system on chip (SoC). Furthermore, software/hardware codesigns using FPGAs are becoming very popular as a solution to certain complex problems, especially those that involve intensive computations [3].

For all the previously discussed reasons, Xilinx System Generator (XSG) brought the idea of compiling an FPGA program starting from a high-level Simulink model, rather than programming in a low-level language [1]. It is important to bear in mind, when working with System Generator, that an FPGA offers freedom in implementation of different functions for signal processing. There is a possibility, for example, to define data path widths throughout the designed system or to use a variety of individual data processors (e.g. multiply-accumulate engines), all depending on the requirements and needs. On the other hand, System Generator provides the level of abstraction that allows the user to primarily focus on the algorithm and not the programming language.

9.1.1 Xilinx System Generator

XSG [1, 15] is a Xilinx tool that allows coupling with Mathworks Simulink models to be adapted for FPGA-design. For a variety of hardware operations, XSG provides a series of Simulink blocks for implementation. These blocks can be used to simulate the hardware system's functionality with Simulink. The design of most DSP applications includes a floating data point format [1]. Although this is relatively easy to incorporate models in many computer systems using applications such as Simulink, the difficulty of the application of floating-point arithmetic makes it more complicated in the hardware environment. For portable DSP systems, these problems increase where the restricting constraints of the system design. The XSG uses a fixed-point format to represent all numerical values. System Generator offers multiple blocks to convert the data supplied by the program (in our case Simulink) and the hardware (System Generator blocks) in the simulation environment.

The Simulink environment for higher level [16, 17] is expanded by Xilinx 's unique Blockset [3, 18]. For the use of a XSG, no previous experience with Xilinx FPGAs or RTL design methods is needed. The Matlab-Simulink high-level graphical interface makes it very easier to work with in comparison to the other software for hardware description [19]. The Xilinx-specific Simulink DSP Block contains more than 90 building blocks. These blocks include not just common components, such as adders, multipliers, and registers, but also complex building blocks including FFTs, filters, and memories. DSP blocks are also included. These blocks are used to generate optimized results for a chosen target system by Xilinx Logi-CORE™ IP.

Downstream implementation of the Simulink environment for the generation of a .bit FPGA programming file are automated – synthesizing, placing, and route. In order to implement JTAG hardware co-simulation, I/O planning and Clock planning can also be performed. After this, a netlist is generated and a draft for the model and programming file in VHDL/Verilog is created. The module is checked for behavioral syntax check, synthesized, and then it can be implemented on FPGA. The XSG itself has also the possibility to generate User constraints file (UCF), Test bench, and Test vectors for testing architecture [3]. In order to optimize FPGA resources, System Generator enables some features such as System Resource Estimation, Hardware Co-Simulation [20], and accelerated simulation through hardware in the loop co-simulation, which can be used to increase the simulation performance [17].

As previously stated, the designs are made using Xilinx-specific Blockset using DSP-friendly Simulink modeling environment [18]. The XSG library includes over 90 DSP building blocks that allow faster prototyping and design from a high-level programming stand point [1, 3, 21]. There are also blocks such as the M code and Black Box, which allow the user to program in MATLAB M code or C code, and Verilog to simplify integration with already existing projects or customized block behavior [3, 21]. Figure 9.1 shows the design flow of hardware implementation of XSG model.

The prerequisite of using System Generator is to have compatible versions of Matlab and System Generator, which will be explained later, as well as that the System Generator token available along with Xilinx has to be configured in Matlab to be able to use the Blocksets [22].

System Generator can be useful for many purposes, out of which the main one could be:

- to test the algorithm to explore its possibilities without hardware translation;
- to use a System Generator design as part of a bigger design;
- to implement a complete design in FPGA hardware.

This part of the chapter summarizes all three possibilities [23].

Algorithm Exploration
System Generator is particularly suitable for exploring algorithms, designs prototyping, and model analyses. This is achieved with the aim to get a sense of design concepts, and to predict the cost and efficiency of the hardware implementation. The work is preparatory and the designs do not have to be translated to hardware. In this environment, key parts of the architecture are installed without caring about fine detail or information. MATLAB M-code and Simulink blocks are

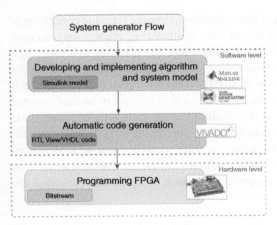

Figure 9.1 Design flow of Xilinx System Generator use in implementing an algorithm on hardware.

stimulants for simulation and outcomes analysis. A rough understanding of the cost of hardware production is provided by the resource estimate [1, 23].

Hardware generation tests may indicate hardware speeds that are possible. System Generator encourages step-by-step changes to be done, meaning that some of the architecture can be ready for hardware execution, while others remain abstract and high-level. System Generator hardware co-simulation facilities are particularly beneficial for optimizing portions of a design [1, 23].

Implementing Part of a Larger Design

System Generator is also used to build a design that is only a component of a bigger design. System Generator, for example, is an excellent environment for introducing data paths and testing, but not ideal for complex and strict timing requirements. It might be useful in this case to incorporate design parts via System Generator, to deploy other components externally and then to merge the components in a greater design.

A common solution to this flow is to construct a full HDL wrapper that represents an entire design and to use the System Generator part as a component. The parts of the architecture of the non-System Generator may also be components in the wrapper or can be instantiated directly in the wrapper [1, 23].

Implementing a Complete Design

Most of the time, all the necessary components for a design are available inside System Generator. To build this design, pressing the Generate button and the System Generator will instruct to convert the design into HDL and write the files necessary to process the HDL using downstream tools. The necessary files are:

- HDL that implements the design itself.
- A clock wrapper that encloses the design.
- An HDL testbench that encloses the clock wrapper. The testbench allows comparison of the outcomes of the Simulink simulations with those of a logic simulator.
- Files and scripts of projects that allow different synthesis tools like XST and Symplify.
- Files that enable System Generator HDL to be used as a project in the Project Navigator [1, 23].

9.1.2 Image Processing on FPGAs Using XSG

A variety of methods are used in different strategies for image processing. Some of them such as acquisition, enhancement, restoration, segmentation, and analysis are usually part of every image processing algorithm [9]. The images, once they are inside the computer system, are represented by the matrices and theoretically, all operations applied on matrices, can be applied on images too – addition, subtraction, multiplication, and division [3]. General processing techniques include removing the noise and performing enhancement [11], transformations to extract features for pattern recognition, compressing for storage and retrieval, or transmission via a computer network or a communication system [3].

Mohammed et al. studied the processing of digital image, especially point processing algorithms [3]. In the co-simulation of both software and hardware, Hanisha et al. examined the image denoising algorithm by using XSG [24]. Using Matlab-Simulink coupled with XSG tools, Kumar et al. examined various image processing operations to transform the image into suitable form for further processing on FPGAs [25]. A standard image processing pipeline was used: file reading, image preprocessing, main image processing techniques, post-processing, and visual output representation [25].

Back to the processing of images in real time, it is impractical and also time consuming to write thousands of lines in low-level language. Therefore, software environment for hardware description [1, 2, 21] is created to be supported by a tool XSG [2] coupled with a Simulink graphical interface in Matlab. While software implementations of various image processing techniques are appropriate for general use, they must be incorporated in hardware in order to satisfy real-time applications [3]. Moreover, processing speed of software versions is not sufficient when it comes to meeting the real-time needs, because a significant amount of information has to be processed in a limited period of time (meaning that throughput is higher) [26]. In comparison to their corresponding software, FPGA-based application-specific hardware design offers higher processing speeds, with the benefits of parallel architecture [26].

Some research on implementation of different image processing algorithms on FPGA using the XSG tool have already been carried out [27]. Christe et al. proposed FPGA architecture for MRI image analysis and tumor characterization. Experimental results for different filtering algorithms are presented in this work using the XSG [16]. It was shown that with a chip XC3S500E-FG320, only 50% of the total resources were used. Que et al. performed the reconstruction of the image from computer tomography and its FPGA-based hardware implementation [28]. This algorithm is implemented on Virtex-2 processor using both Simulink and XSG block for both hardware/software.

Sudeep and Majumdar investigated FPGA-based image edge detection architecture [29]. Sobel, Prewitt, and Robert algorithms are implemented in this architecture using XSG on a Spartan-3A FPGA device. Harinarayan et al. researched feature extraction of digital aerial images on FPGA-based architecture [30]. The Virtex-4 processor is used to implement XSG blocks. Edge detection methods such as Sobel and Prewitt are used to perform feature extraction.

Shanshan and Xiaohong introduced the FPGA image edge detection architecture using the XSG tool [31]. This architecture was mainly used to identify the edge features of the images of the vehicle. This system is implemented on Spartan-3E Board and the summary of resource utilization is also reported. Basu et al. outlined the design and deployment of a real-time DSP program using coupled Matlab-Simulink and XSG [32]. The configuration is implemented on the Spartan-3A Board. The implementation of the FPGA image processing system with the help of XSG is also investigated by Said et al. [33]. This design implements Sobel algorithm for edge detection. The design was implemented on the Spartan 3A and Virtex-5 FPGA devices and their utilization summary is compared. FPGA-based implementation of conventional edge detectors such as Sobel, Prewitt, Robert, and Laplacian of Gaussian (LoG) are already investigated for implementation using XSG [29, 33]. Since XSG-based edge detection based on Simpler Gabor Wavelet hardware implementation was not investigated up to that point, Sujatha et al. implemented it on FPGA and compared with other edge detection methods [27].

Although some of the mentioned algorithms were already implemented on FPGA using XSG, very rarely have the papers investigated the application on medical images. Among those, Šušteršič et al. have studied the use of XSG in the medical image processing [34] and this chapter represents the extension of that work. Therefore, it is very interesting to see how these algorithms are behaving in the case of implementation on different medical images. All this leads to the motivation of this chapter to implement several image processing algorithms as part of a complex medical and biomedical applications using Matlab/Simulink coupled with XSG tool. In this chapter, Vivado System Generator 2019.1 was used, and Matlab version was 2018b; however, the requirements and steps necessary to build any model, independent of the versions are also described in Table 9.2.

9.2 Building a Simple Model Using XSG

System Generator allows device-specific hardware designs to be built directly in a flexible, high-level system modeling environment. Signals in System Generator can be in fixed-point format (signed or unsigned) and transformed into appropriate signal types. Blocks react to their surroundings, automatically adjusting the results they produce to the chosen hardware [1].

Data flow models, traditional hardware design languages (i.e. VHDL, Verilog), and MATLAB programming language functions can be used alongside, simulated, and synthesized together to a working hardware. The results of the System Generator simulation are bit and cycle-accurate, which ensures that the effects shown

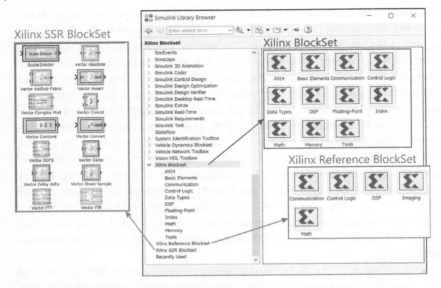

Figure 9.2 Xilinx System Generator library. *Source:* The MathWorks, Inc.

in the simulation precisely match the results found on the hardware. System Generator simulations are much faster to produce rather than conventional HDL simulators, and results are simpler to interpret [1].

The Simulink Xilinx Library is a library of blocks that can be connected to the Simulink block editor to construct fully functional systems (Figure 9.2). System Generator blocksets are in that sense used as any other Simulink blocksets. The blocks include abstractions of logical, mathematical, memory, and DSP functions that can be used to construct complex signal processing systems. There are also blocks that provide interfaces to other tools (i.e. FDATool, ModelSim) as well as code generation from System Generator [1].

The Xilinx Blockset represents a family of libraries that include basic System Generator blocks. Some of the blocks are low-level and provide the access to device-specific hardware. Other blocks are more high-level and their purpose is to implement advanced designs (i.e. signal processing). Those blocks that have the wide application (i.e. I/O blocks) belong to several libraries. Every block is part of the Index library given in Table 9.1.

Prerequisites

In order to be able to build any model using XSG, Matlab and Vivado versions need to be compatible. A full list of supported versions is given in Table 9.2 [22].

Table 9.1 Xilinx Blockset library description [1].

Library Name	Description
Index	Every block in the Xilinx Blockset
Basic elements	Elements Standard building blocks for digital logic
Communication	Forward error correction and modulator blocks, commonly used in digital communications systems
Control Logic	Blocks for control circuitry and state machines
Data Types	Blocks that convert data types (includes gateways)
DSP	Digital signal processing (DSP) blocks
Math	Blocks that implement mathematical functions
Memory	Blocks that implement and access memories
Shared Memory	Blocks that implement and access Xilinx shared memories
Tools	"Utility" blocks, e.g. code generation (System Generator block), resource estimation, HDL co-simulation, etc.

Table 9.2 Supported Matlab and XSG versions.

System Generator version	Supported Matlab release
2020.1	R2020a, R2019b, R2019a
2019.2	R2019b, R2019a, R2018b, R2018a
2019.1	R2019a, R2018b, R2018a
2018.3	R2018a, R2017b, R2017a
2018.2	R2018a, R2017b, R2017a
2018.1	R2017b, R2017a
2017.4	R2017b, R2017a, R2016b, R2016a
2017.3	R2017a, R2016b, R2016a
2017.2	R2017a, R2016b, R2016a
2017.1	R2017a, R2016b, R2016a
2016.4	R2016b, R2016a, R2015b, R2015a
2016.3	R2016b, R2016a, R2015b, R2015a
2016.2	R2015b, R2015a, R2014b, R2014a
2016.1	R2015b, R2015a, R2014b, R2014a

In order to configure MATLAB to the Vivado® Design Suite, the following steps should be performed:

Windows system

- Select Start > All Programs > Xilinx Design Tools > Vivado 2019.x > System Generator > System Generator 2019.x MATLAB Configurator
- Click the check box of the version of MATLAB compatible for configuration and then click OK

Linux systems

- Launching System Generator under Linux is handled using a shell script called *sysgen* located in the *<Vivado install dir>/bin* directory. Before launching this script, MATLAB executable must be found in Linux system's *$PATH* environment variable for Linux system. When *sysgen* script is executed, it will launch the first MATLAB executable found in *$PATH* and attach System Generator to that session of MATLAB. Also, the *sysgen* shell script supports all the options that MATLAB supports and all options can be passed as command line arguments to the *sysgen* script [23].

System Generator can be started using Start > All Programs >Xilinx Design Tools > Vivado 2019.x > System Generator. In the Simulink library browser, there is a list of all different Toolboxes installed within MATLAB, among which the XSG components will be shown and divided into three subcategories:

- Xilinx Blockset
- Xilinx Reference Blockset
- Xilinx SSR Blockset

The category Xilinx Blockset contains all the basic blocks, which can be used in many different applications, out of which some will be shown in this chapter. As in general Simulink, a new model can be created via File> New> Model.

Basic XSG blocks, used in almost any application are explained below, alongside with their icons given in Figure 9.3.

- **Xilinx System Generator Block**: One of the most important blocks used in any System Generator model is the System Generator Block (Figure 9.3a). This block is found in the Simulink category: Xilinx Blockset>Basic Elements>System Generator. The System Generator block in the new model window should be placed as shown in Figure 9.20.
- **Input/Output Gateways**: These blocks belong to System Generator blocks that are used to convert data inside Simulink in the floating-point format to fixed-point format that is used inside the hardware system and modeled by System Generator (Figure 9.3b). In the end of processing, output gateway is used again

Figure 9.3 Schematic icons for some of the widely used XSG elements. (a) Xilinx System Generator Block, (b) Input/Output Gateways, (c) Multipliers, (d) Constants, (e) Adder/Subtractor

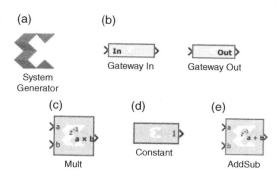

to convert the system output back to floating point. In that sense, there are two types of gateways inside System Generator:

– **Input Gateways**: are used to convert floating point data into fixed-point format. Input gateway properties define in detail the fixed-point format with its three important properties:

- Output type: the values that can be assigned to this property are *boolean* (single bit data representation), *two's complement* data representation, and *unsigned* data representation.
- Number of bits: Number of bits that are used to represent the data. The resolution of the system is directly defined by the number of bits.
- Binary point: represents the position of the binary point in fixed-point format.

– **Output Gateways**: are used to convert the fixed-point format back into the floating-point format. Output gateways are able to automatically detect the fixed-point format from the system output and therefore do not require any other modification.

• **Multipliers**: Most of the models require some multiplication. XSG Blockset provides several blocks for arithmetic operations, among which multiplication is one of them (Figure 9.3c). Multiplier block can be placed using Xilinx Blockset>Math>Multblock. Double click on the Multblock will enable the change of its properties. Two basic options for output precision are to be found in its basic properties: *full* and *user defined*. In the *full precision* option, the multiplier block uses the input fixed-point format to determine the format of the output. In the *user-defined* precision mode, the designer specifies a different format. This means that the designer needs to specify a rounding method for excessive data values. Additionally, in the Implementation Menu, the user has the possibility to optimize speed/area using the dedicated Multipliers embedded in an FPGA or use the LUTs instead.

- **Constants**: Constants are necessary part of every design. Constants are usually implemented using hardwired configurations. They can be found in Xilinx Blockset>Basic Elements>Constant (Figure 9.3d). Double click on the constant block will enable the change of constant properties. If fixed-point (Signed (2's comp)) is used, the user should pay attention to have enough bits to represent the number (i.e. number of bits 16 would allow to represent fraction numbers between the +7 and −8).
- **Adder/Subtractor**: The adders/subtractors are almost a necessity in every model. These blocks can be found in Xilinx Blockset>Math>AddSub (Figure 9.3e). The AddSub block performs addition or subtraction operations using two operands. The fixed-point format of the output is determined based on the format of the inputs.

More detailed explanation on building a simple model and Hardware/Software Co-Simulation is given in [1, 23]. In this chapter, we will focus on implementation of medical image processing algorithms using XSG.

9.3 Medical Image Processing Using XSG

9.3.1 Image Pre- and Post-Processing

In software co-simulation, all other Xilinx blocks are integrated between two blocks – Gateway In and Gateway Out, which serve as input and output for the H/W design, respectively [35]. As Xilinx blocks operate in a fixed-point format, there is a necessity to translate real signals, written with a floating point, to a fixed point, and vice versa. For that purpose, Gateway In and Gateway Out blocks act as translators for conversion of signals into different formats [2, 21, 35].

For every image processing design, blocks like Image Source and Image Viewer will be used to read the source file and later to view the output. Preprocessing and post-processing image units are the same independently of the image processing purpose and are constructed using traditional Simulink block sets [3, 24, 34]. Therefore, preprocessing subsystem should consist of a variety of mentioned blocks – *File image, Transpose, Transform 2D to 1D, Frame translation, and Unbuffer* (Figure 9.4).

Since most MRI/CT images are grayscale images and therefore represented as M-by-N arrays, where the number of rows and the number of columns are M and N, respectively, *Image from file* block is used to read the image of interest as matrix [25]. In case an image is color image, its representation is M-by-N-by-P array, where M and N are rows and columns, respectively, on each color plane. Block *Transpose* performs the mathematic transpose operation of an M-by-N

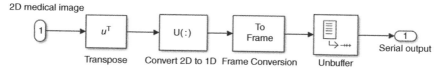

2D medical image

Transpose   Convert 2D to 1D Frame Conversion   Unbuffer

Serial output

Figure 9.4   Image preprocessing module.

matrix to obtain the output N-by-M matrix. Block *Convert 2D to 1D* serves to transform a 2D image matrix to a 1D sequence as the name itself reads. *Frame conversion* gives two options: setting the sampling mode to frame based or sample based and helping prepare the data to unbuffer. *Unbuffer* block performs unbuffering of M-by-N input into a 1-by-N output, where unbuffering is performed row wise [34].

Image post-processing subsystem is again made up of several blocks — *Data type conversion, Buffer, Convert 1D to 2D, Transpose, and Video viewer* (Figure 9.5).

These blocks help the image to be reconstructed based on the obtained 1D array. *Data type conversion* serves to convert the signal to multiple formats, where unsigned integer is used. *Buffer* Block implements frame-based processing by redistributing the data in each column input to generate an output of a different frame size. The buffering of a signal with a greater frame size would result in an output that would be slower than the input frame rate [25]. *Block Convert 1D to 2D* is used to transform the 1D array back to the 2D image matrix to be able to view the image on the screen (using *Video Viewer* block) [34].

The next section of the chapter performs implementation of several algorithms for image processing on medical images, which can serve as a guidance to more complex analyses of images applied in biomedical fields.

9.3.2   Algorithms for Image Preprocessing

9.3.2.1   Algorithm for Negative Image

Transforming an image into negative has many applications in medical imaging, where inverted image is used for further processing. Inverted image can be obtained in several ways, among which an XOR function block/simple Inverter block/Add sub block are just some ways. The algorithm for obtaining grayscale image negative using XOR is given in Figure 9.6.

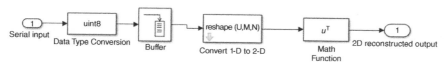

Serial input
Data Type Conversion   Buffer   Convert 1-D to 2-D   Math
                                                            Function

2D reconstructed output

Figure 9.5   Image post-processing module.

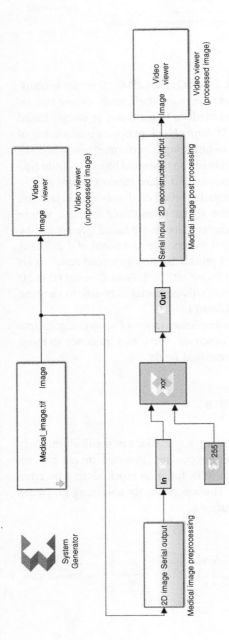

Figure 9.6 Simulink – System Generator DSP model for creating negative image.

Other ways to obtain negative image is using Addsub block by subtracting the constant 255 from each pixel value to generate finally the negation of an input image. The same result would be obtained when an Inverter block is used.

9.3.2.2 Algorithm for Image Contrast Stretching

Contrast of an image is the distribution of light and dark pixels in an image. To enhance the contrast on an image, histogram should be stretched on a range between 0 and 255, meaning to fill the full dynamic range of the image. However, if there is at least one pixel with value 0 or 255, contrast stretching does not make sense. The new pixel value is calculated using the equation [36]:

$$g(x,y) = \frac{f(x,y) - f_{min}}{f_{max} - f_{min}} \bullet \left(2^{bpp} - 1\right) \tag{9.1}$$

This formula requires finding the minimum f_{min} and maximum f_{max} pixel intensity multiplied by levels of gray. In our case, the image is 8 bit, so levels of gray are $2^8 = 256$. In the case of the used image with sagittal view of the spine disc, $f_{min} = 44$ and $f_{max} = 255$, obtaining the Subsystem for image enhancement as in Figure 9.7. In the case of the used image with lungs infected with COVID-19, $f_{min} = 20$ and $f_{max} = 209$; therefore, the constant and multiplier values are 44 and 1.349, respectively.

When visualized in the histogram, that is equivalent to expanding or compressing the histogram around the midpoint value [21, 25].

9.3.2.3 Image Edge Detection

The edge of the image is defined as a point where there is a sharp difference in the image, in the sense of where pixel positions have a sudden shift in brightness, i.e. a discontinuity in gray level values. For this purpose, edge detection in image processing means recognizing and localizing regions in which such brightness transitions occur [21, 35, 37]. Commonly, edge detection techniques are applied using software, but with the advent of Very Large Scale Integration (VLSI) technologies, the hardware implementation of edge detectors has played an important role in real-time applications [35].

Gradient-based methods are often used for the identification of edges. They search for the edge by finding the maximum and the minimum in the first derivative of the image. The key edge detection operator is the gradient operation of the matrix field, which defines the variance between different pixels. The edge detection operator is then determined by constructing a matrix centered around the pixel that is chosen as the center of the matrix area. If the value of this matrix region is above the threshold, the middle pixel is treated as the edge [38]. In the sense of mathematical operations, edge detection in image processing means

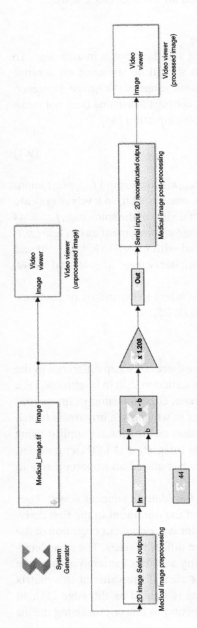

Figure 9.7 Simulink – System Generator DSP model for image contrast stretching.

a masking operation with an acceptable filter mask, meaning the input image is convolved with some of aforementioned filter masks [38].

The use of edge detection algorithms reduces the amount of details that need to be processed and clears information that are less relevant without modifying any of the image properties [26]. As a consequence, several researchers have researched edge detection algorithms for a number of purposes. Fuad et al. published a comparative survey of the edge detection techniques and implemented Canny Edge Detection Algorithm using the XSG on the Nexys3 Board [35]. Kabir et al. have researched another algorithm for the identification of edges – Sobel edge detection in hardware [26]. Although various edge detection methods are available as software implementations such as Roberts, Prewitt, and Sobel method, software versions are suitable for general use. According to the motivation of this chapter, we further study several gradient-based edge detection methods and implement them on medical images in hardware using the XSG.

For a given image I, in the process of calculating horizontal and vertical gradient of each pixel $I(x, y)$, Eqs. (9.2) and (9.3) are applied [33, 34]:

$$G_x(x,y) = \text{horizontal\_mask} * I(x,y) \tag{9.2}$$

$$G_y(x,y) = \text{vertical\_mask} * I(x,y) \tag{9.3}$$

In each pixel (x,y), the horizontal and vertical gradients are combined as in Eq. (9.4), in order to obtain the gradient value:

$$G(x,y) = \sqrt{G_x(x,y)^2 + G_y(x,y)^2} \tag{9.4}$$

Approximation of the magnitude is calculated using the formula:

$$G(x,y) = |G_x(x,y)| + |G_y(x,y)| \tag{9.5}$$

where G_x and G_y are the gradients in the x- and y-directions, respectively. The orientation of the gradient is calculated using the formula:

$$\theta(x,y) = \arctan\left(\frac{G_y(x,y)}{G_x(x,y)}\right) \tag{9.6}$$

In many real-world implementations, the Canny approach is dominant due to its strong localization efficiency and the ability to remove major edges very well. However, the Canny algorithm consists of comprehensive preprocessing (i.e. smoothing) and post-processing (i.e. Non-Maximum Suppression [NMS]). It is also more computationally complex than other gradient-based edge detection algorithms such as Roberts, Prewitt, and Sobel [35, 39, 40]. As a result, we will concentrate on several gradient-based edge detection methods and compare their efficiency, resource utilization, performance time, etc.

Robert method

The basis for Robert method is the use of the Robert operator [38]. It is a type of differential operator that approximates the gradient of an image using a discrete differentiation that calculates the sum of the squares of the differences between the diagonally adjacent pixels of two 2×2 kernels (masks) given by Eq. (9.7) [2, 34].

$$G_x = \begin{bmatrix} 1 & 0 \\ 0 & -1 \end{bmatrix} \quad G_y = \begin{bmatrix} 0 & 1 \\ -1 & 0 \end{bmatrix} \tag{9.7}$$

Masks are separately applied to the input image in order to generate different measurements of the gradient component in the x and y directions (G_x and G_y). Local gradients use convoluted mask coefficients for each pixel. If we define G_x and G_y as horizontal and vertical gradient filter masks, the gradient calculation formulas in both directions are [38, 41] as follows:

$$|G_x| = |p1 - p4| \text{ and } |G_y| = |p2 - p3| \tag{9.8}$$

They are then combined together to find the absolute magnitude of the gradient at each pixel using Eq. (9.5). The model for edge detection using the XSG with implemented Robert method, separately shown for horizontal and vertical calculations is given in Figures 9.8 and 9.9, respectively. The masks are than added and forwarded to the thresholding subsystem (explained later in this chapter).

Prewitt method

The basis for Prewitt method is that the operator uses the central difference. This method is considered to be better in certain aspects than Roberts. Image is convolved with the kernel in horizontal and vertical direction with the kernels given by Eq. (9.9) [2, 34].

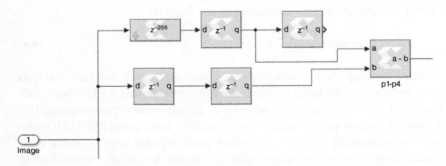

Figure 9.8 Robert method – System Generator DSP sub-model for horizontal mask in edge detection.

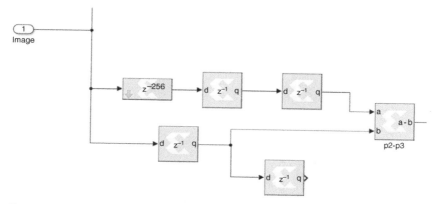

Figure 9.9 Robert method - System Generator DSP sub-model for vertical mask in edge detection.

$$G_x = \begin{bmatrix} -1 & 0 & 1 \\ -1 & 0 & 1 \\ -1 & 0 & 1 \end{bmatrix} \quad G_y = \begin{bmatrix} 1 & 1 & 1 \\ 0 & 0 & 0 \\ -1 & -1 & -1 \end{bmatrix} \tag{9.9}$$

Prewitt edge detection uses moving masks over the image to calculate the gradient of the image. If we define G_x and G_y as horizontal and vertical gradient filter masks, equations for gradient measurement are as Eq. (9.10) [38, 41]:

$$|G_x| = |(p3 + p6 + p9) - (p1 + p4 + p7)|$$
$$|G_y| = |(p1 + p2 + p3) - (p7 + p8 + p9)| \tag{9.10}$$

Prewitt's method is less susceptible to noise due to its kernel [2]. The methodology for the identification of edges using the XSG with implemented Prewitt method is given in Figures 9.10 and 9.11.

Sobel method

The Sobel operator is similar to the Prewitt operator, and has a basis in central difference, but larger weights are added to the central pixels and the mask is therefore as Eq. (9.11) [2, 34]:

$$G_x = \begin{bmatrix} -1 & 0 & 1 \\ -2 & 0 & 2 \\ -1 & 0 & 1 \end{bmatrix} \quad G_y = \begin{bmatrix} 1 & 2 & 1 \\ 0 & 0 & 0 \\ -1 & -2 & -1 \end{bmatrix} \tag{9.11}$$

These kernels are generated to adapt as much as possible to edges that are vertical and horizontal relatively to the pixel grid, where one kernel is used for each direction x and y. These kernels can be used separately on the input image to allow

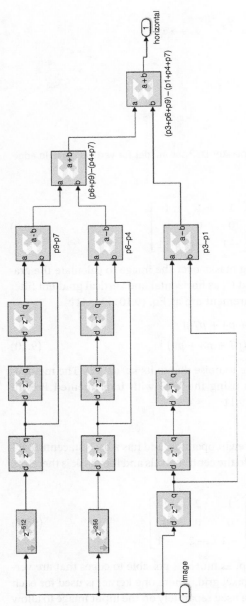

Figure 9.10   Prewitt method – System Generator DSP sub-model for horizontal mask in edge detection.

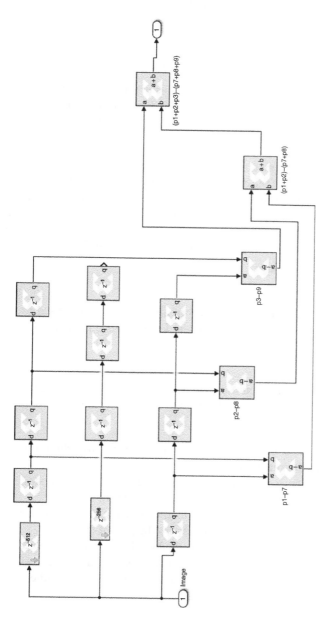

Figure 9.11 Prewitt method – System Generator DSP sub-model for vertical mask in edge detection.

for independent calculations of the gradient components in each direction [38]. Similarly as other methods, moving mask in Sobel edge detection goes over the image and calculates the gradient of the image. If G_x and G_y are defined as masks for horizontal and vertical gradient kernels, gradient calculation equations are given as [38, 41]:

$$|G_x| = |(p3 + 2*p6 + p9) - (p1 + 2*p4 + p7)|$$
$$|G_y| = |(p1 + 2*p2 + p3) - (p7 + 2*p8 + p9)| \tag{9.12}$$

The Sobel method improves noise reduction and is one of the simpler operators with very good results [33]. The methodology for the detection of edges using the XSG with implemented Sobel algorithm is given in Figures 9.12 and 9.13.

Thresholding method

Once the edge detection algorithm is finished, it is important to evaluate the strength of the edges, since some of the detected edges may not be the actual edges. We will achieve this by using thresholding. Thresholding is performed in such a way that each pixel with a value greater than the threshold value is marked as a strong edge and the edge pixels with the values smaller than the threshold are discarded, which means that the following equation is applied [25, 34, 35]:

$$F(x,y) = \begin{cases} 255, & \text{if } f(x,y) > T \\ 0, & \text{otherwise} \end{cases} \tag{9.13}$$

where T is the threshold assigned. In the context of application, Mux is used to compare and replace the pixels with the new values. The explained methodology using XSG is given in Figure 9.14.

Canny method

Canny method is one of the more accurate edge detection methods among edge detection techniques based on intensity gradients [42]. The Canny Edge Detection method is divided into several sub-steps:

- Gaussian smoothing: noise is eliminated using Gaussian filter over the input image (using a Gaussian discrete kernel).
- Calculate Intensity Gradients: identifies the image areas with the strongest intensity gradients (using a kernel of Sobel, Prewitt, or Robert).
- Non-Maximum Suppression: NMS is applied to thin out the edges. We are attempting to delete unnecessary pixels that may not be part of the edges.
- Thresholding with Hysteresis: Hysteresis or double thresholding involves accepting the pixels as edges if the intensity gradient magnitude overcomes the upper threshold. The pixels are rejected as edges, if the intensity gradient value is below the lower threshold [42].

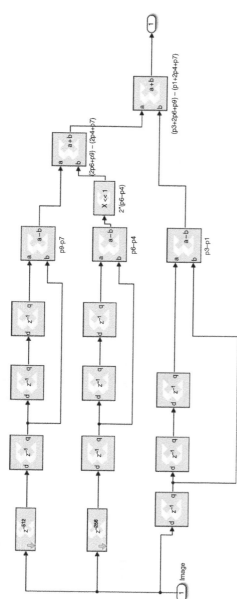

Figure 9.12 Sobel method – System Generator DSP sub-model for horizontal mask in edge detection.

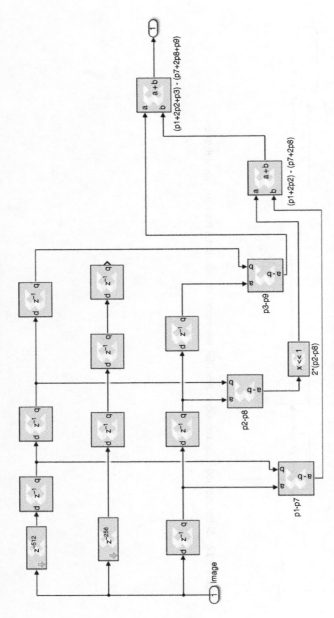

Figure 9.13 Sobel method – System Generator DSP sub-model for vertical mask in edge detection.

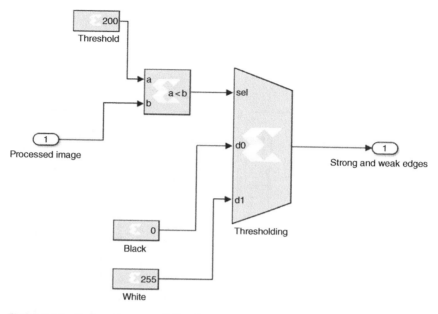

Figure 9.14 System Generator DSP sub-model for thresholding method.

The methodology for the detection of edges using the XSG with implemented Canny algorithm is given in Figure 9.15. Different subsystems are then detailed in Figures 9.16–9.19.

Filter is applied on each of the 5×5 regions in the image and convolved with that region (with the pixel of interest as the center cell called anchor). The result of the convolution is the new intensity value for all the pixels. The block in XSG that performs this operation has a total of nine applications, out of which we choose Gaussian filtering. In that sense, it acts as a low-pass filter and removes unwanted high-frequency elements – noise [42].

After removing the noise using the Gaussian smoothing, the next step is to calculate the intensity gradients. Any of the discussed kernels – Robert, Prewitt, or Sobel – can be used, but usually the adopted kernel is Sobel. We would not give the detailed description of Sobel horizontal and vertical filters, as they were already explained in this chapter. In order to calculate the resultant magnitude, square root operation is necessary. However, the cost to realize the square root operation in hardware is high, therefore low-cost hardware implementation equation based on square approximation is adopted to compute the magnitude [25]. This formula is given in Eq. (9.5).

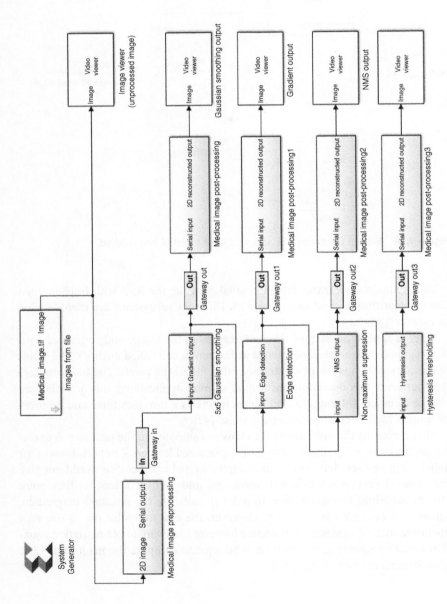

Figure 9.15 System Generator DSP model for Canny Edge detection.

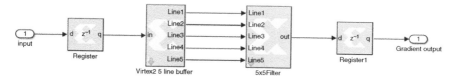

Figure 9.16 Canny method – System Generator DSP sub-model for Gaussian smoothing.

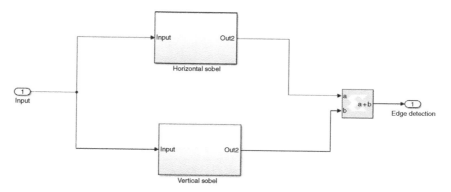

Figure 9.17 Canny method – System Generator DSP sub-model for edge detection (includes Sobel vertical and horizontal kernels).

Once the gradient magnitude has been calculated, NMS is performed. This means that the whole image is scanned in order to remove the pixels that might not be the part of the edges. This is done in such a way that pixels that are local maxima are found in the direction of the gradient (gradient direction is perpendicular to edges). An example of this is to look at the three pixels that are next to each other: pixels a, b, and c. Let pixel b have larger intensity than both a and c, where pixels a and c are in the gradient direction of b. As a result, pixel b is marked as an edge. Otherwise, if pixel b was not a local maximum, it would be set to 0, meaning that it would not be considered an edge pixel.

This block (designed as shown in Figure 9.18) provides edge thinning. It compares all neighboring pixel values to the central pixel and suppresses it to zero value if it is not maximum. Otherwise, it is forwarded to the next block [42].

NMS does not give the perfect result, as some edges might still actually be noise. In order to solve this, canny method applies thresholding to remove the weakest edges and keep the strongest. All of this is achieved using high and low threshold. These thresholds values are calculated based on gradient magnitude histogram of the image. If the gradient magnitude is greater than the high threshold value, the pixel is marked as a strong edge. If the gradient magnitude is lower than the low threshold value, the pixel is marked as a weak edge. The high threshold value is

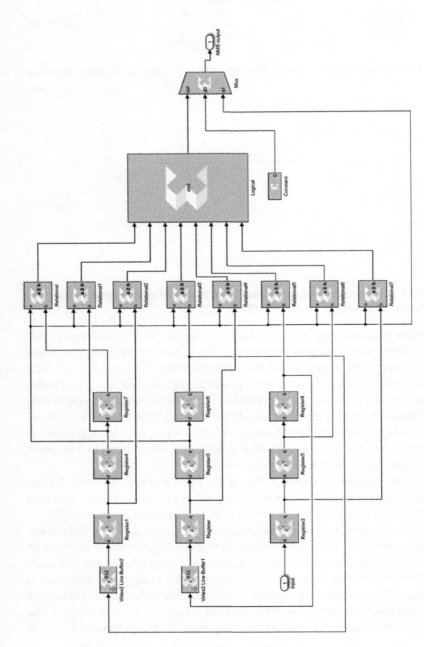

Figure 9.18 Canny method – System Generator DSP sub-model for Non-Maximum Suppression.

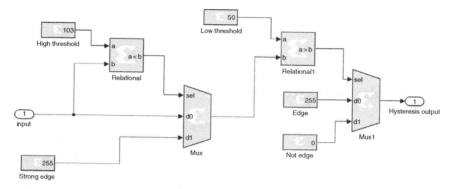

Figure 9.19 Canny method – System Generator DSP sub-model for Hysteresis thresholding.

chosen based on P1 percentage of the total value of pixel intensities. The lower threshold value is P2 percentage of the total value of pixel intensities. Although the percentage values are usually calculated using the cumulative distribution function (CDF), empirically the values of P1 are P2 are chosen as 20 and 40% of the pixel intensities, respectively [42]. Figure 9.19 gives the design of the described subsystem.

9.3.3 Hardware Co-Simulation

The model has been software co-simulated to check that it satisfies the specifications of the selected hardware platform. After software simulation, the System Generator token has to be configured to allow the model to be compiled into hardware [3]. System Generator token dialog box has to be configured with the necessary settings for compilation, synthesis, and clocking. Firstly, compilation Target needs to be chosen by setting up the preferred FPGA platform. Secondly, synthesis tool helps in synthesizing the design (settings are hardware language, creating test bench and design that is synthesized and implemented) and clocking tab (settings are period of the system clock (in nanoseconds) and pin location for hardware clock) [2, 3, 24].

Once the block is configured with the adequate settings, a bit stream (.bit) file is generated. After the bit stream file is generated, hardware co-simulation target is selected and a block will be generated. For our purposes, Artix 7 AC701 Evaluation Platform 1.0 (with the chip Artix7 xc7a200t-1sbg484) is chosen to test the algorithms. This block can be now added to the existing design and both software and hardware the blocks now coexist. The complete design with the hardware and software co-simulation subsystems for the edge detection using Sobel method is presented in Figure 9.20. All the other designs can be hardware co-simulated using the same process.

9.4 Results and Discussion

The model can generally accept and process any medical image. All input images in this chapter were grayscale, size 256×256, 8-bit representation. The model can also work with other dimensions of the images, although the parts of the design dependent on the size of the image have to be adapted. Also, it should be emphasized that the simulation time has to be greater than product of dimensions. Therefore, for all the simulations in this chapter, time was set to 70 000, which is greater than 65 536.

For comparative study in implementation of the algorithm for negative, two images are used – *spine_axial.tiff* and *lungs.tiff*. The results of creating negative image are given in Figure 9.21, one for the example of axial view of the spine discus (Figure 9.21a) and second for the lungs infected with COVID-19 (Figure 9.21b). As it can be seen, negative image represents total inversion, where light areas became dark and vice versa. Reversing the intensity levels of an image produces the equivalent of a photographic negative. The implementation of this type of processing is suited when white or gray detail enhancement in dark regions of an image is needed, especially in those cases where black areas are dominant. Also, image negatives may contribute to localizing regions of interest better and therefore in setting up an adequate diagnosis.

If the image was color image, the same procedure for creating negative image would be applied separately on R, G, and B color components. In this case, negative color image is color-reversed, where red areas become cyan, greens become magenta, and blues become yellow.

For comparative study in implementation of the algorithm for negative, two images are used – *spine_sagittal.tiff* and *lungs.tiff*. The results of creating negative image are given in Figure 9.22, one for the example of sagittal view of the spine discus (Figure 9.22a) and second for the lungs infected with COVID-19 (Figure 9.22b).

It should be noted that in the case where there is no enough contrast in an image (meaning there were both pixels with values 0 and 255), the implementation of Eq. (9.1) would not make sense. Therefore, contrast stretching makes only sense if pixel intensities fall under narrower histogram, with no values close to 0 and 255.

If the image was color image, the same procedure for enhancing contrast in an image would be applied separately on R, G, and B color components. This algorithm is suitable for use in more complex processing.

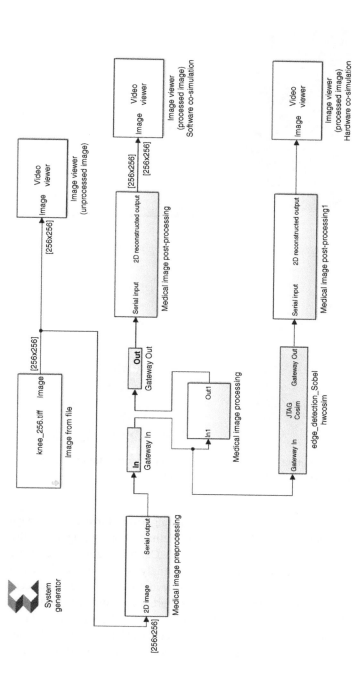

Figure 9.20 Combined software and hardware co-simulation model for the edge detection (implementation of Sobel method).

(a)

(b)

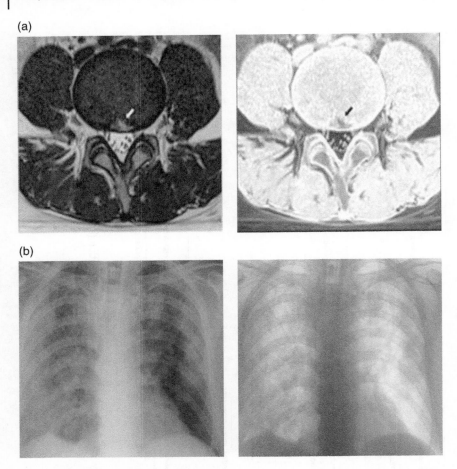

Figure 9.21 Original and output image after application of algorithm for image negative: (a) axial view of spine discus and (b) lungs infected with COVID-19.

For comparative study in implementation of the algorithm for negative, image *knee.tiff* is used. Gradient-based edge detection methods Robert (Figure 9.23b), Prewitt (Figure 9.23b), Sobel (Figure 9.23d), and Canny (Figure 9.23e) were applied and their detection was compared. Visual analysis of results is shown in Figure 9.23, and it can be seen that Sobel and Canny methods achieve more accurate detection. In contrast, the calculation using Robert or Prewitt kernel is simpler and less resource demanding in comparison to the other kernels, since only adder–subtractor logic is used. It should be also emphasized that Canny outputs thinner edges, due to its hysteresis thresholding. The threshold set in all simulations can be

(a)

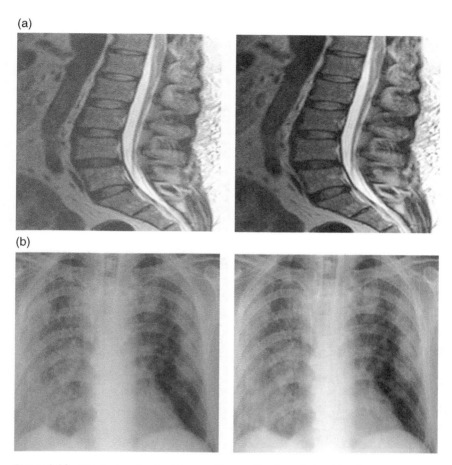

(b)

Figure 9.22 Original and output image after application of algorithm for contrast stretching: (a) sagittal view of spine discus and (b) lungs infected with COVID-19.

changed and varied in order to increase signal-to-noise-ratio and achieve best results (Figure 9.23a).

It should be emphasized that Canny method, due to several additional steps in edge detection Gaussian smoothing (Figure 9.24a), calculation of Intensity Gradients (Figure 9.24b), NMS (Figure 9.24c), and Thresholding with Hysteresis (Figure 9.24d), could be plotted after each of the subprocesses (Figure 9.24).

It is noticeable that the presence of noise threshold dependent and some false edges may be detected. With the threshold that was varied, Sobel tends to detect more false edges, and the performance comparable with that of Prewitt method.

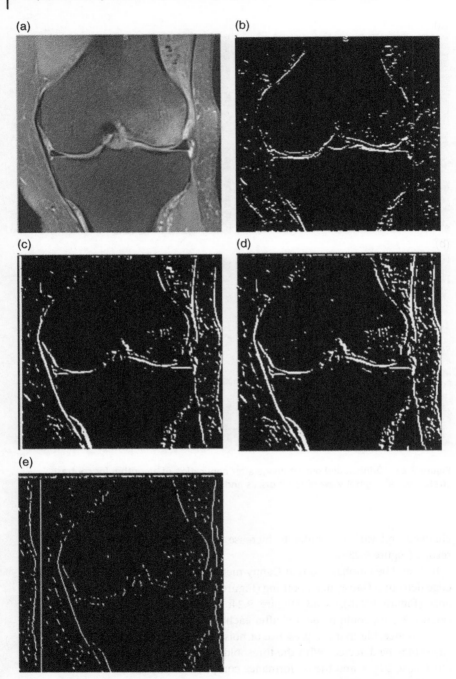

Figure 9.23 Original image (a) compared with output images after application of algorithm for edge detection using Robert (b), Prewitt (c), Sobel (d), and Canny (e) methods.

(a)

(b)

(c)

(d)

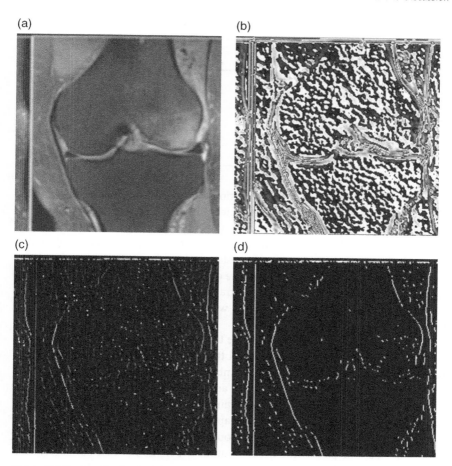

Figure 9.24 Canny edge detection sub-steps: Gaussian smoothing (a), calculation of Intensity Gradients (b), Non-Maximum Suppression (c), and Thresholding with Hysteresis (d).

The same algorithms can be tested with different medical images, which would lead to the conclusion that Robert method is more efficient in images with lower contrast, in which cases the presence of noise is reduced. In the sense of number of edges, Sobel outputs maximum edges whereas Robert provides minimum edges. This is due to the fact that Sobel operator emphasizes on central pixels resulting in thicker edges in comparison to the Robert method. In contrast, Canny output achieves thinner edges.

All the designs are also synthesized and implemented using Xilinx Vivado 2019.1. Table 9.3 shows a comparative analysis of the investigated edge detection

Table 9.3 Post synthesis resource utilization summary.

Edge detection method	LUTs	DSPs	BRAMs	Registers
Robert method	75	0	1	112
Prewitt method	223	0	1.5	266
Sobel method	223	0	1.5	266
Canny method	686	5	7	1024
available	13460	740	365	269200

methods, with respect to resource utilization. The resource utilization speaks a lot about the complexity of the algorithm, meaning that from this analysis it can be also seen that Robert is the simplest and Canny is the most complex algorithm, out of those investigated. Classical gradient-based operators typically have low latency and high throughput and are used in systems where image processing has to be performed in real time [41]. They are the option when applications where precision is not a big concern (i.e. security, tracking, and barcode reader). However, additional techniques need to be added when processing biomedical images, in order to improve precision.

Table 9.4 provides a comparison of the timing paths for the investigated gradient-based edge detection techniques. All the algorithms have passed the Timing path analysis without any violations. Number of paths reported in Timing analysis for Robert method is 8, both for Prewitt and Sobel 19, and Canny 50.

Thus, in applications where medium-range contrast images are of primary concern, Sobel and Prewitt tend to be more appropriate. Robert is the best choice for faster performance with less device resources. As an optimal compromise, the attention should be paid to the desired output – if we do not care about resource utilization in terms that complexity is not a problem, Canny algorithm will satisfy the needs. However, if the resource and timing consumption is the priority, Robert

Table 9.4 Post synthesis timing paths.

Edge detection method	Max_Slack (ns)	Min_Slack (ns)	Max_Delay (ns)	Min_Delay (ns)	Number of paths
Robert	9.118	7.122	2.771	0.635	8
Prewitt	9.118	7.122	2.771	0.635	19
Sobel	9.118	7.122	2.771	0.635	19
Canny	8.699	7.122	2.771	1.332	50

would be the chosen algorithm. In between lies Sobel and Prewitt with balance complexity and accuracy.

9.5 Conclusions

Due to short time-to-market constraints, the need for fast prototyping using the tools such as XSG or Matlab Simulink is now very significant. The XSG tool makes it possible to construct complex designs and implement algorithms using friendly environment with processing units represented by blocks. This tool not only supports the software simulation of different algorithms, but it is also very useful in FPGA hardware synthesization to maximally exploit the hardware parallelism, robustness, and speed. Simulink/XSG characteristics that give them advantages over other solutions are fast time to market for image processing algorithms (shortening the production period from algorithm design to implementation on hardware); easy-to-use interface (easily arranged and separated into subsystems); flexible modeling and simulation (relatively easy debugging process); easier software to hardware transfers (the generator produces all necessary files for the Xilinx FPGAs, with additional possibility to monitor and control the utilization of resources).

In this chapter, image contrast stretching, edge detection, and thresholds are implemented using coupled Simulink-XSG tool, in order to be prepared for FPGA. Algorithms are compared, both visually and quantitatively (resulting accuracy, device utilization, power, minimum period, and maximum frequency). The results suggest that Robert method is, due to its simplicity, most efficient when it comes to device utilization; however, other algorithms, such as Prewitt, Sobel, and Canny are more accurate. The result given in this chapter proves that the proposed hardware implementation of different algorithms gives optimal results for (bio)medical images. Thus, this proposed architecture is very well suited for real-time medical image processing applications.

The main advantages of implementing different algorithms on a FPGA is full use of large memory, embedded multipliers, and the programming the application-specific hardware that offers greater processing speed. The methodology used in this chapter can be expanded further to be used in more complex real-time medical image processing systems.

Acknowledgments

This research is funded by Serbian Ministry of Education, Science, and Technological Development [451-03-68/2020-14/200107 (Faculty of Engineering, University of Kragujevac)]. This research was also supported by the CEI project "Use of

Regressive Artificial Intelligence (AI) and Machine Learning (ML) Methods in Modelling of COVID-19 spread – COVIDAi."

References

1 Xilinx, (2009). *System Generator for DSP User Guide*, USA: Xilinx.
2 Reddy, G.B. and Anusudha, K., (2016). Implementation of image edge detection on FPGA using XSG. Nagercoil, India.
3 Mohammed, A., Rachid, E., and Laamari, H. (2014). High level FPGA modeling for image processing algorithms using Xilinx System Generator. *International Journal of Computer Science and Telecommunications* 5 (6): 1–8.
4 Ownby, M. and Mahmoud, W.H. (2003). *A design methodology for implementing DSP with Xilinx/sup/spl reg//System Generator for Matlab/sup/spl reg.* Morgantown, WV, USA, USA.
5 Vukicevic, A.M., Stojadinovic, M., Radovic, M. et al. (2016). Automated development of artificial neural networks for clinical purposes: Application for predicting the outcome of choledocholithiasis surgery. *Computers in Biology and Medicine* 75: 80–89.
6 Šušteršič, T., Milovanović, V., Ranković, V., and Filipović, N. (2020). A comparison of classifiers in biomedical signal processing as a decision support system in disc hernia diagnosis. *Computers in Biology and Medicine* 125 (103978).
7 Šušteršič, T., Ranković, V., Peulić, M., and Peulić, A. (2019). An early disc herniation identification system for advancement in the standard medical screening procedure based on bayes theorem. *IEEE Journal of Biomedical and Health Informatics* 24 (1): 151–159.
8 Hasan, S., Yakovlev, A. and Boussakta, S. (2010). *Performance efficient FPGA implementation of parallel 2-D MRI image filtering algorithms using Xilinx System Generator.* Newcastle upon Tyne, UK.
9 Othman, M.F.B., Abdullah, N. and Rusli, N.A.B.A. (2010). *An overview of MRI brain classification using FPGA implementation.* Penang, Malaysia.
10 Tana, H., Sazish, A.N., Ahmad, A. et al., (2010). *Efficient FPGA implementation of a wireless communication system using bluetooth connectivity.* Paris, France.
11 Moses, C.J., Selvathi, D. and Rani, S.S. (2010). *FPGA implementation of an efficient partial volume interpolation for medical image registration.* Ramanathapuram, India.
12 Šušteršič, T. and Peulić, A. (2019). Implementation of face recognition algorithm on field programmable gate array (FPGA). *Journal of Circuits, Systems and Computers* 28 (8): 1950129.
13 Šušteršič, T., Vulović, A., Filipović, N., and Peulić, A. (2016). *FPGA implementation of face recognition algorithm.* In: *Pervasive Computing Paradigms for Mental Health* (eds. N. Oliver, S. Serino, A. Matic, et al.), 93–99. Cham: Springer.

14 Šušteršič, T., Vulovic, A., Trifunovic, N. et al. (2019). Face recognition using maxeler dataflow. In: *Exploring the DataFlow Supercomputing Paradigm* (eds. V. Milutinovic and M. Kotlar), 171–196. Cham: Springer.

15 Elamaran, V., Aswini, A., Niraimathi, V., and Kokilavani, D. (2012). FPGA implementation of audio enhancement using adaptive LMS filters. *Journal of Artificial Intelligence* **5** (4): 221.

16 Christe, S.A., Vignesh, M., and Kandaswamy, A. (2011). An efficient FPGA implementation of MRI image filtering and tumor characterization using Xilinx system generator. *International Journal of VLSI Design and Communication Systems* **2**: 95–109.

17 Karthigaikumar, P., Kirubavathy, K.J., and Baskaran, K. (2011). FPGA based audio watermarking—covert communication. *Microelectronics Journal* **42** (5): 778–784.

18 Raut, N.P. and Gokhale, A.V. (2013). FPGA implementation for image processing algorithms using Xilinx System Generator. *IOSR Journal of VLSI and Signal Processing (IOSR-JVSP)* **2** (4): 26–36.

19 Moreo, A.T., Lorente, P.N., Valles, F.S. et al. (2005). Experiences on developing computer vision hardware algorithms using Xilinx System Generator. *Microprocessors and Microsystems* **29** (8-9): 411–419.

20 Saidani, T., Dia, D., Elhamzi, W. et al. (2009). Hardware co-simulation for video processing using xilinx system generator. London, U.K.

21 Gupta, A., Vaishnav, H., and Garg, H. (2015). Image processing using Xilinx System Generator (XSG) in FPGA. *International Journal of Research and Scientific Innovation* **2** (9): 119–125.

22 Xilinx, (2019). *Compatibilty versions of system generator for DSP with versions of Vivado design tools.* https://www.xilinx.com/support/answers/55830.html (accessed 25 January 2019).

23 Design, M.B.D. (2012). Vivado Design Suite Reference Guide

24 Hanisha, V. and Kumari, G.V. (2016). Hardware implementation of image denoise filters using system generator. *International Journal of VLSI System Design and Communication System* **4** (10): 940–946.

25 Kumar, K.A. and Kumar, M.V. (2014). Implementation of image processing lab using Xilinx System Generator. *Advances in Image and Video Processing* **2** (5): 27–35.

26 Kabir, S. and Alam, A.A. (2014). Hardware design and simulation of sobel edge detection algorithm. *International Journal of Image, Graphics and Signal Processing* **6** (5): 10.

27 Sujatha, C. and Selvathi, D. (2014). Hardware implementation of image edge detection using Xilinx System Generator. *Asian journal of scientific research* **7** (2): 188.

28 Que, Z., Zhu, Y., Wang, X. et al. (2010). Implementing medical CT algorithms on stand-alone FPGA based systems using an efficient workflow with SysGen and simulink. *Proceeding of the IEEE International Conference on Computer and*

Information Technology (CIT 2010), Bradford, UK: (29 June to 1 July 2010), pp. 2391–2396.

29 Sudeep, K.C. and Majumdar, J. (2011). A novel architecture for real time implementation of edge detectors on FPGA. *International Journal of Computer Science Issues (IJCSI)* **8** (1): 193.

30 Harinarayan, R.R., Pannerselvam, M.M. and Tripathi, D. K. (2011). Feature extraction of digital aerial images by FPGA based implementation of edge detection algorithms. *Proceedings of the International Conference on Emerging Trends in Electrical and Computer Technology*, Tamil Nadu, March 23-24

31 Shanshan, Z. and Xiaohong, W. (2010). Vehicle image edge detection algorithm hardware implementation on FPGA. *2010 International Conference on Computer Application and System Modeling (ICCASM 2010)*.

32 Basu, A., Das, T.S., and Sarkar, S.K. (2012). SysGen architecture for visual information hiding framework. *International Journal of Emerging Technology and Advanced Engineering* **2**: 32–40.

33 Said, Y., Saidani, T., Smach, F. and Atri, M., (2012). *Real time hardware co-simulation of edge detection for video processing system*. Yasmine Hammamet, Tunisia.

34 Šušteršič, T., Milovanović, V, Ranković, V. et al., (2019). Medical image processing using Xilinx system generator. *International Conference on Computational Bioengineering*. Springer, Cham, pp. 104–116.

35 Fuad, K.A.A. and Rizvi, S.M. (2015). Hardware software co-simulation of Canny edge detection algorithm. *International Journal of Computer Applications* **122** (19): 7–12.

36 Tutorials_point, (2019). *Histogram stretching*. https://www.tutorialspoint.com/dip/histogram_stretching.htm (accessed 25 April 2019).

37 Gonzalez, R.C. and Woods, R.E. (2008). *Digital Image Processing*, 3e. New Jersey: Prentice Hall.

38 Yasri, I., bin Hamid, N.H. and Yap, V.V., (2009). *An FPGA implementation of gradient based edge detection algorithm design*. Kota Kinabalu, Malaysia.

39 Xu, Q., Chakrabarti, C. and Karam, L.J., (2011). *A distributed Canny edge detector and its implementation on FPGA*. Sedona, AZ, USA.

40 Xu, Q., Varadarajan, S., Chakrabarti, C., and Karam, L.J. (2014). A distributed canny edge detector: algorithm and FPGA implementation. *IEEE Transactions on Image Processing* **23** (7): 2944–2960.

41 Mahalle, A.G. and Shah, A.M. (2018). Hardware co-simulation of classical edge detection algorithms using Xilinx System Generator. *International Research Journal of Engineering and Technology (IRJET)* **5** (1): 912–916.

42 Kumar, M. and Singh, S. (2020). Hardware and software implementation of canny edge detection algorithm. *Journal of Information and Computational Science* **10** (4): 415–426.

Index

Computational Modeling and Simulation Examples in Bioengineering, First Edition. Edited by Nenad D. Filipovic. © 2022 The Institute of Electrical and Electronics Engineers, Inc. Published 2022 by John Wiley & Sons, Inc.

IEEE Press Series in Biomedical Engineering

The focus of our series is to introduce current and emerging technologies to biomedical and electrical engineering practitioners, researchers, and students. This series seeks to foster interdisciplinary biomedical engineering education to satisfy the needs of the industrial and academic areas. This requires an innovative approach that overcomes the difficulties associated with the traditional textbooks and edited collections.

Series Editor, Metin Akay, University of Houston, TX

Printed and bound by CPI Group (UK) Ltd, Croydon, CR0 4YY

16/04/2025

14658530-0001